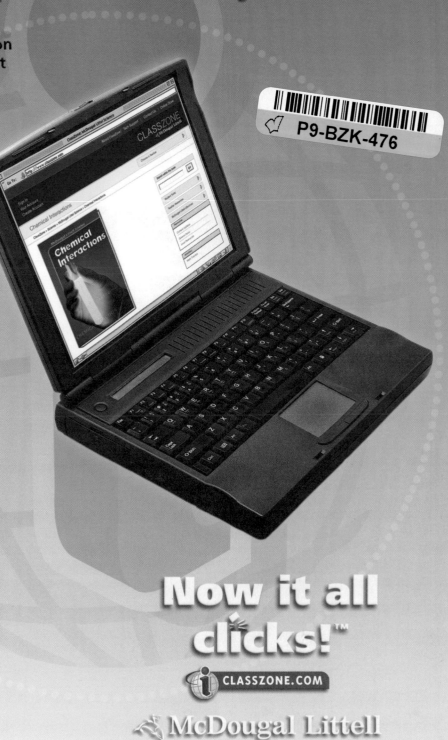

TEACHER'S EDITION

McDougal Littell Science

Chemical Interactions

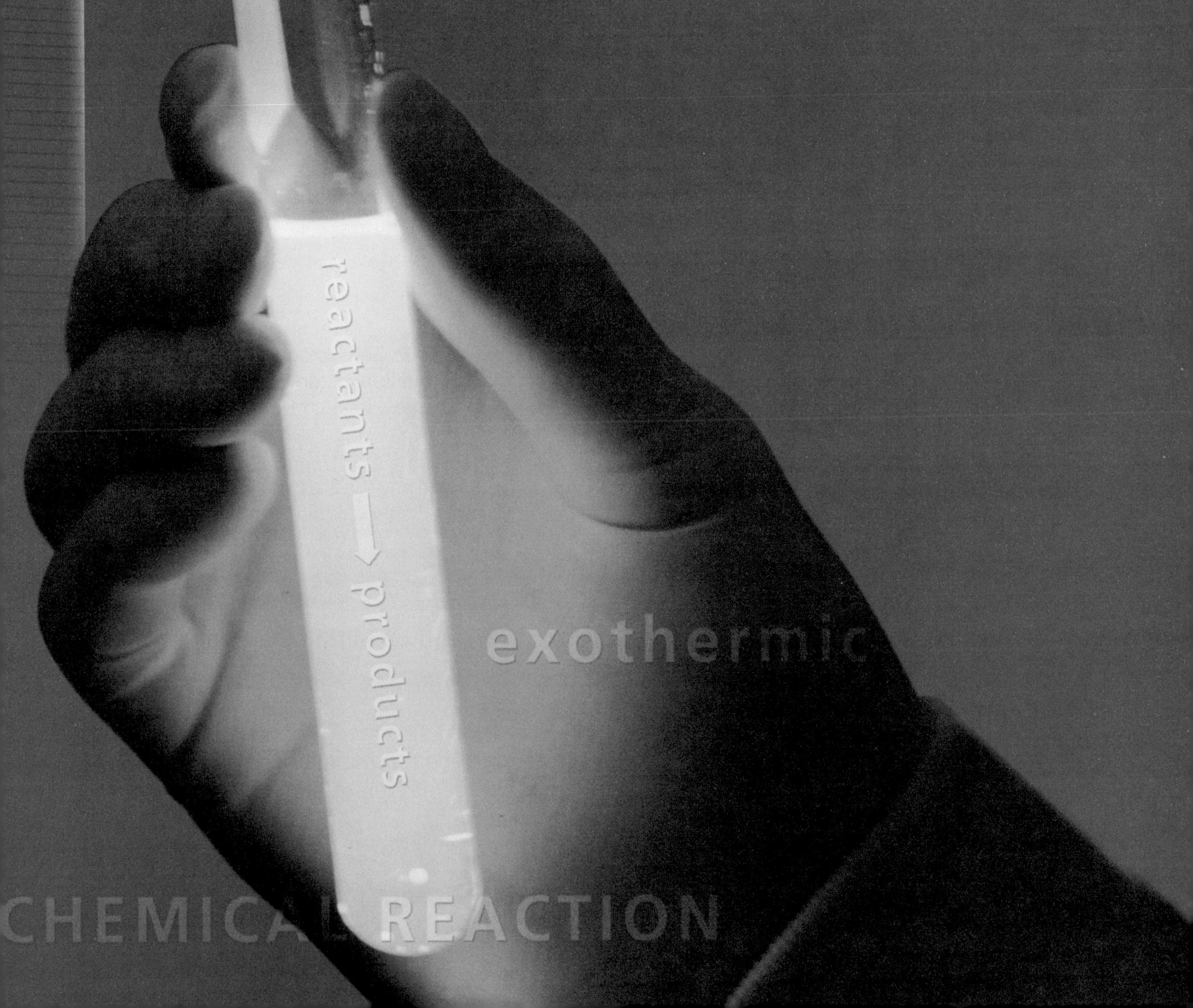
reactants ⟶ products
exothermic
CHEMICAL REACTION

Credits
5B, 65B, 65C, 107B Illustrations by Stephen Durke; **107C** © Photodisc/Getty Images;
© Greg Pease/Stock Connection/PictureQuest; © Stockbyte; © S. Feld/Robertstock.com;
Jellinek & Sampson, London/Bridgeman Art Library; **143B** Illustration by Stephen Durke.

Acknowledgements
Excerpts and adaptations from National Science Education Standards by the National Academy of
Sciences. Copyright © 1996 by the National Academy of Sciences. Reprinted with permission from the
National Academies Press, Washington, D.C.

ISBN: 0-618-33439-4   1 2 3 4 5 6 7 8 VJM 08 07 06 05 04

Internet Web Site: http://www.mcdougallittell.com

McDougal Littell Science

Effective Science Instruction Tailored for Middle School Learners

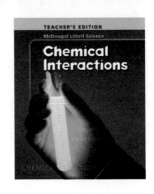

Chemical Interactions
Teacher's Edition
Contents

Consultants and Reviewers

Science Consultants

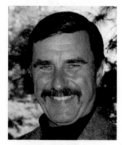

Chief Science Consultant

James Trefil, Ph.D. is the Clarence J. Robinson Professor of Physics at George Mason University. He is the author or co-author of more than 25 books, including *Science Matters* and *The Nature of Science.* Dr. Trefil is a member of the American Association for the Advancement of Science's Committee on the Public Understanding of Science and Technology. He is also a fellow of the World Economic Forum and a frequent contributor to *Smithsonian* magazine.

Rita Ann Calvo, Ph.D. is Senior Lecturer in Molecular Biology and Genetics at Cornell University, where for 12 years she also directed the Cornell Institute for Biology Teachers. Dr. Calvo is the 1999 recipient of the College and University Teaching Award from the National Association of Biology Teachers.

Kenneth Cutler, M.S. is the Education Coordinator for the Julius L. Chambers Biomedical Biotechnology Research Institute at North Carolina Central University. A former middle school and high school science teacher, he received a 1999 Presidential Award for Excellence in Science Teaching.

Instructional Design Consultants

Douglas Carnine, Ph.D. is Professor of Education and Director of the National Center for Improving the Tools of Educators at the University of Oregon. He is the author of seven books and over 100 other scholarly publications, primarily in the areas of instructional design and effective instructional strategies and tools for diverse learners. Dr. Carnine also serves as a member of the National Institute for Literacy Advisory Board.

Linda Carnine, Ph.D. consults with school districts on curriculum development and effective instruction for students struggling academically. A former teacher and school administrator, Dr. Carnine also co-authored a popular remedial reading program.

Donald Steely, Ph.D. serves as principal investigator at the Oregon Center for Applied Science (ORCAS) on federal grants for science and language arts programs. His background also includes teaching and authoring of print and multimedia programs in science, mathematics, history, and spelling.

Sam Miller, Ph.D. is a middle school science teacher and the Teacher Development Liaison for the Eugene, Oregon, Public Schools. He is the author of curricula for teaching science, mathematics, computer skills, and language arts.

Vicky Vachon, Ph.D. consults with school districts throughout the United States and Canada on improving overall academic achievement with a focus on literacy. She is also co-author of a widely used program for remedial readers.

Content Reviewers

John Beaver, Ph.D.
Ecology
Professor, Director of Science Education Center
College of Education and Human Services
Western Illinois University
Macomb, IL

Donald J. DeCoste, Ph.D.
Matter and Energy, Chemical Interactions
Chemistry Instructor
University of Illinois
Urbana-Champaign, IL

Dorothy Ann Fallows, Ph.D., MSc
Diversity of Living Things, Microbiology
Partners in Health
Boston, MA

Michael Foote, Ph.D.
The Changing Earth, Life Over Time
Associate Professor
Department of the Geophysical Sciences
The University of Chicago
Chicago, IL

Lucy Fortson, Ph.D.
Space Science
Director of Astronomy
Adler Planetarium and Astronomy Museum
Chicago, IL

Elizabeth Godrick, Ph.D.
Human Biology
Professor, CAS Biology
Boston University
Boston, MA

Isabelle Sacramento Grilo, M.S.
The Changing Earth
Lecturer, Department of the Geological Sciences
Montana State University
Bozeman, MT

David Harbster, MSc
Diversity of Living Things
Professor of Biology
Paradise Valley Community College
Phoenix, AZ

Richard D. Norris, Ph.D.
Earth's Waters
Professor of Paleobiology
Scripps Institution of Oceanography
University of California, San Diego
La Jolla, CA

Donald B. Peck, M.S.
*Motion and Forces; Waves, Sound, and Light;
Electricity and Magnetism*
Director of the Center for Science Education (retired)
Fairleigh Dickinson University
Madison, NJ

Javier Penalosa, Ph.D.
Diversity of Living Things, Plants
Associate Professor, Biology Department
Buffalo State College
Buffalo, NY

Raymond T. Pierrehumbert, Ph.D.
Earth's Atmosphere
Professor in Geophysical Sciences (Atmospheric Science)
The University of Chicago
Chicago, IL

Brian J. Skinner, Ph.D.
Earth's Surface
Eugene Higgins Professor of Geology and Geophysics
Yale University
New Haven, CT

Nancy E. Spaulding, M.S.
Earth's Surface, The Changing Earth, Earth's Waters
Earth Science Teacher (retired)
Elmira Free Academy
Elmira, NY

Steven S. Zumdahl, Ph.D.
Matter and Energy, Chemical Interactions
Professor Emeritus of Chemistry
University of Illinois
Urbana-Champaign, IL

Susan L. Zumdahl, M.S.
Matter and Energy, Chemical Interactions
Chemistry Education Specialist
University of Illinois
Urbana-Champaign, IL

Safety Consultant

Juliana Texley, Ph.D.
Former K–12 Science Teacher and School Superintendent
Boca Raton, FL

English Language Advisor

Judy Lewis, M.A.
Director, State and Federal Programs for reading proficiency
and high risk populations
Rancho Cordova, CA

Research-Based Solutions for Your Classroom

The distinguished program consultant team and a thorough, research-based planning and development process assure that *McDougal Littell Science* supports all students in learning science concepts, acquiring inquiry skills, and thinking scientifically.

Standards-Based Instruction

Concepts and skills were selected based on careful analysis of national and state standards.

- National Science Education Standards
- Project 2061 Benchmarks for Science Literacy
- Comprehensive database of state science standards

Standards and Benchmarks

Each chapter in **Chemical Interactions** covers some of the learning goals that are described in the *National Science Education Standards* (NSES) and the Project 2061 *Benchmarks for Science Literacy*. Selected content and skill standards are shown below in shortened form. The following National Science Education Standards are covered on pages xii–xxvii, in Frontiers in Science, and in Timelines in Science, as well as in chapter features and laboratory investigations: Understandings About Scientific Inquiry (A.9), Understandings About Science and Technology (E.6), Science and Technology in Society (F.5), Science as a Human Endeavor (G.1), Nature of Science (G.2), and History of Science (G.3).

Content Standards

1 Atomic Structure and the Periodic Table

National Science Education Standards

B.1.b	Substances react chemically in characteristic ways with other substances to form new substances.
B.1.c	There are more than 100 known elements that combine to produce compounds.

Project 2061 Benchmarks

4.D.1	The atoms of any element are alike but are different from atoms of other elements.
4.D.6	There are groups of elements that have similar properties, including • highly reactive metals • less-reactive metals • highly reactive nonmetals • some almost completely nonreactive gases
10.F.2	Scientists are still learning about the basic kinds of matter and how they combine.
10.G.1	The discovery that minerals containing uranium darken photographic film led to the idea of radioactivity.
10.G.2	Scientists Marie Curie and Pierre Curie isolated the elements radium and polonium.

2 Chemical Bonds and Compounds

National Science Education Standards

B.1.b	Substances react chemically in characteristic ways with other substances to form new substances.

Project 2061 Benchmarks

4.D.1	Atoms may stick together in molecules or may be packed together in large arrays.
10.F.2	Scientists are still learning about the basic kinds of matter and how they combine.
11.B.1	Models are often used to explore processes that • happen too slowly, too quickly, or are on too small a scale to observe directly • are on too great a scale to be studied experimentally • are potentially dangerous
11.C.5	Symmetry or a lack of symmetry may determine the properties of many objects.

3 Chemical Reactions

National Science Education Standards

B.1.b	In chemical reactions, the total mass is conserved.
B.3.e	In most chemical reactions, energy is transferred into or out of a system.

x Unit: Chemical Interactions

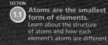

Atomic Structure and the Periodic Table

the BIG idea

A substance's atomic structure determines its physical and chemical properties.

You can't zoom in any closer than this! The picture is an extremely close-up view of nickel. How do things look different the closer you get to them?

Key Concepts

SECTION
1.1 Atoms are the smallest form of elements.
Learn about the structure of atoms and how each element's atoms are different.

SECTION
1.2 Elements make up the periodic table.
Learn how the periodic table of the elements is organized.

SECTION
1.3 The periodic table is a map of the elements.
Learn more about the groups of elements in the periodic table.

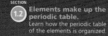

Internet Preview

CLASSZONE.COM
Chapter 1 online resources: Content Review, Simulation, Visualization, three Resource Centers, Math Tutorial, Test Practice

D 6 Unit: Chemical Interactions

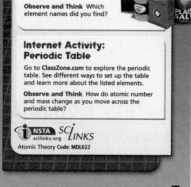

Observe and Think Which element names did you find?

Internet Activity: Periodic Table

Go to **ClassZone.com** to explore the periodic table. See different ways to set up the table and learn more about the listed elements.

Observe and Think How do atomic number and mass change as you move across the periodic table?

NSTA
scilinks.org
SCiLINKS
Atomic Theory Code: MDL022

Chapter 1: Atomic Structure and the Periodic Table 7 D

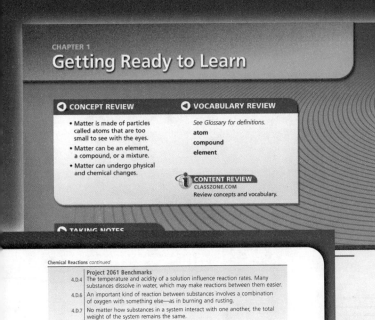

Getting Ready to Learn

◐ CONCEPT REVIEW

- Matter is made of particles called atoms that are too small to see with the eyes.
- Matter can be an element, a compound, or a mixture.
- Matter can undergo physical and chemical changes.

◐ VOCABULARY REVIEW

See Glossary for definitions.

atom

compound

element

ⓘ CONTENT REVIEW
CLASSZONE.COM
Review concepts and vocabulary.

◐ TAKING NOTES

Chemical Reactions *continued*

Project 2061 Benchmarks

4.D.4	The temperature and acidity of a solution influence reaction rates. Many substances dissolve in water, which may make reactions between them easier.
4.D.6	An important kind of reaction between substances involves a combination of oxygen with something else—as in burning and rusting.
4.D.7	No matter how substances in a system interact with one another, the total weight of the system remains the same.
10.F.3	The work of scientist Antoine Lavoisier led to the modern science of chemistry.
10.F.4	Lavoisier tested the concept of conservation of matter by measuring the substances involved in burning.

4 Solutions

National Science Education Standards

B.1.a	A substance has characteristic properties, such as density, a boiling point, and solubility, all of which do not depend on the amount of the sample.
B.1.c	Chemical elements do not break down during normal reactions involving • heating • electric current • acids
D.1.h	The atmosphere is a mixture of nitrogen, oxygen, and small amounts of gases such as water vapor.

Project 2061 Benchmarks

4.D.4	The temperature and acidity of a solution influence reaction rates. Many substances dissolve in water, which may make reactions between them easier.

5 Carbon in Life and Materials

National Science Education Standards

B.1.c	There are more than 100 known elements that combine to produce compounds.
C.1.a	Living systems demonstrate the complementary nature of structure and function.

Project 2061 Benchmarks

4.D.6	Carbon and hydrogen are essential elements of living matter.

Process and Skill Standards

National Science Education Standards

A.1	Identify questions that can be answered through investigation.
A.2	Design and conduct a scientific investigation.
A.3	Use appropriate tools and techniques to gather and interpret data.
A.4	Use evidence to describe, predict, explain, and model.
A.5	Use critical thinking to find relationships between results and interpretations.
A.7	Communicate procedures, results, and conclusions.
E.1	Identify a problem to be solved.
E.2	Design a solution or product.
E.3	Implement the proposed solution.
E.4	Evaluate the solution or design.

Project 2061 Benchmarks

12.B.1	Find what percentage one number is of another.
12.B.2	Use, interpret, and compare numbers in several equivalent forms, such as integers, fractions, decimals, and percents.
12.C.3	Using appropriate units, use and read instruments that measure length, volume, weight, time, rate, and temperature.
12.D.1	Use tables and graphs to organize information and identify relationships.
12.D.2	Read, interpret, and describe tables and graphs.
12.D.4	Understand information that includes different types of charts and graphs, including circle charts, bar graphs, line graphs, data tables, diagrams, and symbols.

Standards and Benchmarks xi

Effective Instructional Strategies

McDougal Littell Science incorporates strategies that research shows are effective in improving student achievement. These strategies include

- Notetaking and nonlinguistic representations (Marzano, Pickering, and Pollock)

- A focus on big ideas (Kameenui and Carnine)

- Background knowledge and active involvement (Project CRISS)

Robert J. Marzano, Debra J. Pickering, and Jane E. Pollock, *Classroom Instruction that Works; Research-Based Strategies for Increasing Student Achievement* (ASCD, 2001)

Edward J. Kameenui and Douglas Carnine, *Effective Teaching Strategies that Accommodate Diverse Learners* (Pearson, 2002)

Project CRISS (Creating Independence through Student Owned Strategies)

VOCA...

proton
neutro...
nucleu...
electro...
atomi...
atomi...
isotop...
ion p...

All matter is made of atoms.

Think of all the substances you see and touch every day. Are all of these substances the same? Obviously, the substances that make up this book you're reading are quite different from the substances in the air around you. So how many different substances can there be? This is a question people have been asking for thousands of years.

About 2400 years ago, Greek philosophers proposed that everything on Earth was made of only four basic substances—air, water, fire, and earth. Everything else contained a mixture of these four substances. As time went on, chemists came to realize that there had to be more than four basic substances. Today chemists know that about 100 basic substances, or elements, account for everything we see and touch. Sometimes these elements appear by themselves. Most often, however, these elements appear in combination with other elements to make new substances. In this section, you'll learn about the atoms of the elements that make up the world and how these atoms differ from one another.

READING TIP
The word *element* is related to *elementary,* which means "basic."

Chapter 1: **Atomic Structure and the Periodic Table 9** **D**

Comprehensive Research, Review, and Field Testing

An ongoing program of research and review guided the development of *McDougal Littell Science.*

- Program plans based on extensive data from classroom visits, research surveys, teacher panels, and focus groups

- All pupil edition activities and labs classroom-tested by middle school teachers and students

- All chapters reviewed for clarity and scientific accuracy by the Content Reviewers listed on page T5

- Selected chapters field-tested in the classroom to assess student learning, ease of use, and student interest

Content Organized Around Big Ideas

Each chapter develops a big idea of science, helping students to place key concepts in context.

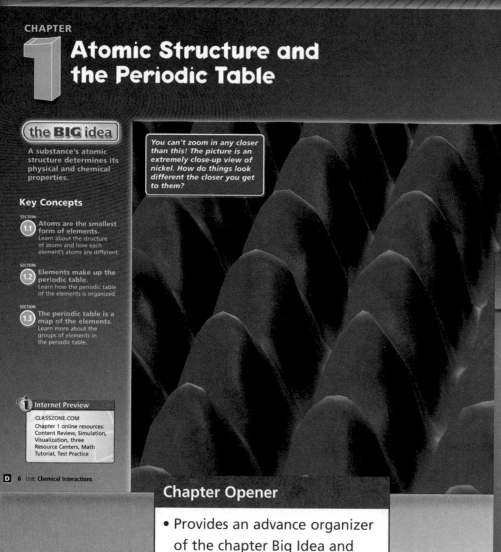

CHAPTER

1 Atomic Structure and the Periodic Table

the BIG idea

A substance's atomic structure determines its physical and chemical properties.

You can't zoom in any closer than this! The picture is an extremely close-up view of nickel. How do things look different the closer you get to them?

Key Concepts

SECTION
1.1 Atoms are the smallest form of elements.
Learn about the structure of atoms and how each element's atoms are different.

SECTION
1.2 Elements make up the periodic table.
Learn how the periodic table of the elements is organized.

SECTION
1.3 The periodic table is a map of the elements.
Learn more about the groups of elements in the periodic table.

Internet Preview

CLASSZONE.COM
Chapter 1 online resources:
Content Review, Simulation, Visualization, three Resource Centers, Math Tutorial, Test Practice

D 6 Unit: Chemical Interactions

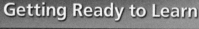

EXPLORE the **BIG** idea

That's Far!

Place a baseball in the middle of a large field. Hold a dime and count off the number of steps from the baseball to the edge of the field. If the baseball were an atom's nucleus and the dime an electron, you would need to go about 6000 steps to walk the distance between the nucleus and the electrons.

Observe and Think How far were you able to go? How much farther would you need to go to model the proportion of an atom? What does this tell you about atomic structure?

Element Safari

Locate the following products in your home or in a grocery store: baking soda, vinegar, cereal flakes, and antacid tablets. You may examine

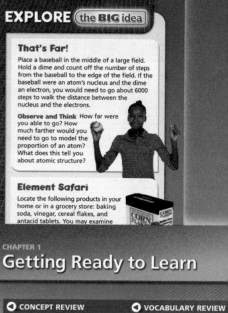

CHAPTER 1

Getting Ready to Learn

○ CONCEPT REVIEW

• Matter is made of particles called atoms that are too small to see with the eyes.

• Matter can be an element, a compound, or a mixture.

• Matter can undergo physical and chemical changes.

○ VOCABULARY REVIEW

See Glossary for definitions.

atom
compound
element

CONTENT REVIEW
CLASSZONE.COM
Review concepts and vocabulary.

▶ TAKING NOTES

MAIN IDEA WEB

Write each new blue heading in a box. Then write notes in boxes around the center box that give important terms and details about that blue heading.

VOCABULARY STRATEGY

Write each new vocabulary term in the center of a frame game diagram. Decide what information to frame it with. Use examples, descriptions, parts, sentences that use the term in context, or pictures. You can change the frame to fit each term.

See the Note-Taking Handbook on pages R45–R51.

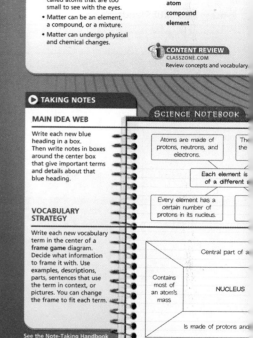

SCIENCE NOTEBOOK

Atoms are made of protons, neutrons, and electrons.

Each element is of a different

Every element has a certain number of protons in its nucleus.

Central part of a

Contains most of an atom's mass

NUCLEUS

Is made of protons and

D 8 Unit: Chemical Interactions

Chapter Opener

• Provides an advance organizer of the chapter Big Idea and Key Concepts

• Connects the Big Idea to the real world through an engaging photo and related question

Visual Summary

- Summarizes Key Concepts using both text and visuals
- Reinforces the connection of Key Concepts to the Big Idea

Section Opener

- Highlights the Key Concept
- Connects new learning to prior knowledge
- Previews important vocabulary

---

the **BIG** idea

A substance's atomic structure determines its physical and chemical properties.

 CONTENT REVIEW
CLASSZONE.COM

KEY CONCEPTS SUMMARY

1.1 Atoms are the smallest form of elements.

- All matter is made of the atoms of approximately 100 elements.
- Atoms are made of protons, neutrons, and electrons.
- Different elements are made of different atoms.
- Atoms form ions by gaining or losing electrons.

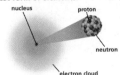

nucleus   proton

neutron

electron cloud

VOCABULARY
proton p. 11
neutron p. 11
nucleus p. 11
electron p. 11
atomic number p. 12
atomic mass number p. 12
isotope p. 12
ion p. 14

1.2 Elements make up the periodic table.

- Elements can be organized by similarities.
- The periodic table organizes the atoms of the elements by properties and atomic number.

Na Mg   Al Si P S Cl

Groups of elements have similar properties.
Elements in a period have varying properties.

VOCABULARY
atomic mass p. 17
periodic table p. 18
group p. 22
period p. 22

1.3 The periodic table is a map of the elements.

- The periodic table has distinct regions.
- Most elements are metals.
- Nonmetals and metalloids have a wide range of properties.
- Some atoms can change their identity through radioactive decay.

Reviewing Vocabulary

Describe how the vocabulary terms in the following pairs are related to each other. Explain the relationship in a one- or two-sentence answer. Underline each vocabulary term in your answer.

1. isotope, nucleus

2. atomic mass, atomic number

3. electron, proton

4. atomic number, atomic mass number

5. group, period

6. metals, nonmetals

7. radioactivity, half-life

Reviewing Key Concepts

Multiple Choice Choose the letter of the best answer.

8. The central part of an atom is called the
   - a. electron
   - b. nucleus
   - c. proton
   - d. neutron

9. The electric charge on a proton is
   - a. positive
   - b. negative
   - c. neutral
   - d. changing

10. The number of protons in the nucleus is the
    - a. atomic mass
    - b. isotope
    - c. atomic number
    - d. half-life

Thinking Critically

The table below lists some properties of six elements. Use the information and your knowledge of the properties of elements to answer the next three questions.

Element	Appearance	Density (g/cm³)	Conducts Electricity
A	dark purple crystals	4.93	no
B	shiny silvery solid	0.97	yes
C	shiny silvery solid	22.65	yes
D	yellow powder	2.07	no
E	shiny gray solid	5.32	semiconductor
F	shiny bluish solid	8.91	yes

23. **ANALYZE** Based on the listed properties, identify each of the elements as a metal, nonmetal, or metalloid.

24. **APPLY** Which would weigh more: a cube of element A or a same-sized cube of element D?

25. **HYPOTHESIZE** Which element(s) do you think you might find in electronic devices? Why?

26. **HYPOTHESIZE** The thyroid gland, located in your throat, secretes hormones. In 1924 iodine was added to table salt. As more and more Americans used iodized salt, the number of cases of thyroid diseases decreased. Write a hypothesis that explains the observed decrease in thyroid-related diseases.

27. **INFER** How does the size of a beryllium (Be) atom compare with the size of an oxygen (O) atom?

28. **PREDICT** Although noble gases do not naturally react with other elements, xenon and krypton have been made to react with halogens such as chlorine in laboratories. Wh... most likely to react with the...

D 36 Unit: Chemical Interactions

Below is an element square from the periodic table. Use it to answer the next two questions.

80
Hg
Mercury
200.59

29. **CALCULATE** One of the more common isotopes of mercury is mercury-200. How many protons and neutrons are in the nucleus of mercury-200?

30. **INFER** Cadmium occupies the square directly above mercury on the periodic table. Is a cadium atom larger or smaller than a mercury atom?

31. **CALCULATE** An isotope has a half-life of 40 minutes. How much of a 100-gram sample would remain unchanged after two hours?

32. **APPLY** When a uranium atom with 92 protons and 146 neutrons undergoes radioactive decay, it produces a particle that consists of two protons and two neutrons from its nucleus. Into which element is the uranium atom transformed?

the **BIG** idea

33. **ANALYZE** Look again at the photograph on pages 6–7. Answer the question again, using what you have learned in the chapter.

34. **DRAW CONCLUSIONS** Suppose you've been given the ability to take apart and assemble atoms. How could you turn lead into gold?

35. **ANALYZE** Explain how the structure of an atom determines its place in the periodic table.

UNIT PROJECTS

If you are doing a unit project, make a folder for your project. Include in your folder a list of the resources you will need, the date on which the project is due, and a schedule to track your progress. Begin gathering data.

The Big Idea Questions

- Help students connect their new learning back to the Big Idea
- Prompt students to synthesize and apply the Big Idea and Key Concepts

---

KEY CONCEPT

1.1 Atoms are the smallest form of elements.

◀ **BEFORE, you learned**
- All matter is made of atoms
- Elements are the simplest substances

▶ **NOW, you will learn**
- Where atoms are found and how they are named
- About the structure of atoms
- How ions are formed from atoms

VOCABULARY
proton p. 11
neutron p. 11
nucleus p. 11
electron p. 11
atomic number p. 12
atomic mass number p. 12
isotope p. 12
ion p. 14

EXPLORE The Size of Atoms

How small can you cut paper?

PROCEDURE

1. Cut the strip of paper in half. Cut one of these halves in half.

2. Continue cutting one piece of paper in half as many times as you can.

WHAT DO YOU THINK?
- How many cuts were you able to make?
- Do you think you could keep cutting the paper forever? Why or why not?

MATERIALS
- strip of paper about 30 centimeters long
- scissors

All matter is made of atoms.

Think of all the substances you see and touch every day. Are all of these substances the same? Obviously, the substances that make up this book you're reading are quite different from the substances in the air around you. So how many different substances can there be? This is a question people have been asking for thousands of years.

About 2400 years ago, Greek philosophers proposed that everything on Earth was made of only four basic substances—air, water, fire, and earth. Everything else contained a mixture of these four substances. As time went on, chemists came to realize that there had to be more than four basic substances. Today chemists know that about 100 basic substances, or elements, account for everything we see and touch. Sometimes these elements appear by themselves. Most often, however, these elements appear in combination with other elements to make new substances. In this section, you'll learn about the atoms of the elements that make up the world and how these atoms differ from one another.

READING TiP
The word element is related to elementary, which means "basic."

Chapter 1: Atomic Structure and the Periodic Table 9 **D**

Many Ways to Learn

Because students learn in so many ways, *McDougal Littell Science* gives them a variety of experiences with important concepts and skills. Text, visuals, activities, and technology all focus on Big Ideas and Key Concepts.

Integrated Technology

- Interaction with Key Concepts through Simulations and Visualizations

- Easy access to relevant Web resources through Resource Centers and SciLinks

- Opportunities for review through Content Review and Math Tutorials

Hands-on Learning

- Activities that reinforce Key Concepts

- Skill Focus for important inquiry and process skills

- Multiple activities in every chapter, from quick Explores to full-period Chapter Investigations

The rates of chemical reactions can vary.

Most chemical reactions take place when particles of reactants collide with enough force to react. Chemical reactions can occur at different rates. Striking a match causes a very quick chemical reaction, while the rusting of an iron nail may take months. However, the rate of a reaction can be changed. For instance, a nail can be made to rust more quickly. Three physical factors—concentration, surface area, and temperature—and a chemical factor—a catalyst—can greatly affect the rate of a chemical reaction.

Concentration

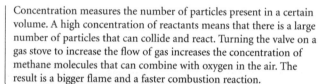

VISUALIZATION
CLASSZONE.COM

Observe how changing the concentration of a reactant can change the rate of a reaction.

Concentration measures the number of particles present in a certain volume. A high concentration of reactants means that there is a large number of particles that can collide and react. Turning the valve on a gas stove to increase the flow of gas increases the concentration of methane molecules that can combine with oxygen in the air. The result is a bigger flame and a faster combustion reaction.

Surface Area

Suppose one of the reactants in a chemical reaction is present as a single large piece of material. Particles of the second reactant cannot get inside the large piece, so they can react only with particles on the surface. To make the reaction go faster, the large piece of material could be broken into smaller pieces before the reaction starts.

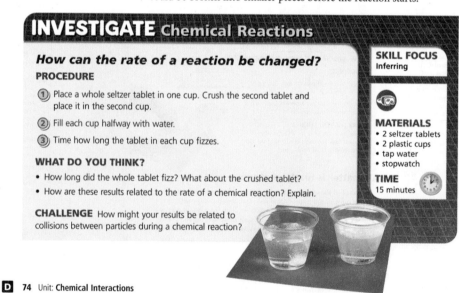

INVESTIGATE Chemical Reactions

How can the rate of a reaction be changed?

PROCEDURE

1. Place a whole seltzer tablet in one cup. Crush the second tablet and place it in the second cup.
2. Fill each cup halfway with water.
3. Time how long the tablet in each cup fizzes.

WHAT DO YOU THINK?

- How long did the whole tablet fizz? What about the crushed tablet?
- How are these results related to the rate of a chemical reaction? Explain.

CHALLENGE How might your results be related to collisions between particles during a chemical reaction?

SKILL FOCUS
Inferring

MATERIALS
- 2 seltzer tablets
- 2 plastic cups
- tap water
- stopwatch

TIME
15 minutes

Breaking a large piece of material into smaller parts increases the surface area of the material. All of the inner material has no surface when it is inside a larger piece. Each time the large piece is broken, however, more surfaces are exposed. The amount of material does not change, but breaking it into smaller parts increases its surface area. Increasing the surface area increases the rate of the reaction.

CHECK YOUR READING Why does a reaction proceed faster when the reactants have greater surface areas?

Temperature

The rate of a reaction can be increased by making the particles move faster. The result is that more collisions take place per second and occur with greater force. The most common way to make the particles move faster is to add energy to the reactants, which will raise their temperature.

Many chemical reactions during cooking go very slowly, or do not take place at all, unless energy is added to the reactants. Too much heat can make a reaction go too fast, and food ends up burned. Chemical reactions can also be slowed or stopped by decreasing the temperature of the reactants. Again, think about cooking. The reactions that take place during cooking can be stopped by removing the food from the heat source.

REMINDER
Temperature is the average amount of kinetic energy of the particles in a substance.

Particles and Reaction Rates		
Changes in Reactants	**Normal Reaction Rate**	**Increased Reaction Rate**
Concentration An increase in concentration of the reactants increases the number of particles that can interact.		
Surface area An increase in the surface area of the reactants increases the number of particles that can interact.		
Temperature Adding energy makes particles move faster and increases temperature. The increase in motion allows reactants to collide and react more frequently.		

Differentiated Instruction

A full spectrum of resources for differentiating instruction supports you in reaching the wide range of learners in your classroom.

1.1 INSTRUCT

Mathematics Connection

Explain to students that not all pie charts (or other charts that display gathered information) are completely secure in their data. Deviations might occur.

Address Misconceptions

IDENTIFY Ask: If you had a microscope powerful enough to see atoms in pure gold, what would the atoms look like? Ask students to sketch them. If students draw atoms of various shapes and sizes with no space between them, they hold the misconception that atoms of the same element are solid bits that vary in size and shape and are jammed together.

CORRECT Have students imagine a wall covered with evenly spaced black dots. Then have them imagine backing away from the wall: the dots would appear smaller and closer together as students get farther away. Eventually, the wall would look solid black. Say that the black dots are like the atoms in matter, which merge into the form of the matter—solid, liquid, or gas.

REASSESS Ask: If you could see the atoms in a grain of sand, what would they look like? *dots with space around them*

Technology Resources

Visit **ClassZone.com** for background on common student misconceptions.

 MISCONCEPTION DATABASE

Ongoing Assessment

Recognize where atoms of common elements are found and how they are named.

Ask: In terms of percentages, which has more oxygen, Earth's crust or a human body? *a human body*

CHECK YOUR READING Answer: hydrogen

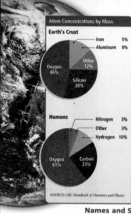

Types of Atoms in Earth's Crust and Living Things

Atoms of the element hydrogen account for about 90 percent of the total mass of the universe. Hydrogen atoms make up only about 1 percent of Earth's crust, however, and most of those hydrogen atoms are combined with oxygen atoms in the form of water. The graph on the left shows the types of atoms in approximately the top 100 kilometers of Earth's crust.

The distribution of the atoms of the elements in living things is very different from what it is in Earth's crust. Living things contain at least 25 types of atoms. Although the amounts of these atoms vary somewhat, all living things—animals, plants, and bacteria—are composed primarily of atoms of oxygen, carbon, hydrogen, and nitrogen. As you can see in the lower graph on the left, oxygen atoms account for more than half your body's mass.

 A

CHECK YOUR READING What is the most common element in the universe?

Names and Symbols of Elements

 B

Elements get their names in many different ways. Magnesium, for example, was named for the region in Greece known as Magnesia. Lithium comes from the Greek word *lithos*, which means "stone." Neptunium was named after the planet Neptune. The elements einsteinium and fermium were named after scientists Albert Einstein and Enrico Fermi.

Each element has its own unique symbol. For some elements, the symbol is simply the first letter of its name.

hydrogen (H)   sulfur (S)   carbon (C)

The symbols for other elements use the first letter plus one other letter of the element's name. Notice that the first letter is capitalized but the second letter is not.

aluminum (Al)   platinum (Pt)   cadmium (Cd)   zinc (Zn)

The origins of some symbols, however, are less obvious. The symbol for gold (Au), for example, doesn't seem to have anything to do with the element's name. The symbol refers instead to gold's name in Latin, *aurum*. Lead (Pb), iron (Fe), and copper (Cu) are a few other elements whose symbols come from Latin names.

Each element is made of a different atom.

In the early 1800s British scientist John Dalton proposed that each element is made of tiny particles called atoms. Dalton stated that all of the atoms of a particular element are identical but are different from atoms of all other elements. Every atom of silver, for example, is similar to every other atom of silver but different from an atom of iron.

Dalton's theory also assumed that atoms could not be divided into anything simpler. Scientists later discovered that this was not exactly true. They found that atoms are made of even smaller particles.

 RESOURCE CENTER
CLASSZONE.COM
Learn more about the atom.

The Structure of an Atom

A key discovery leading to the current model of the atom was that atoms contain charged particles. The charge on a particle can be either positive or negative. Particles with the same type of charge repel each other—they are pushed apart. Particles with different charges attract each other—they are drawn toward each other.

Atoms are composed of three types of particles—electrons, protons, and neutrons. A **proton** is a positively charged particle, and a **neutron** is an uncharged particle. The neutron has approximately the same mass as a proton. The protons and neutrons of an atom are grouped together in the atom's center. This combination of protons and neutrons is called the **nucleus** of the atom. Because it contains protons, the nucleus has a positive charge. **Electrons** are negatively charged particles that move around outside the nucleus.

C

D

VOCABULARY Remember to make a frame for *neutron*, *proton*, and *electron* and for other vocabulary terms.

The Atomic Model

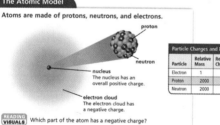

Atoms are made of protons, neutrons, and electrons.

proton

neutron

nucleus
The nucleus has an overall positive charge.

electron cloud
The electron cloud has a negative charge.

Particle Charges and Mass		
Particle	Relative Mass	Relative Charge
Electron	1	−1
Proton	2000	+1
Neutron	2000	0

READING VISUALS Which part of the atom has a negative charge?

DIFFERENTIATE INSTRUCTION

? More Reading Support

A What element makes up more than half of your body's mass? *oxygen*

B From where do elements get their names? *from people, places, Greek words*

English Learners Some English learners may not be familiar with how pie charts display information. Use a simpler example than the "Atoms Concentrations by Mass" charts. For example, create a pie chart displaying the numbers of different hair colors among students in the class.

Inclusion Help students with hearing impairments identify the names of elements presented on this page. Write each element's name on the board; then pronounce it distinctly for the students.

DIFFERENTIATE INSTRUCTION

? More Reading Support

C What particles are grouped in the atom's nucleus? *protons and neutrons*

D What particles move around outside the nucleus? *electrons*

Advanced Challenge students to find out about quarks—smaller, more fundamental particles of atoms. Since quarks' existence is still hypothetical, have students report on the evidence used to support their existence.

R Challenge and Extension, p. 19

English Learners Have English learners write each vocabulary term and its definition on separate index cards, or put the terms on the Science Word Wall so students have quick and easy reminders to refer to during the lesson.

Teacher's Edition

- More Reading Support for below-level readers
- Strategies for below-level and advanced learners, English learners, and inclusion students

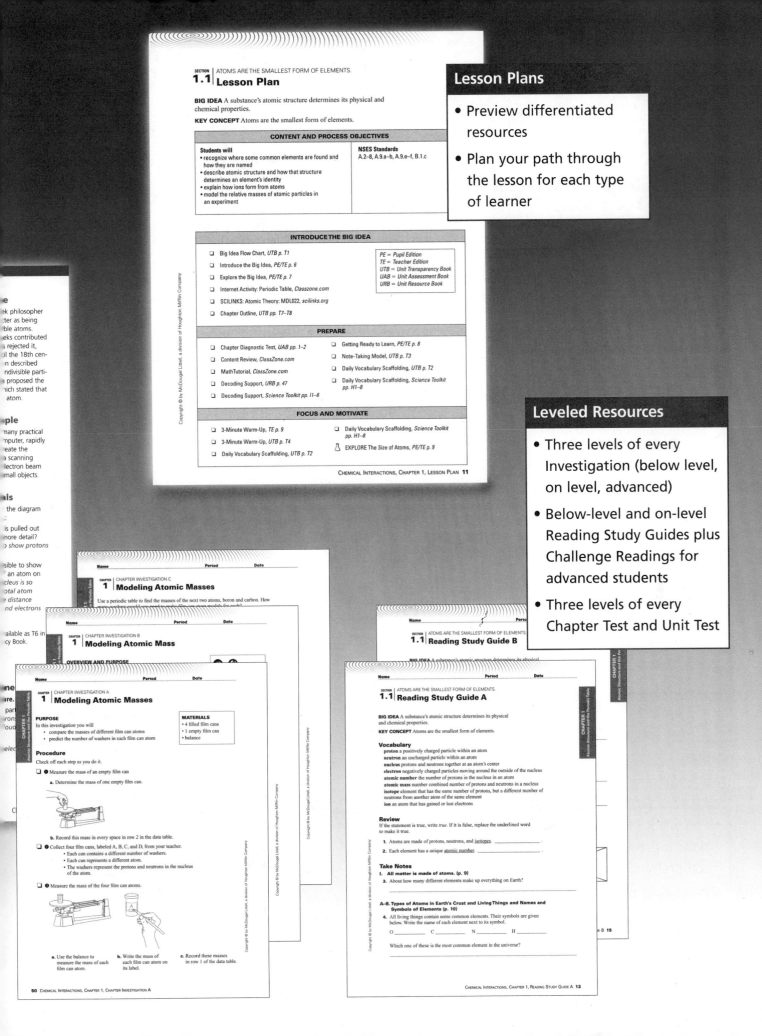

Lesson Plans

- Preview differentiated resources
- Plan your path through the lesson for each type of learner

Leveled Resources

- Three levels of every Investigation (below level, on level, advanced)
- Below-level and on-level Reading Study Guides plus Challenge Readings for advanced students
- Three levels of every Chapter Test and Unit Test

SECTION ATOMS ARE THE SMALLEST FORM OF ELEMENTS.

1.1 Lesson Plan

BIG IDEA A substance's atomic structure determines its physical and chemical properties.

KEY CONCEPT Atoms are the smallest form of elements.

CONTENT AND PROCESS OBJECTIVES

Students will	NSES Standards
• recognize where some common elements are found and how they are named • describe atomic structure and how that structure determines an element's identity • explain how ions form from atoms • model the relative masses of atomic particles in an experiment	A.2–8, A.9.a–b, A.9.e–f, B.1.c

INTRODUCE THE BIG IDEA

- ☐ Big Idea Flow Chart, *UTB p. T1*
- ☐ Introduce the Big Idea, *PE/TE p. 6*
- ☐ Explore the Big Idea, *PE/TE p. 7*
- ☐ Internet Activity: Periodic Table, *Classzone.com*
- ☐ SCILINKS: Atomic Theory: MDL022, *scilinks.org*
- ☐ Chapter Outline, *UTB pp. T7–T8*

| PE = Pupil Edition |
| TE = Teacher Edition |
| UTB = Unit Transparency Book |
| UAB = Unit Assessment Book |
| URB = Unit Resource Book |

PREPARE

- ☐ Chapter Diagnostic Test, *UAB pp. 1–2*
- ☐ Content Review, *ClassZone.com*
- ☐ MathTutorial, *ClassZone.com*
- ☐ Decoding Support, *URB p. 47*
- ☐ Decoding Support, *Science Toolkit pp. I1–6*
- ☐ Getting Ready to Learn, *PE/TE p. 8*
- ☐ Note-Taking Model, *UTB p. T3*
- ☐ Daily Vocabulary Scaffolding, *UTB p. T2*
- ☐ Daily Vocabulary Scaffolding, *Science Toolkit pp. H1–8*

FOCUS AND MOTIVATE

- ☐ 3-Minute Warm-Up, *TE p. 9*
- ☐ 3-Minute Warm-Up, *UTB p. T4*
- ☐ Daily Vocabulary Scaffolding, *UTB p. T2*
- ☐ Daily Vocabulary Scaffolding, *Science Toolkit pp. H1–8*
- ☐ EXPLORE The Size of Atoms, *PE/TE p. 9*

Copyright © by McDougal Littell, a division of Houghton Mifflin Company

---

Name _____ Period _____ Date _____

CHAPTER INVESTIGATION C

1 Modeling Atomic Masses

Use a periodic table to find the masses of the next two atoms, boron and carbon. How

---

Name _____ Period _____ Date _____

CHAPTER INVESTIGATION B

1 Modeling Atomic Mass

OVERVIEW AND PURPOSE

Name _____ Period _____ Date _____

CHAPTER INVESTIGATION A

1 Modeling Atomic Masses

PURPOSE
In this investigation you will
- compare the masses of different film can atoms
- predict the number of washers in each film can atom

MATERIALS
- 4 filled film cans
- 1 empty film can
- balance

Procedure
Check off each step as you do it.

☐ ➊ Measure the mass of an empty film can.

   a. Determine the mass of one empty film can.

   b. Record this mass in every space in row 2 in the data table.

☐ ➋ Collect four film cans, labeled A, B, C, and D, from your teacher.
- Each can contains a different number of washers.
- Each can represents a different atom.
- The washers represent the protons and neutrons in the nucleus of the atom.

☐ ➌ Measure the mass of the four film can atoms.

   a. Use the balance to measure the mass of each film can atom.

   b. Write the mass of each film can atom on its label.

   c. Record these masses in row 1 of the data table.

Copyright © by McDougal Littell, a division of Houghton Mifflin Company

---

Name _____ Period _____ Date _____

SECTION ATOMS ARE THE SMALLEST FORM OF ELEMENTS.

1.1 Reading Study Guide A

BIG IDEA A substance's atomic structure determines its physical and chemical properties.

KEY CONCEPT Atoms are the smallest form of elements.

Vocabulary
proton a positively charged particle within an atom
neutron an uncharged particle within an atom
nucleus protons and neutrons together at an atom's center
electron negatively charged particles moving around the outside of the nucleus
atomic number the number of protons in the nucleus in an atom
atomic mass number combined number of protons and neutrons in a nucleus
isotope element that has the same number of protons, but a different number of neutrons from another atom of the same element
ion an atom that has gained or lost electrons

Review
If the statement is true, write *true*. If it is false, replace the underlined word to make it true.

1. Atoms are made of protons, neutrons, and <u>isotopes</u>. _____ .

2. Each element has a unique <u>atomic number</u>. _____ .

Take Notes
I. **All matter is made of atoms. (p. 9)**

3. About how many different elements make up everything on Earth?

_____

A–B. Types of Atoms in Earth's Crust and Living Things and Names and Symbols of Elements (p. 10)

4. All living things contain some common elements. Their symbols are given below. Write the name of each element next to its symbol.

O _____ C _____ N _____ H _____

Which one of these is the most common element in the universe?

_____

Copyright © by McDougal Littell, a division of Houghton Mifflin Company

CHAPTER 1 Atomic Structure and the Periodic Table

SECTION ATOMS ARE THE SMALLEST FORM OF ELEMENTS.

1.1 Reading Study Guide B

BIG IDEA A substance's atomic structure determines its physical

---

Effective Assessment

McDougal Littell Science incorporates a comprehensive set of resources for assessing student knowledge and performance before, during, and after instruction.

Diagnostic Tests

- Assessment of students' prior knowledge
- Readiness check for concepts and skills in the upcoming chapter

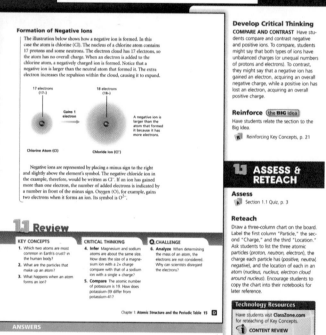

Formation of Negative Ions

The illustration below shows how a negative ion is formed. In this case the atom is chlorine (Cl). The nucleus of a chlorine atom contains 17 protons and some neutrons. The electron cloud has 17 electrons, so the atom has no overall charge. When an electron is added to the chlorine atom, a negatively charged ion is formed. Notice that a negative ion is larger than the neutral atom that formed it. The extra electron increases the repulsion within the cloud, causing it to expand.

17 electrons (17–)   Gains 1 electron   18 electrons (18–)

A negative ion is larger than the atom that formed it because it has more electrons.

Chlorine Atom (Cl)   Chloride Ion (Cl⁻)

Negative ions are represented by placing a minus sign to the right and slightly above the element's symbol. The negative chloride ion in the example, therefore, would be written as Cl⁻. If an ion has gained more than one electron, the number of added electrons is indicated by a number in front of the minus sign. Oxygen (O), for example, gains two electrons when it forms an ion. Its symbol is O²⁻.

1.1 Review

KEY CONCEPTS
1. Which two atoms are most common in Earth's crust? in the human body?
2. What are the particles that make up an atom?
3. What happens when an atom forms an ion?

CRITICAL THINKING
4. **Infer** Magnesium and sodium atoms are about the same size. How does the size of a magnesium ion with a 2+ charge compare with that of a sodium ion with a single + charge?
5. **Compare** The atomic number of potassium is 19. How does potassium-39 differ from potassium-41?

CHALLENGE
6. **Analyze** When determining the mass of an atom, the electrons are not considered. Why can scientists disregard the electrons?

Chapter 1: Atomic Structure and the Periodic Table 15 **D**

ANSWERS

1. oxygen and silicon; oxygen and carbon
2. protons, neutrons, and electrons
3. It gains or loses an electron.
4. The magnesium ion is smaller.
5. Potassium-41 has two more neutrons.
6. Scientists can disregard electrons because they have such a small mass.

Develop Critical Thinking
COMPARE AND CONTRAST Have students compare and contrast negative and positive ions. To compare, students might say that both types of ions have unbalanced charges (or unequal numbers of protons and electrons). To contrast, they might say that a negative ion has gained an electron, acquiring an overall negative charge, while a positive ion has lost an electron, acquiring an overall positive charge.

Reinforce (the **BIG** idea)
Have students relate the section to the Big Idea.
Reinforcing Key Concepts, p. 21

1.1 ASSESS & RETEACH

Assess
Section 1.1 Quiz, p. 3

Reteach
Draw a three-column chart on the board. Label the first column "Particle," the second "Charge," and the third "Location." Ask students to list the three atomic particles (*proton*, *neutron*, *electron*), the charge each particle has (*positive*, *neutral*, *negative*), and the location of each in an atom (*nucleus*, *nucleus*, *electron cloud around nucleus*). Encourage students to copy the chart into their notebooks for later reference.

Technology Resources
Have students visit ClassZone.com for reteaching of Key Concepts.
⟳ CONTENT REVIEW
⟳ CONTENT REVIEW CD-ROM

Chapter 1 15 **D**

Ongoing Assessment

- Check Your Reading questions for student self-check of comprehension
- Consistent Teacher Edition prompts for assessing understanding of Key Concepts

---

Reviewing Vocabulary

Describe how the vocabulary terms in the following pairs are related to each other. Explain the relationship in a one- or two-sentence answer. Underline each vocabulary term in your answer.

1. isotope, nucleus
2. atomic mass, atomic number
3. electron, proton
4. atomic number, atomic mass number
5. group, period
6. metals, nonmetals
7. radioactivity, half-life

Reviewing Key Concepts

Multiple Choice *Choose the letter of the best answer.*

8. The central part of an atom is called the
   a. electron   c. proton
   b. nucleus   d. neutron

9. The electric charge on a proton is
   a. positive   c. neutral
   b. negative   d. changing

10. The number of protons in the nucleus is the
    a. atomic mass   c. atomic number
    b. isotope   d. half-life

11. Nitrogen has atomic number 7. An isotope of nitrogen containing seven neutrons would be
    a. nitrogen-13   c. nitrogen-15
    b. nitrogen-14   d. nitrogen-16

12. How does the size of a negative ion compare to the size of the atom that formed it?
    a. It's smaller.
    b. It's larger.
    c. It's the same size.
    d. It varies.

13. The modern periodic table is organized by
    a. size of atom
    b. atomic mass
    c. number of neutrons
    d. atomic number

14. Elements in a group have
    a. a wide range of chemical properties
    b. the same atomic radius
    c. similar chemical properties
    d. the same number of protons

15. Elements in a period have
    a. a wide range of chemical properties
    b. the same atomic radius
    c. similar chemical properties
    d. the same number of protons

16. From left to right in a period, the size of atoms
    a. increases   c. remains the same
    b. decreases   d. shows no pattern

17. The elements in Group 1 of the periodic table are commonly called the
    a. alkali metals   c. alkaline earth metals
    b. transition metals   d. rare earth metals

18. The isotope nitrogen-13 has a half-life of 10 minutes. If you start with 40 grams of this isotope, how many grams will you have left after 20 minutes?
    a. 10   c. 20
    b. 15   d. 30

Short Answer *Write a short answer to each question. You may need to consult a periodic table.*

19. Rubidium forms the positive ion Rb⁺. Is this ion larger or smaller than the neutral atom? Explain.

20. How can you find the number of neutrons in the isotope nitrogen-16?

21. Explain how density varies across and up and down the periodic table.

22. Place these elements in order from least reactive to most reactive: nickel (Ni), xenon (Xe), lithium (Li). How did you determine the order?

Section and Chapter Reviews

- Focus on Key Concepts and critical thinking skills
- A full range of question types and levels of thinking

Leveled Chapter and Unit Tests

- Three levels of test for every chapter and unit
- Same Big Ideas, Key Concepts, and essential skills assessed on all levels

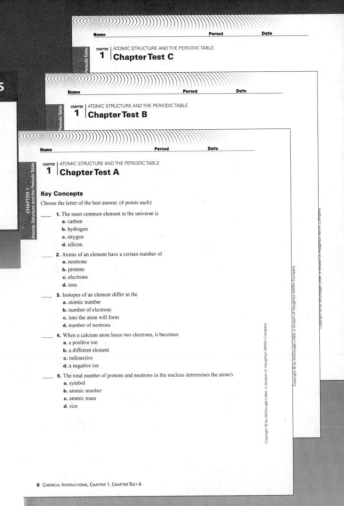

Name   Period   Date

CHAPTER **1** | ATOMIC STRUCTURE AND THE PERIODIC TABLE
Chapter Test C

Name   Period   Date

CHAPTER **1** | ATOMIC STRUCTURE AND THE PERIODIC TABLE
Chapter Test B

Name   Period   Date

CHAPTER **1** | ATOMIC STRUCTURE AND THE PERIODIC TABLE
Chapter Test A

Key Concepts

Choose the letter of the best answer. (4 points each)

_____ 1. The most common element in the universe is
     a. carbon
     b. hydrogen
     c. oxygen
     d. silicon

_____ 2. Atoms of an element have a certain number of
     a. neutrons
     b. protons
     c. electrons
     d. ions

_____ 3. Isotopes of an element differ in the
     a. atomic number
     b. number of electrons
     c. ions the atom will form
     d. number of neutrons

_____ 4. When a calcium atom loses two electrons, it becomes
     a. a positive ion
     b. a different element
     c. radioactive
     d. a negative ion

_____ 5. The total number of protons and neutrons in the nucleus determines the atom's
     a. symbol
     b. atomic number
     c. atomic mass
     d. size

6 CHEMICAL INTERACTIONS, CHAPTER 1, CHAPTER TEST A

Thinking Critically

The table below lists some properties of six elements. Use the information and your knowledge of the properties of elements to answer the next three questions.

Element	Appearance	Density (g/cm³)	Conducts Electricity
A	dark purple crystals	4.93	no
B	shiny silvery solid	0.97	yes
C	shiny silvery solid	22.65	yes
D	yellow powder	2.07	no
E	shiny gray solid	5.32	semiconductor
F	shiny bluish solid	8.91	yes

23. ANALYZE Based on the listed properties, identify each of the elements as a metal, nonmetal, or metalloid.

24. APPLY Which would weigh more: a cube of element A or a same-sized cube of element D?

25. HYPOTHESIZE Which element(s) do you think you might find in electronic devices? Why?

26. HYPOTHESIZE The thyroid gland, located in your throat, secretes hormones. In 1924 iodine was added to table salt. As more and more Americans used iodized salt, the number of cases of thyroid diseases decreased. Write a hypothesis that explains the observed decrease in thyroid-related diseases.

27. INFER How does the size of a beryllium (Be) atom compare with the size of an oxygen (O) atom?

28. PREDICT Although noble gases do not naturally react with other elements, xenon and krypton have been made to react with halogens such as chlorine in laboratories. Why are the halogens most likely to react with the noble gases?

Below is an element square from the periodic table. Use it to answer the next two questions.

80
Hg
Mercury
200.59

29. CALCULATE One of the more common isotopes of mercury is mercury-200. How many protons and neutrons are in the nucleus of mercury-200?

30. INFER Cadmium occupies the square directly above mercury on the periodic table. Is a cadium atom larger or smaller than a mercury atom?

31. CALCULATE An isotope has a half-life of 40 minutes. How much of a 100-gram sample would remain unchanged after two hours?

32. APPLY When a uranium atom with 92 protons and 146 neutrons undergoes radioactive decay, it produces a particle that consists of two protons and two neutrons from its nucleus. Into which element is the uranium atom transformed?

the BIG idea

33. ANALYZE Look again at the photograph on pages 6–7. Answer the question again, using what you have learned in the chapter.

34. DRAW CONCLUSIONS Suppose you've been given the ability to take apart and assemble atoms. How could you turn lead into gold?

35. ANALYZE Explain how the structure of an atom determines its place in the periodic table.

Rubrics

- Rubrics in Teacher Edition for all extended response questions
- Rubrics for all Unit Projects
- Alternative Assessment with rubric for each chapter
- A wide range of additional rubrics in the Science Toolkit

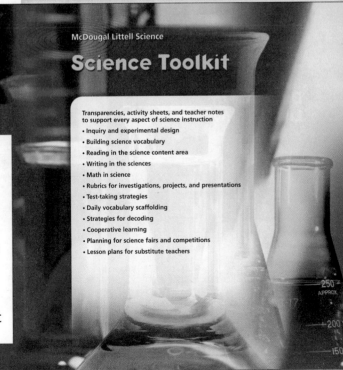

McDougal Littell Science

Science Toolkit

Transparencies, activity sheets, and teacher notes to support every aspect of science instruction

- Inquiry and experimental design
- Building science vocabulary
- Reading in the science content area
- Writing in the sciences
- Math in science
- Rubrics for investigations, projects, and presentations
- Test-taking strategies
- Daily vocabulary scaffolding
- Strategies for decoding
- Cooperative learning
- Planning for science fairs and competitions
- Lesson plans for substitute teachers

McDougal Littell Science Modular Series

McDougal Littell Science lets you choose the titles that match your curriculum. Each module in this flexible 15-book series takes an in-depth look at a specific area of life, earth, or physical science.

- Flexibility to match your curriculum
- Convenience of smaller books
- Complete Student Resource Handbooks in every module

Life Science Titles

A ▶ Cells and Heredity
1. The Cell
2. How Cells Function
3. Cell Division
4. Patterns of Heredity
5. DNA and Modern Genetics

B ▶ Life Over Time
1. The History of Life on Earth
2. Classification of Living Things
3. Population Dynamics

C ▶ Diversity of Living Things
1. Single-Celled Organisms and Viruses
2. Introduction to Multicellular Organisms
3. Plants
4. Invertebrate Animals
5. Vertebrate Animals

D ▶ Ecology
1. Ecosystems and Biomes
2. Interactions Within Ecosystems
3. Human Impact on Ecosystems

E ▶ Human Biology
1. Systems, Support, and Movement
2. Absorption, Digestion, and Exchange
3. Transport and Protection
4. Control and Reproduction
5. Growth, Development, and Health

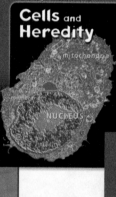

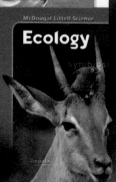

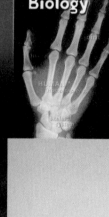

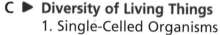

Earth Science Titles

A ▶ Earth's Surface
1. Views of Earth Today
2. Minerals
3. Rocks
4. Weathering and Soil Formation
5. Erosion and Deposition

B ▶ The Changing Earth
1. Plate Tectonics
2. Earthquakes
3. Mountains and Volcanoes
4. Views of Earth's Past
5. Natural Resources

C ▶ Earth's Waters
1. The Water Planet
2. Freshwater Resources
3. Ocean Systems
4. Ocean Environments

D ▶ Earth's Atmosphere
1. Earth's Changing Atmosphere
2. Weather Patterns
3. Weather Fronts and Storms
4. Climate and Climate Change

E ▶ Space Science
1. Exploring Space
2. Earth, Moon, and Sun
3. Our Solar System
4. Stars, Galaxies, and the Universe

Physical Science Titles

A ▶ Matter and Energy
1. Introduction to Matter
2. Properties of Matter
3. Energy
4. Temperature and Heat

B ▶ Chemical Interactions
1. Atomic Structure and the Periodic Table
2. Chemical Bonds and Compounds
3. Chemical Reactions
4. Solutions
5. Carbon in Life and Materials

C ▶ Motion and Forces
1. Motion
2. Forces
3. Gravity, Friction, and Pressure
4. Work and Energy
5. Machines

D ▶ Waves, Sound, and Light
1. Waves
2. Sound
3. Electromagnetic Waves
4. Light and Optics

E ▶ Electricity and Magnetism
1. Electricity
2. Circuits and Electronics
3. Magnetism

Teaching Resources

A wealth of print and technology resources help you adapt the program to your teaching style and to the specific needs of your students.

Book-Specific Print Resources

Unit Resource Book provides all of the teaching resources for the unit organized by chapter and section.
- Family Letters
- *Scientific American Frontiers* Video Guide
- Unit Projects
- Lesson Plans
- Reading Study Guides (Levels A and B)
- Spanish Reading Study Guides
- Challenge Readings
- Challenge and Extension Activities
- Reinforcing Key Concepts
- Vocabulary Practice
- Math Support and Practice
- Investigation Datasheets
- Chapter Investigations (Levels A, B, and C)
- Additional Investigations (Levels A, B, and C)
- Summarizing the Chapter

Unit Assessment Book contains complete resources for assessing student knowledge and performance.
- Chapter Diagnostic Tests
- Section Quizzes
- Chapter Tests (Levels A, B, and C)
- Alternative Assessments
- Unit Tests (Levels A, B, and C)

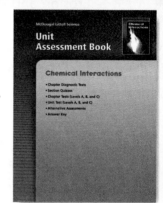

Unit Transparency Book includes instructional visuals for each chapter.
- Three-Minute Warm-Ups
- Note-Taking Models
- Daily Vocabulary Scaffolding
- Chapter Outlines
- Big Idea Flow Charts
- Chapter Teaching Visuals

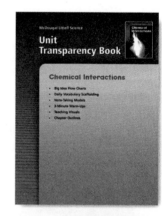

Unit Lab Manual

Unit Note-Taking/Reading Study Guide

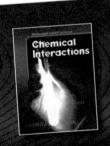

McDougal Littell Science

Unit Resource Book

Chemical Interactions

- Family Letters (English and Spanish)
- *Scientific American Frontiers* Video Guides
- Unit Projects (with Rubrics)
- Lesson Plans
- Reading Study Guides (Levels A and B and Spanish)
- Challenge Activities and Readings
- Reinforcing Key Concepts
- Vocabulary Practice and Decoding Support
- Math Support and Practice
- Investigation Datasheets
- Chapter Investigations (Levels A, B, and C)
- Additional Investigations (Levels A, B, and C)

Program-Wide Print Resources

Process and Lab Skills

Problem Solving and Critical Thinking

Standardized Test Practice

Science Toolkit

City Science

Visual Glossary

Multi-Language Glossary

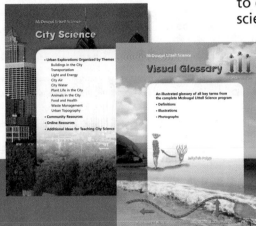

English Learners Package

Scientific American Frontiers Video Guide

How Stuff Works Express This quarterly magazine offers opportunities to explore current science topics.

Technology Resources

Scientific American Frontiers **Video Program**
Each specially-tailored segment from this award-winning PBS series correlates to a unit; available on VHS and DVD

Audio CDs Complete chapter texts read in both English and Spanish

Lab Generator CD-ROM
A searchable database of all activities from the program plus additional labs for each unit; edit and print your own version of labs

Test Generator CD-ROM

eEdition CD-ROM

EasyPlanner CD-ROM

Content Review CD-ROM

Power Presentations CD-ROM

Online Resources

   ClassZone.com

   Content Review Online

   eEdition Plus Online

   EasyPlanner Plus Online

   eTest Plus Online

Correlation to National Science Education Standards

This chart provides an overview of how the five Physical Science modules of *McDougal Littell Science* address the National Science Education Standards.

A Matter and Energy
B Chemical Interactions
C Motion and Forces
D Waves, Sound, and Light
E Electricity and Magnetism

A. Science as Inquiry

	Book, Chapter, and Section
A.1– A.8 Abilities necessary to do scientific inquiry Identify questions for investigation; design and conduct investigations; use evidence; think critically and logically; analyze alternative explanations; communicate; use mathematics.	All books (pp. R2–R44), All Chapter Investigations, All Think Science features
A.9 Understandings about scientific inquiry Different kinds of investigations for different questions; investigations guided by current scientific knowledge; importance of mathematics and technology for data gathering and analysis; importance of evidence, logical argument, principles, models, and theories; role of legitimate skepticism; scientific investigations lead to new investigations.	All books (pp. xxii–xxv) A3.1, B2.2, C4.2, D3.2, E3.1

B. Physical Science

	Book, Chapter, and Section
B.1 Properties and changes of properties in matter Physical properties; substances, elements, and compounds; chemical reactions.	A1.1, A1.2, A1.3, A1.4, A2.1, A2.2, B1, B3.2, B4.1, B4.2, B4.3, B5.1, B5.3, C3.4
B.2 Motions and forces Position, speed, direction of motion; balanced and unbalanced forces.	C1.1, C1.2, C1.3, C2.1, C2.3, C3.1, C3.2, C3.3, C3.4, C4.1, E5.1
B.3 Transfer of energy Energy transfer; forms of energy; heat and light; electrical circuits; sun as source of Earth's energy.	A3.1, A3.2, A3.3, A4.1, A4.2, A4.3, B3.3, C4.2, D1.1, D3.3, D3.4, D4.1, D4.2, D4.3, E1.1, E1.2, E1.3, E2.1, E2.2

C. Life Science

	Book, Chapter, and Section
C.1 Structure and function in living systems Systems; structure and function; levels of organization; cells and cell activities; specialization; human body systems; disease.	B1.1 (Connecting Sciences), B5.2, C5.2 (Connecting Sciences)

D. Earth and Space Science

	Book, Chapter, and Section
D.1 Earth's changing atmosphere	B4.2 (Connecting Sciences)
D.3 Earth in the solar system Sun, planets, asteroids, comets; regular and predictable motion and day, year, phases of the moon, and eclipses; gravity and orbits; sun as source of energy for earth; cause of seasons.	A3.2, A3.3, C3.1, D3.3

E. Science and Technology

	Book, Chapter, and Section
E.1– E.5 Abilities of technological design Identify problems; design a solution or product; implement a proposed design; evaluate completed designs or products; communicate the process of technological design.	A2.3, A3.1, A4.3, B (p. 5), B4.2, C1.2, C2.1, C3.2, C5.3, D2.4, D3.1, D3.3, D4.4, E2.3
E.6 Understandings about science and technology Similarities and differences between scientific inquiry and technological design; contributions of people in different cultures; reciprocal nature of science and technology; nonexistence of perfectly designed solutions; constraints, benefits, and unintended consequences of technological designs.	All books (pp. xxvi–xxvii) All books (Frontiers in Science, Timelines in Science) A.1.2, A3.3, B3.4, B3.1, B3.3, C5.3, D4.4, E1.2, E1.3

F.	Science in Personal and Social Perspectives	Book, Chapter, and Section
F.1	**Personal health** Exercise; fitness; hazards and safety; tobacco, alcohol, and other drugs; nutrition; STDs; environmental health	B4.2, B5.2
F.2	**Populations, resources, and environments** Overpopulation and resource depletion; environmental degradation.	A3.1
F.3	**Natural hazards** Earthquakes, landslides, wildfires, volcanic eruptions, floods, storms; hazards from human activity; personal and societal challenges.	B3.2, D3.2, E1.2
F.4	**Risks and benefits** Risk analysis; natural, chemical, biological, social, and personal hazards; decisions based on risks and benefits.	B4.3
F.5	**Science and technology in society** Science's influence on knowledge and world view; societal challenges and scientific research; technological influences on society; contributions from people of different cultures and times; work of scientists and engineers; ethical codes; limitations of science and technology.	All books (Timelines in Science) A1.2, A3.2, A3.3, B3.4, B4.4, C5.3, D2, D3.2, D3.3, D4.4, E1.2, E1.3

G.	History and Nature of Science	Book, Chapter, and Section
G.1	**Science as a human endeavor** Diversity of people w.orking in science, technology, and related fields; abilities required by science	All books (pp. xxii–xxv; Frontiers in Science)
G.2	**Nature of science** Observations, experiments, and models; tentative nature of scientific ideas; differences in interpretation of evidence; evaluation of results of investigations, experiments, observations, theoretical models, and explanations; importance of questioning, response to criticism, and communication.	B1.2, B2.1, B2.3, E3.2
G.3	**History of science** Historical examples of inquiry and relationships between science and society; scientists and engineers as valued contributors to culture; challenges of breaking through accepted ideas.	All books (Frontiers in Science; Timelines in Science) B1.2, B3.2, C2.1, D2.4

Correlations to Benchmarks

This chart provides an overview of how the five Physical Science modules of *McDougal Littell Science* address the National Science Education Standards.

A Matter and Energy
B Chemical Interactions
C Motion and Forces
D Waves, Sound, and Light
E Electricity and Magnetism

1. The Nature of Science	Book, Chapter, and Section
	The Nature of Science (pp. xxii–xxv); E2.3; Think Science Features: A3.1, B2.2, C2.1, C4.2, D3.2, E3.1; Scientific Thinking Handbook (pp. R2–R9); Lab Handbook (pp. R10–R35)

3. The Nature of Technology	Book, Chapter, and Section
	The Nature of Technology (pp. xxvi–xxvii); A3.3, B4.4, D4.4, E1, E2.3, E3.2, E3.3, E3.4; Timelines in Science Features

4. The Physical Setting	Book, Chapter, and Section
4.B THE EARTH	A3.1, A4.3, C3.1
4.D STRUCTURE OF MATTER	
4.D.1 All matter is made of atoms; atoms of any element are alike but different from atoms of other elements; different arrangements of atoms into groups compose all substances.	A1.2, A1.3, B1.1, B2.1, B2.2
4.D.2 Equal volumes of different substances usually have different weights.	A2.1, A2.3
4.D.3 Atoms and molecules are perpetually in motion; increased temperature means greater average energy of motion; states of matter: solids, liquids, gases.	A1.2, A4.1
4.D.4 Temperature and acidity of a solution influence reaction rates. Many substances dissolve in water, which may facilitate reactions between them.	B3.1, B4.2, B4.3
4.D.5 Greek philospheres' scientific ideas about elements; most elements tend to combine with others, so few elements are found in their pure form.	B1.1
4.D.6 Groups of elements have similar properties; oxidation; some elements, like carbon and hydrogen, don't fit into any category and are essential elements of living matter.	B1.2, B1.3, B3.1, B5.1, B5.2
4.D.7 Conservation of matter: the total weight of a closed system remains the same because the total number of atoms stays the same regardless of how they interact with one another.	B3.2
4.E ENERGY TRANSFORMATIONS	
4.E.1 Energy cannot be created or destroyed, but only changed from one form into another.	A3.2, C4.2
4.E.2 Most of what goes on in the universe involves energy transformations.	A3, A4.2, A4.3
4.E.3 Heat can be transferred through materials by the collisions of atoms or across space by radiation; convection currents transfer heat in fluid materials.	A4.2, A4.3
4.E.4 Energy appears in many different forms, including heat energy, chemical energy, mechanical energy, and gravitational energy.	A3.1, A4.2, A4.3, C4.2
4.F MOTION	
4.F.1 Light from the Sun is made up of many different colors of light; objects that give off or reflect light have a different mix of colors.	D3.3, D3.4
4.F.2 Something can be "seen" when light waves emitted or reflected by it enter the eye.	D4.1, D4.3

4.F.3 An unbalanced force acting on an object changes its speed or direction of motion, or both. If the force acts toward a single center, the object's path may curve into an orbit around the center.	C2.1, C2.2, C3.1
4.F.4 Vibrations in materials set up wavelike disturbances (such as sound) that spread away from the source; waves move at different speeds in different materials.	D1, D2.1, D2.2, D3.1, D3.4
4.F.5 Human eyes respond to only a narrow range of wavelengths of electromagnetic radiation—visible light. Differences of wavelengths within that range are perceived as differences in color.	D3.2, D3.4, D4.3
4.G FORCES OF NATURE	
4.G.1 Objects exerts gravitational forces on one another, but these forces depend on the mass and distance of objects, and may be too small to detect.	C3.1
4.G.2 The Sun's gravitational pull holds Earth and other planets in their orbits; planets' gravitational pull keeps their moons in orbit around them.	C3.1
4.G.3 Electric currents and magnets can exert a force on each other.	E3.1, E3.2, E3.3

5. The Living Environment	**Book, Chapter, and Section**
5.E Flow of Matter and Energy	B5.2
8. The Designed World	A2.1, A2.3, A3.3, B3.4, B4.4, B5.3, C5.3, E2.3, E3.2
9. The Mathematical World	All Math in Science Features, E2.3
10. Historical Perspectives	B1, B2, B3.2, C1.1, C2, D4.4

12. Habits of Mind	**Book, Chapter, and Section**
12.A VALUES AND ATTITUDES	Think Science Features: A3.1, B2.2, C2.1, C4.2, D3.2, E3.1
12.B Computation and Estimation	All Math in Science Features, Lab Handbook (pp. R10–R35)
12.C Manipulation and Observation	All Investigates and Chapter Investigations
12.D Communication Skills	All Chapter Investigations, Lab Handbook (pp. R10–R35)
12.E Critical-Response Skills	Think Science Features: A3.1, B2.2, C2.1, C4.2, D3.2, E3.1; Scientific Thinking Handbook (pp. R2–R9)

Planning the Unit

The Pacing Guide provides suggested pacing for all chapters in the unit as well as the two unit features shown below.

Frontiers in Science

- Features cutting-edge research as an engaging point of entry into the unit
- Connects to an accompanying *Scientific American Frontiers* video and viewing guide
- Introduces three options for unit projects.

FRONTIERS in Science

VIDEO SUMMARY

"Endangered Wonder Drug," a segment of the *Scientific American Frontiers* series that aired on PBS stations, traces the development of the cancer-fighting drug Taxol. The drug comes from the bark of the increasingly rare Pacific yew tree. Scientists are trying to find alternative sources in nature and in the laboratory. Taxol has been clinically tested as a treatment for women with ovarian cancer. If the drug succeeds as a cancer treatment and the demand increases, the natural supply of Taxol will soon be used up. One alternative is to grow yew trees in nurseries; another is to extract Taxol from the needles rather than the bark. Chemists are also working to make synthetic Taxol in large quantities. The ultimate solution may be a hybrid of natural and synthetic chemical compounds.

National Science Education Standards

A.9.a–d Understandings About Scientific Inquiry
E.6.a–f Understandings About Science and Technology
F.5.a–e Science and Technology in Society
G.1.a–b Science as a Human Endeavor
G.2.a Nature of Science

ADDITIONAL RESOURCES

Technology Resources

- **Scientific American Frontiers Video:** *Endangered Wonder Drug:* 11-minute video segment that introduces the unit.
- **ClassZone.com** CAREER LINK, Chemist

D 2 Unit: **Chemical Interactions**

Guide student viewing and comprehension of the video:
Video Teaching Guide, pp. 1–2; Video Viewing Guide, p. 3; Video Wrap-Up, p. 4

Scientific American Frontiers Video Guide, pp. 43–46

Unit projects procedures and rubrics:
Unit Projects, pp. 5–10

FRONTIERS in Science

Medicines from Nature

Where have people found medicines?

View the "Endangered Wonder Drug" segment of your Scientific American Frontiers video to see how chemicals found in nature can improve the health of people.

In Brazil, extracts from plants are used to treat Parkinson's Disease to arthritis.

Finding Natural Remedies

In the 1960s, people were searching desperately for cancer-fighting agents. Scientists tested over 35,000 which came from the bark of the Pacific yew. These have strong effects on the body. The tests in the bark stopped the growth of cancer eventually derived the drug Taxol from the yew tree.

Natural medicines are much more than the sapi karta leaves that the Kuna people in increase creativity. Many powerful medicines are based on compounds found in nature. But even though these natural compounds may be very effective at treating diseases, they can be limited in supply and can have harmful side effects. Organic chemists must find ways to make these compounds safer and produce them in greater amounts.

Modeling the Molecule

To make a compound, a chemist must know what its molecule looks like, atom by atom. Many useful drugs have structures that contain many atoms arranged in complicated ways. The chemist must know exactly how many atoms of each kind are in the molecule and how they are arranged. One atom in the wrong place might mean that the drug won't work the way it should.

Out of 110,000 tested compounds, very few have led to cancer-fighting drugs. Ask: What do you think are important personal qualities for a chemist who is looking for new medicines? *Sample answer: optimism and determination*

Technology Design

Point out that some natural remedies are produced in tiny quantities by plants or animals. Ask: What is the first step to finding out whether enough of a useful chemical can be synthesized? *analyze its composition and structure*

DIFFERENTIATE INSTRUCTION

More Reading Support
A What disease can be treated with extracts from yew bark? *cancer*
B How do chemists solve the problem of the small quantities of natural medicines? *manufacture more*

Advanced Have students research natural remedies that have been used for so long that most people forget they were once extracted from plants, such as the compound in aspirin, a drug that originally came from willow trees.

Frontiers in Science 3 **D**

TIMELINES in Science

FOCUS

Set Learning Goals
Students will
- Examine how the concept of atomic structure has changed over the years.
- Learn about the tools used to study atoms and subatomic particles.
- Model the discovery of the atomic nucleus.

National Science Education Standards

A.9.a–g Understandings About Scientific Inquiry
E.6.a–c Understandings About Science and Technology
F.5.a–e, F.5.g Science and Technology in Society
G.1.a–b Science as a Human Endeavor
G.2.a Nature of Science
G.3.a–c History of Science

INSTRUCT

Point out that the top half of the timeline

Timelines in Science

- Traces the history of key scientific discoveries
- Highlights interactions between science and technology.

TIMELINES in Science

THE STORY OF ATOMIC STRUCTURE

About 2500 years ago, certain Greek thinkers proposed that all matter consisted of extremely tiny particles called atoms. The sizes and shapes of different atoms, they reasoned, was what determined the properties of a substance. This early atomic theory, however, was not widely accepted. Many at the time found these tiny, invisible particles difficult to accept.

What everyone could observe was that all substances were liquid, solid, or gas; light or heavy; hot or cold. Everything, they thought, must then be made of only a few basic substances, or elements. They reasoned these elements must be water, air, fire, and earth. Different substances contained different amounts of each of these four substances.

The timeline shows a few of the major events that led scientists to accept the idea that matter is made of atoms and agree on the basic structure of atoms. With the revised atomic theory, scientists were able to explain how elements could be basic but different.

EVENTS

1600   1620   1640   1660

1661
Boyle Challenges Concept of the Four Elements
British chemist Robert Boyle proposes that more than four simple substances exist. Boyle also concludes that all matter is made of very tiny particles he calls corpuscles.

1808
John Dalton Says: "Bring Back the Atom!"
English chemist John Dalton revives the ancient Greek idea that all matter is made of atoms. Dalton claims that each element has its own type of atom and that the atoms combine in fixed and predictable ratios with one another in different substances.

1808
Humphry Davy Shocks Chemistry
English chemist Humphry Davy applies an electric current to different materials. He discovers that many materials once thought to be elements break apart into even simpler materials. Davy succeeds in isolating the elements sodium, calcium, strontium, and barium.

1897
It's Smaller Than the Atom!
English physicist Joseph John Thomson discovers the electron—the first sub-atomic particle to be identified. Thomson concludes that these tiny particles have a negative charge. Thomson will later propose that atoms are made of a great many of these negative particles floating in a sea of positive charge. Thomson suggests that each atom resembles a dish of pudding with raisins in it. The electrons are the raisins and the pudding the positive charge in which they float.

1800   1820   1840   1860   1880

APPLICATIONS AND TECHNOLOGY

TECHNOLOGY
Collecting and Studying Gases
Throughout the 1600s, scientists tried to study gases but had difficulty collecting them. English biologist Stephen Hales designed an apparatus to collect gases. The "pneumatic trough" was a breakthrough in chemistry because it allowed scientists to collect and study gases for the first time. The pneumatic trough was later used by such chemists as Joseph Black, Henry Cavendish, and Joseph Priestley to study the gases that make up the air we breathe. The work of these scientists showed that air was made of more than a single gas.

D 104 Chemical Interactions

TECHNOLOGY
Chemistry and Electric Charge
In 1800 Italian physicist Alessandro Volta announced that he had produced an electric current from a pile, or battery, of alternating zinc and silver discs. Volta's invention was important for the study of atoms and elements in two ways. First, the fact that the contact of two different metals could produce an electric current (electric charge must be part of matter. Second, the powerful electric current produced by the batteries enabled chemists to break apart many other substances, showing that there were more elements than previously thought.

Scientific Process
Ask: What old theory did Thomson's atomic model prove false? *Atoms are the smallest form of matter.*

Technology
1815 Davy and other scientists were using batteries to produce electricity for experiments, but otherwise batteries weren't used by most people. Ask: Why do you think ordinary people weren't using electricity at that time? *Batteries were large, expensive, and difficult to maintain in working order. Also, uses had to be developed for batteries before they could be used.*

Language Arts Connection
Mary Shelley wrote one of the first science fiction novels, *Frankenstein*, about a scientist who created a monster. He used electricity to give it life. Electricity was a new and exciting subject when Shelley wrote her novel.

DIFFERENTIATE INSTRUCTION

Below Level Point out the structure of the timeline. Tell students that events get closer to the present as you read the line from left to right. Discuss how one discovery often leads to another, farther to the right. Ask: How did electricity, produced by Volta in 1800, lead to Davy's isolation of elements? *Electricity gave investigators a new tool to use to try to break down substances.*

DIFFERENTIATE INSTRUCTION

Advanced Have students research batteries today. They should chart what types are available, what they are used for, and the chemical reaction that produces electricity.

Chemical Interactions Pacing Guide

The following pacing guide shows how the chapters in *Chemical Interactions* can be adapted to fit your specific course needs.

	TRADITIONAL SCHEDULE (DAYS)	BLOCK SCHEDULE (DAYS)
Frontiers in Science: Medicines from Nature	1	0.5
Chapter 1 Atomic Structure and the Periodic Table		
1.1 Atoms are the smallest form of elements.	2	1
1.2 Elements make up the periodic table.	2	1
1.3 The periodic table is a map of the elements.	3	1.5
Chapter Investigation	1	0.5
Chapter 2 Chemical Bonds and Compounds		
2.1 Elements combine to form compounds.	2	1
2.2 Chemical bonds hold compounds together.	2	1
2.3 Substances' properties depend on their bonds.	3	1.5
Chapter Investigation	1	0.5
Chapter 3 Chemical Reactions		
3.1 Chemical reactions alter arrangements of atoms.	2	1
3.2 The masses of reactants and products are equal.	2	1
3.3 Chemical reactions involve energy changes.	2	1
3.4 Life and industry depend on chemical reactions.	3	1.5
Chapter Investigation	1	0.5
Timelines in Science: The Story of Atomic Structure	1	0.5
Chapter 4 Solutions		
4.1 A solution is a type of mixture.	2	1
4.2 The amount of solute that dissolves can vary.	2	1
4.3 Solutions can be acidic, basic, or neutral.	2	1
4.4 Metal alloys are solid mixtures.	3	1.5
Chapter Investigation	1	0.5
Chapter 5 Carbon in Life and Materials		
5.1 Carbon-based molecules have many structures.	2	1
5.2 Carbon-based molecules are life's building blocks.	2	1
5.3 Carbon-based molecules are in many materials.	3	1.5
Chapter Investigation	1	0.5
Total Days for Module	**46**	**23**

Planning the Chapter

Complete planning support precedes each chapter.

Previewing Content

- Section-by-section science background notes
- Common Misconceptions notes

Previewing Content

SECTION

1.1 Atoms are the smallest form of elements. pp. 9–16

1. All matter is made of atoms.
All matter is made up of atoms of about 100 elements, or basic substances. Hydrogen is the most abundant element in the universe; oxygen is the most abundant element in Earth's crust. Every element has a unique name and symbol. Names and symbols of the elements come from many sources.

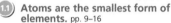

SECT

1.

1. E
M
th
M
e

CHAPTER

1 Atomic Structure and the Periodic Table

Physical Science
UNIFYING PRINCIPLES

PRINCIPLE 1	PRINCIPLE 2	PRINCIPLE 3	PRINCIPLE 4
Matter is made of particles too small to see.	Matter changes form and moves from place to place.	Energy change form to anoth cannot be cre destroyed.	

Unit:
Chemical
Interactions
BIG IDEAS

CHAPTER 1 Atomic Structure and the Periodic Table A substance's atomic structure determines its physical and chemical properties.	CHAPTER 2 Chemical Bonds and Compounds The properties of compounds depend on their atoms and chemical bonds.	CHAPTER 3 Chemical Reactions Chemical reactions form new substances by breaking and making chemical bonds.	C S V c s C

CHAPTER 1 KEY CONCEPTS

SECTION **1.1**	SECTION **1.2**
Atoms are the smallest form of elements.	Elements make up the periodic tab
1. All matter is made of atoms.	**1.** Elements can be organized by similarities.
2. Each element is made of a different atom.	**2.** The periodic table organizes the ato of the elements by properties and atomic number.
3. Atoms form ions.	

 The Big Idea Flow Chart is available on p. T1 in the **UNIT TRANSPARENCY BOOK**

Previewing Content

SECTION

1.3 The periodic table is a map of the elements. pp. 26–33

1. The periodic table has distinct regions.
Position in the periodic table reveals something about how **reactive** an element is. Elements in Groups 1 and 17 are especially reactive. Elements in Group 18 are the least reactive.

2. Most elements are metals.
Metals are usually shiny, often conduct electricity and heat well, and can be easily shaped and drawn into a wire.
- Alkali metals and alkaline earth metals are at the left of the periodic table and are very reactive.
- Transition metals are near the center of the periodic table and include copper, gold, silver, and iron.
- Rare earth metals are in the top row of the two rows of metals shown outside the main body of the periodic table.
- The two bottom rows are separated from the table to save space.

3. Nonmetals and metalloids have a wide range of properties.
Nonmetals appear on the right side of the periodic table. They include elements with a wide range of properties. Carbon, nitrogen, oxygen, and sulfur are nonmetals, as are the extremely reactive halogens, such as chlorine, and noble, or inert, gases, such as neon.
Metalloids lie between metals and nonmetals in the periodic table. They have characteristics of both. An important use of metalloids is in the making of semiconductors for electronic devices.

4. Some atoms can change their identity.
The atomic nucleus is held together by forces. Sometimes there can be too many or too few neutrons in a nucleus, and so the forces holding it together cannot hold it together properly. To regain its stability, the nucleus will produce particles and eject them. This process is called radioactivity. The identity of radioactive atoms changes when the number of protons change. This is called radioactive decay. Radioactive decay occurs at a steady rate that is characteristic of the particular isotope. The amount of time that it takes for one-half of the atoms in a particular sample to decay is called the half-life of the isotope. The chart shown below illustrates the progress of radioactive decay.

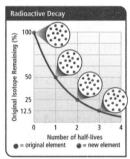

Radioactive Decay

Original Isotope Remaining (%) vs Number of half-lives
- = original element
- = new element

Common Misconceptions

PROPERTIES OF ELEMENTS Students may hold the misconception that a single atom of an element will exhibit the same properties as the element in bulk. Samples of elements exhibit particular, identifiable chemical properties. At the atomic level, properties such as color and texture are meaningless.

TE This misconception is addressed on p. 27.

MISCONCEPTION DATABASE
CLASSZONE.COM Background on student misconceptions

Previewing Chapter Resources

- Section-by-section listing of all print and technology resources
- Suggested pacing
- Correlations to National Science Education Standards

Previewing Chapter Resources

KEY TO ICONS
👁 CD/CD-ROM TE Teacher Edition
🖐 INTERNET PE Pupil Edition R UNIT RESOURCE BOO...

	INTEGRATED TECHNOLOGY		READING AND REINFORCEMENT	ASSESSMENT

CHAPTER 1
Atomic Structure and the Periodic Table

🖐 **CLASSZONE.COM**
- eEdition Plus
- EasyPlanner
- Misconception Database
- Content Review
- Test Practice
- Simulation
- Visualization
- Resource Centers
- Internet Activity: Periodic Table
- Math Tutorial

🖐 **SCILINKS.ORG**
SCI**LINKS**

👁 **CD-ROM**
- eEdition
- EasyPlanner
- Power Presentations
- Content Review
- Lab Generator
- Test Generator

👁 **AUDIO CDS**
- Audio Readings
- Audio Readings in Spanish

- That's Far!
- Element Safari
- Internet Activity: Periodic Table

R **UNIT RESOURCE BOOK**
- Family Letter, p. ix
- Spanish Family Letter, p. x
- Unit Projects, pp. 5–10

🧪 **Lab Generator CD-ROM**
Generate customized labs.

- Frame Game, B26–27
- Main Idea Web, C38–39
- Daily Vocabulary Scaffolding, H1–8

R **UNIT RESOURCE BOOK**
- Vocabulary Practice, pp. 45–46
- Decoding Support, p. 47
- Summarizing the Chapter, pp. 68–69

👁 **Audio Readings CD**
Listen to Pupil Edition.

👁 **Audio Readings in Spanish CD**
Listen to Pupil Edition in Spanish.

PE • Chapter R...
• Standardi...

A **UNIT ASSE...**
- Diagnostic
- Chapter T...
- Alternative

SP A Spanish Cha...

Test Gener...
Generate cu...

🧪 **Lab Genera...**
Rubrics for I...

SECTION 1.1 **Atoms are the smallest form of elements.**

🖐 • RESOURCE CENTER, The Atom, Elements Important to Life
• SIMULATION, Build an Atom

PE • EXPLORE The Size of Atoms, p. 9
• INVESTIGATE Masses of Atomic Particles, p. 13

R **UNIT RESOURCE BOOK**
- Reading Study Guide, A & B, pp. 13–16
- Spanish Reading Study Guide, pp. 17–18
- Challenge and Extension, p. 19
- Reinforcing Key Concepts, p. 21

TE Ongoing As...

PE Section 1.1

A **UNIT ASSE...**
Section 1.1

R **UNIT RESOURCE BOOK**
- Reading Study Guide, A & B, pp. 24–27
- Spanish Reading Study Guide, pp. 28–29
- Challenge and Extension, p. 30
- Reinforcing Key Concepts, p. 31

TE Ongoing As...

PE Section 1.2

A **UNIT ASSES...**
Section 1.2

R **UNIT RESOURCE BOOK**
- Reading Study Guide, A & B, pp. 34–37
- Spanish Reading Study Guide, pp. 38–39
- Challenge and Extension, p. 40
- Reinforcing Key Concepts, p. 42
- Challenge Reading, pp. 43–44

TE Ongoing As...

PE Section 1.3

A **UNIT ASSES...**
Section 1.3

Previewing Labs

EXPLORE (the BIG idea)

🧪 **Lab Generator CD-ROM**
Edit these Pupil Edition and generate alternative.

That's Far! p. 7
Students pace out the relative distance between an electron and a nucleus, demonstrating that most of the volume of an atom is empty space.

TIME 10 minutes
MATERIALS baseball, dime

Element Safari, p. 7
Students look at labels on food packages to learn that elements make up common materials.

TIME 10 minutes
MATERIALS periodic table; packages of (or labels from) baking soda, vinegar, cereal flakes, antacid tablets

Internet Activity: Periodic Table, p. 7
Students explore the organization of the periodic table.

TIME 20 minutes
MATERIALS computer with Internet access

SECTION 1.1
EXPLORE The Size of Atoms, p. 9
Students model the particulate nature of matter by cutting paper into smaller and smaller pieces.

TIME 10 minutes
MATERIALS 30 cm strip of paper, pair of scissors

INVESTIGATE Masses of Atomic Particles, p. 13
Students use common objects to model the relative masses of atomic particles.

TIME 20 minutes
MATERIALS balance; large paper clip; items available in bulk, such as sand, clay, or water

SECTION 1.2
EXPLORE Similarities and Differences of Objects, p. 17
Students classify an assortment of buttons to see how objects can be organized using different criteria.

TIME 15 minutes
MATERIALS assorted buttons like those found on shirts, etc.

CHAPTER INVESTIGATION
Modeling Atomic Masses, pp. 24–25
Students make model atoms from film cans and washers to determine the atom's relative masses.

TIME 40 minutes
MATERIALS 5 film cans, 21 washers, balance

SECTION 1.3
INVESTIGATE Radioactivity, p. 31
Students use pennies to model the half-life of radioactive elements.

TIME 30 minutes
MATERIALS 50 pennies, bag, graph paper

R **Additional INVESTIGATION,** Investigating the Unseen, A, B, & C, pp. 59–67; Teacher Instructions, pp. 329–330

Previewing Labs

- Brief descriptions of all chapter labs and activities
- Time and materials required for each activity

Chapter 1: **Atomic Structure and the Periodic Table 5D** D

Planning the Lesson

Point-of-use support for each lesson provides a wealth of teaching options.

1. Prepare

- Concept and vocabulary review
- Note-taking and vocabulary strategies

2. Focus

- Set Learning Goals
- 3-Minute Warm-up

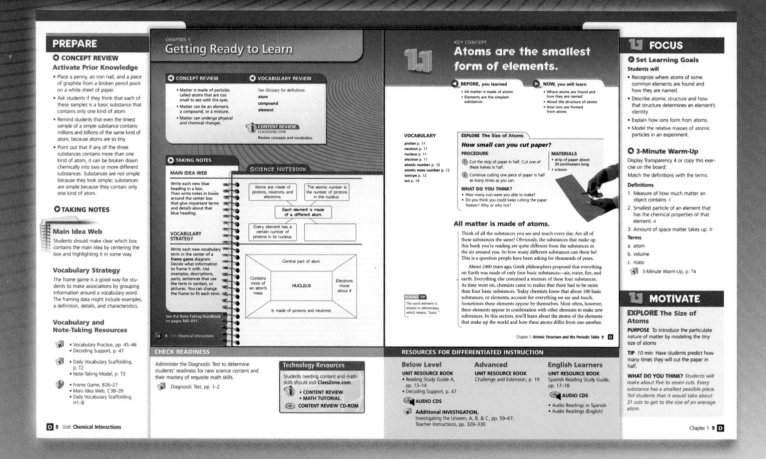

3. Motivate

- Engaging entry into the section
- Explore activity or Think About question

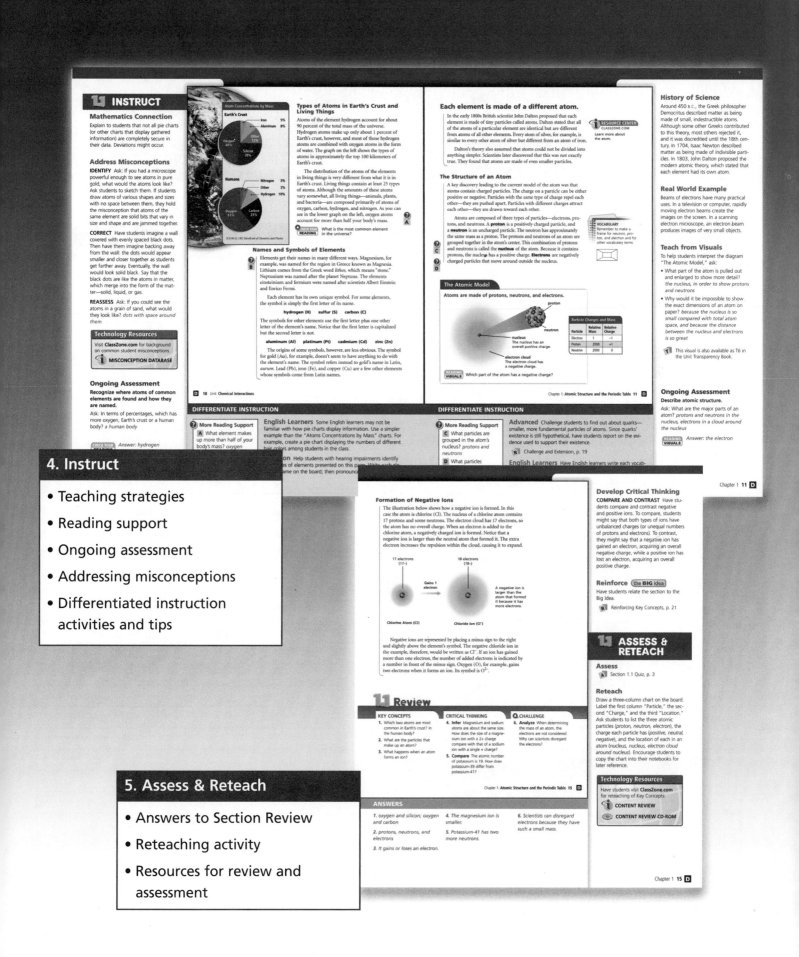

INSTRUCT

Mathematics Connection

Explain to students that not all pie charts (or other charts that display gathered information) in their data. Deviations might occur.

Address Misconceptions

IDENTIFY Ask: If you had a microscope powerful enough to see atoms in pure gold, what would the atoms look like? Ask students to sketch them. If students draw atoms of various shapes and sizes with no space between them, they hold the misconception that atoms of the same element are solid bits that vary in size and shape and are jammed together.

CORRECT Have students imagine a wall covered with evenly spaced black dots. Then have them imagine backing away from the wall: the dots would appear smaller and closer together as students get farther away. Eventually, the wall would look solid black. Say that the black dots are like the atoms in matter, which merge into the form of the matter—solid, liquid, or gas.

REASSESS Ask: If you could see the atoms in a grain of sand, what would they look like? *dots with space around them*

Technology Resources

Visit ClassZone.com for background on common student misconceptions.

MISCONCEPTION DATABASE

Ongoing Assessment

Recognize where atoms of common elements are found and how they are named.

Ask: In terms of percentages, which has more oxygen, Earth's crust or a human body? *a human body*

CHECK YOUR Answer: hydrogen

Atom Concentrations by Mass

Earth's Crust

- Iron 5%
- Aluminum 8%
- Other 12%
- Silicon 28%
- Oxygen 46%

Humans

- Nitrogen 3%
- Other 3%
- Hydrogen 10%
- Oxygen 61%
- Carbon 23%

SOURCE: CRC Handbook of Chemistry and Physics

Types of Atoms in Earth's Crust and Living Things

Atoms of the element hydrogen account for about 90 percent of the total mass of the universe. Hydrogen atoms make up only about 1 percent of Earth's crust, however, and most of those hydrogen atoms are combined with oxygen atoms in the form of water. The graph on the left shows the types of atoms in approximately the top 100 kilometers of atoms in approximately the top 100 kilometers.

The distribution of the atoms of the elements in living things is very different from what it is in Earth's crust. Living things contain at least 25 types of atoms. Although the amounts of these atoms vary somewhat, all living things—animals, plants, and bacteria—are composed primarily of atoms of oxygen, carbon, hydrogen, and nitrogen. As you can see in the lower graph on the left, oxygen atoms account for more than half your body's mass.

CHECK YOUR READING What is the most common element in the universe?

Names and Symbols of Elements

Elements get their names in many different ways. Magnesium, for example, was named for the region in Greece known as Magnesia. Lithium comes from the Greek word *lithos*, which means "stone." Neptunium was named after the planet Neptune. The elements einsteinium and fermium were named after scientists Albert Einstein and Enrico Fermi.

Each element has its own unique symbol. For some elements, the symbol is simply the first letter of its name.

hydrogen (H)   sulfur (S)   carbon (C)

The symbols for other elements use the first letter plus one other letter of the element's name. Notice that the first letter is capitalized but the second letter is not.

aluminum (Al)   platinum (Pt)   cadmium (Cd)   zinc (Zn)

The origins of some symbols, however, are less obvious. The symbol for gold (Au), for example, doesn't seem to have anything to do with the element's name. The symbol refers instead to gold's name in Latin, *aurum*. Lead (Pb), iron (Fe), and copper (Cu) are a few other elements whose symbols come from Latin names.

D 10 Unit: Chemical Interactions

Each element is made of a different atom.

In the early 1800s British scientist John Dalton proposed that each element is made of tiny particles called atoms. Dalton stated that all of the atoms of a particular element are identical but are different from atoms of all other elements. Every atom of silver, for example, is similar to every other atom of silver but different from an atom of iron.

Dalton's theory also assumed that atoms could not be divided into anything simpler. Scientists later discovered that this was not exactly true. They found that atoms are made of even smaller particles.

RESOURCE CENTER
CLASSZONE.COM
Learn more about the atom.

The Structure of an Atom

A key discovery leading to the current model of the atom was that atoms contain charged particles. The charge on a particle can be either positive or negative. Particles with the same type of charge repel each other—they are pushed apart. Particles with different charges attract each other—they are drawn toward each other.

Atoms are composed of three types of particles—electrons, protons, and neutrons. A **proton** is a positively charged particle, and a **neutron** is an uncharged particle. The neutron has approximately the same mass as a proton. The protons and neutrons of an atom are grouped together in the atom's center. This combination of protons and neutrons is called the **nucleus** of the atom. Because it contains protons, the nucleus has a positive charge. **Electrons** are negatively charged particles that move around outside the nucleus.

VOCABULARY
Remember to make a frame for neutron, proton, and electron and for other vocabulary terms.

The Atomic Model

Atoms are made of protons, neutrons, and electrons.

proton

nucleus
The nucleus has an overall positive charge.

neutron

electron cloud
The electron cloud has a negative charge.

Particle	Relative Mass	Relative Charge
Electron	1	−1
Proton	2000	+1
Neutron	2000	0

READING VISUALS Which part of the atom has a negative charge?

Chapter 1: Atomic Structure and the Periodic Table 11 **D**

History of Science

Around 450 B.C., the Greek philosopher Democritus described matter as being made of small, indestructible atoms. Although some other Greeks contributed to this theory, most others rejected it, and it was discredited until the 18th century. In 1704, Isaac Newton described matter as being made of indivisible particles. In 1803, John Dalton proposed the modern atomic theory, which stated that each element had its own atom.

Real World Example

Beams of electrons have many practical uses. In a television or computer, rapidly moving electron beams create the images on the screen. In a scanning electron microscope, an electron beam produces images of very small objects.

Teach from Visuals

To help students interpret the diagram "The Atomic Model," ask:

- What part of the atom is pulled out and enlarged to show more detail? *the nucleus, in order to show protons and neutrons*
- Why would it be impossible to show the exact dimensions of an atom on paper? *because the nucleus is so small compared with total atom space, and because the distance between the electrons and electrons is so great*

This visual is also available as T6 in the Unit Transparency Book.

Ongoing Assessment

Describe atomic structure.

Ask: What are the major parts of an atom? *protons and neutrons in the nucleus, electrons in a cloud around the nucleus*

READING VISUALS Answer: the electron

Chapter 1 11 **D**

DIFFERENTIATE INSTRUCTION

More Reading Support

A What concentration makes up more than half of your body's mass? *oxygen*

English Learners Some English learners may not be familiar with how pie charts display information. Use a simpler example than the "Atoms Concentrations by Mass" charts. For example, create a pie chart displaying the numbers of different hair colors among students in the class.

More Reading Support

B ...

DIFFERENTIATE INSTRUCTION

More Reading Support

C What particles are grouped in the atom's nucleus? *protons and neutrons*

D What particles

Advanced Challenge students to find out about quarks—smaller, more fundamental particles of atoms. Since quarks' existence is still hypothetical, have students report on the evidence used to support their existence.

Challenge and Extension, p. 19

English Learners Have English learners write each vocab-

---

4. Instruct

- Teaching strategies
- Reading support
- Ongoing assessment
- Addressing misconceptions
- Differentiated instruction activities and tips

Formation of Negative Ions

The illustration below shows how a negative ion is formed. In this case the atom is chlorine (Cl). The nucleus of a chlorine atom contains 17 protons and some neutrons. The electron cloud has 17 electrons, so the atom has no overall charge. When an electron is added to the chlorine atom, a negatively charged ion is formed. Notice that the negative ion is larger than the neutral atom that formed it. The extra electron increases the repulsion within the cloud, causing it to expand.

17 electrons
(17−)

Gains 1 electron

18 electrons
(18−)

A negative ion is larger than the atom that formed it because it has more electrons.

Chlorine Atom (Cl)   **Chloride Ion (Cl⁻)**

Negative ions are represented by placing a minus sign to the right and slightly above the element's symbol. The negative chloride ion in the example, therefore, would be written as Cl⁻. If an ion has gained more than one electron, the number of added electrons is indicated by a number in front of the minus sign. Oxygen (O), for example, gains two electrons when it forms an ion. Its symbol is O^{2-}.

Review

KEY CONCEPTS

1. Which two atoms are most common in Earth's crust? in the human body?
2. What are the particles that make up an atom?
3. What happens when an atom forms an ion?

CRITICAL THINKING

4. **Infer** Magnesium and sodium atoms are about the same size. How does the size of a magnesium ion with a 2+ charge compare with that of a sodium ion with a single − charge?
5. **Compare** The atomic number of potassium is 19. How does potassium-39 differ from potassium-41?

CHALLENGE

6. **Analyze** When determining the mass of an atom, the electrons are not considered. Why can scientists disregard the electrons?

Chapter 1: Atomic Structure and the Periodic Table 15 **D**

ANSWERS

1. oxygen and silicon; oxygen and carbon
2. protons, neutrons, and electrons
3. It gains or loses an electron.

4. The magnesium ion is smaller.
5. Potassium-41 has two more neutrons.

6. Scientists can disregard electrons because they have such a small mass.

Develop Critical Thinking

COMPARE AND CONTRAST Have students compare and contrast negative and positive ions. To compare, students might say that both types of ions have unbalanced charges (or unequal numbers of protons and electrons). To contrast, they might say that a negative ion has gained an electron, acquiring an overall negative charge, while a positive ion has lost an electron, acquiring an overall positive charge.

Reinforce (the BIG idea)

Have students relate the section to the Big Idea.

Reinforcing Key Concepts, p. 21

ASSESS & RETEACH

Assess

Section 1.1 Quiz, p. 3

Reteach

Draw a three-column chart on the board. Label the first column "Particle," the second "Charge," and the third "Location." Ask students to list the three atomic particles (*proton, neutron, electron*), the charge each particle has (*positive, neutral, negative*), and the location of each in an atom (*nucleus, nucleus, electron cloud around nucleus*). Encourage students to copy the chart into their notebooks for later reference.

Technology Resources

Have students visit ClassZone.com for reteaching of Key Concepts.

CONTENT REVIEW

CONTENT REVIEW CD-ROM

Chapter 1 15 **D**

---

5. Assess & Reteach

- Answers to Section Review
- Reteaching activity
- Resources for review and assessment

Lab Materials List

The following charts list the consumables, nonconsumables, and equipment needed for all activities. Quantities are per group of four students. Lab aprons, goggles, water, books, paper, pens, pencils, and calculators are assumed to be available for all activities.

Materials kits are available. For more information, please call McDougal Littell at 1-800-323-5435.

Consumables

Description	Quantity per Group	Explore *page*	Investigate *page*	Chapter Investigation *page*
antacid tablet with calcium carbonate	2	125		
ashes	1/2 tsp		95	
bag, paper	1	31		
bag, zip-top sandwich	1			170
baking soda	5 tsp		79	92, 132
ball, Styrofoam, 1"	6		149	
ball, Styrofoam, 3"	2		149	
balloon	1		79	
borax	2 tsp			170
bottle, plastic, 1 liter	1		113	
bottle, plastic, 2 liter with cap	1			170
bottle, plastic, pint	1		79	
bread, cube 1/4"	1		158	
candle, tealight	4	41, 154	95	60
carrot, 1/4" slice	1	154		
coffee filter, cone	1		113	
cornstarch solution, 10%	1–2 mL		158	
cup, plastic clear	27	86, 111, 117, 125	74, 120	60, 92, 132
detergent powder	1 tsp			132
Epsom salts	10 tsp	86		60
flour	1 tsp	111		
food coloring	1 bottle			170
fruit juice	30 mL			132
gelatin, liquid	1–2 mL		158	
glue, white	40 mL			170
hydrogen peroxide	30 mL			92
iodine solution	2 mL		158	
iron filings	20 grams			60
marker, permanent black	1		149	132
marker, water soluble black	1		113	

Description	Quantity per Group	Explore page	Investigate page	Chapter Investigation page
marshmallow, small	1	154		
match, wooden	10	41, 154	95	
nail, iron, 3"	4	69	137	
nail, steel, 3"	3		137	
paper, graph, 8.5" x 11"	2	31		24
paper clip	2		13, 53	
paper towel	2–5			132
pH test strip, universal	7			132
pie plate, aluminum	1	154		
salt, table	2 cups	69, 111	120	132
seltzer tablet	2		74	
shampoo	30 mL			132
soda water	1 1/2 cup	117		132
spoon, plastic	7	69, 86, 111	120	60, 92, 170
stirring stick	1		53	
straw, drinking	5–6			170
string	15 cm		53	
sugar	3 1/2 cups	41	53	60
sugar cube	2		95	
tape, masking	1 roll	56		132
tofu, cube 1/4"	1		158	
toothpick	10		149	
vinegar	4 cups	69, 125	79	92, 132
water, distilled	200 mL			60, 132
wire, copper, uninsulated	45 cm	56		
yeast	1 gram			92

Nonconsumables

Description	Quantity per Group	Explore *page*	Investigate *page*	Chapter Investigation *page*
balance, triple beam	1		13, 79, 120, 137	24, 92
battery, 6 volt	1			60
battery, D cell	1	56		
beaker, 100 mL	2	41	53	60, 92
beaker, 500 mL	2		120	
bolt with matching nut	10		43	
bowl, large clear	1	69		
bowl, small plastic	2			170
button, assortment	20	17		
carbon sample	1	41		
coin, penny	50	31, 69		
electrode, copper	1			60
electrode, zinc	1			60
eyedropper	3		113, 158	
film canister with lid	5			24
funnel	1		79	
graduated cylinder, 100 mL	1	86	137	92, 132
hand lens	1		53	60
jar, baby food with lid	1			170
lid, baby food jar	4		158	
light bulb holder	1	56		60
light bulb, flashlight	1	56		60
measuring spoon, teaspoon	1		79	92, 132, 170
ring stand with ring	1			170
scissors	1	9		170
stopwatch	1	86	74, 158	92, 170
test tube	3	41		60
thermometer	2	86	120	92
tongs	1	154	95	
tongs, test tube	1	41		60
washer, metal 7/16"	21			24
wire lead with alligator clips	3			60

Safety Equipment

Description		Explore *page*	Investigate *page*	Chapter Investigation *page*
gloves			158	

McDougal Littell Science

Chemical Interactions

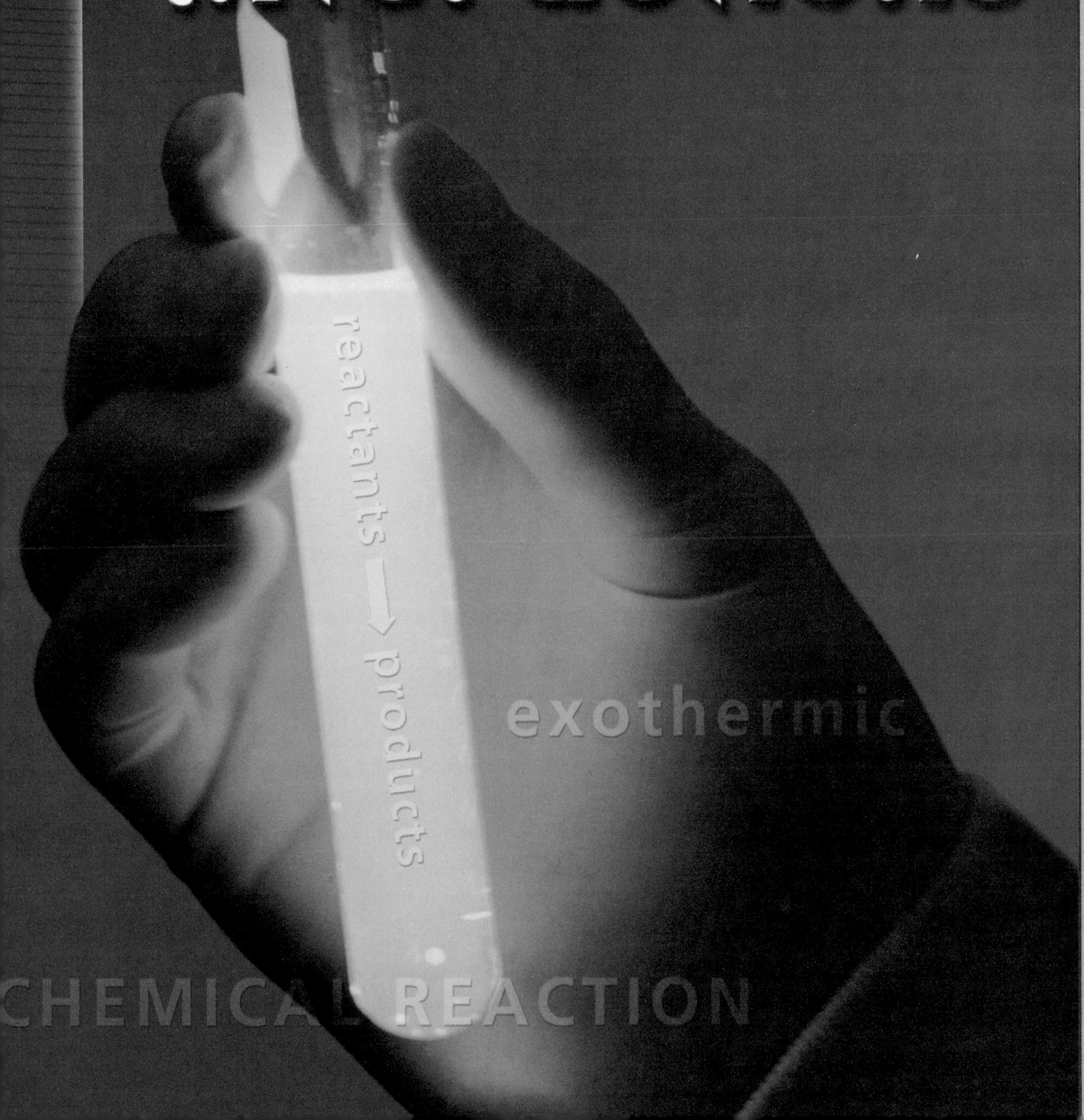

reactants → products
exothermic
CHEMICAL REACTION

PHYSICAL SCIENCE

A ▶ Matter and Energy
B ▶ Chemical Interactions
C ▶ Motion and Forces
D ▶ Waves, Sound, and Light
E ▶ Electricity and Magnetism

LIFE SCIENCE

A ▶ Cells and Heredity
B ▶ Life Over Time
C ▶ Diversity of Living Things
D ▶ Ecology
E ▶ Human Biology

EARTH SCIENCE

A ▶ Earth's Surface
B ▶ The Changing Earth
C ▶ Earth's Waters
D ▶ Earth's Atmosphere
E ▶ Space Science

ISBN: 0-618-33438-6   1 2 3 4 5 6 7 8 VJM 08 07 06 05 04

Internet Web Site: http://www.mcdougallittell.com

Science Consultants

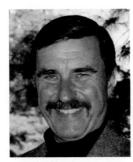

Chief Science Consultant

James Trefil, Ph.D. is the Clarence J. Robinson Professor of Physics at George Mason University. He is the author or co-author of more than 25 books, including *Science Matters* and *The Nature of Science*. Dr. Trefil is a member of the American Association for the Advancement of Science's Committee on the Public Understanding of Science and Technology. He is also a fellow of the World Economic Forum and a frequent contributor to *Smithsonian* magazine.

Rita Ann Calvo, Ph.D. is Senior Lecturer in Molecular Biology and Genetics at Cornell University, where for 12 years she also directed the Cornell Institute for Biology Teachers. Dr. Calvo is the 1999 recipient of the College and University Teaching Award from the National Association of Biology Teachers.

Kenneth Cutler, M.S. is the Education Coordinator for the Julius L. Chambers Biomedical Biotechnology Research Institute at North Carolina Central University. A former middle school and high school science teacher, he received a 1999 Presidential Award for Excellence in Science Teaching.

Instructional Design Consultants

Douglas Carnine, Ph.D. is Professor of Education and Director of the National Center for Improving the Tools of Educators at the University of Oregon. He is the author of seven books and over 100 other scholarly publications, primarily in the areas of instructional design and effective instructional strategies and tools for diverse learners. Dr. Carnine also serves as a member of the National Institute for Literacy Advisory Board.

Linda Carnine, Ph.D. consults with school districts on curriculum development and effective instruction for students struggling academically. A former teacher and school administrator, Dr. Carnine also co-authored a popular remedial reading program.

Donald Steely, Ph.D. serves as principal investigator at the Oregon Center for Applied Science (ORCAS) on federal grants for science and language arts programs. His background also includes teaching and authoring of print and multimedia programs in science, mathematics, history, and spelling.

Sam Miller, Ph.D. is a middle school science teacher and the Teacher Development Liaison for the Eugene, Oregon, Public Schools. He is the author of curricula for teaching science, mathematics, computer skills, and language arts.

Vicky Vachon, Ph.D. consults with school districts throughout the United States and Canada on improving overall academic achievement with a focus on literacy. She is also co-author of a widely used program for remedial readers.

Content Reviewers

John Beaver, Ph.D.
Ecology
Professor, Director of Science Education Center
College of Education and Human Services
Western Illinois University
Macomb, IL

Donald J. DeCoste, Ph.D.
Matter and Energy, Chemical Interactions
Chemistry Instructor
University of Illinois
Urbana-Champaign, IL

Dorothy Ann Fallows, Ph.D., MSc
Diversity of Living Things, Microbiology
Partners in Health
Boston, MA

Michael Foote, Ph.D.
The Changing Earth, Life Over Time
Associate Professor
Department of the Geophysical Sciences
The University of Chicago
Chicago, IL

Lucy Fortson, Ph.D.
Space Science
Director of Astronomy
Adler Planetarium and Astronomy Museum
Chicago, IL

Elizabeth Godrick, Ph.D.
Human Biology
Professor, CAS Biology
Boston University
Boston, MA

Isabelle Sacramento Grilo, M.S.
The Changing Earth
Lecturer, Department of the Geological Sciences
Montana State University
Bozeman, MT

David Harbster, MSc
Diversity of Living Things
Professor of Biology
Paradise Valley Community College
Phoenix, AZ

Richard D. Norris, Ph.D.
Earth's Waters
Professor of Paleobiology
Scripps Institution of Oceanography
University of California, San Diego
La Jolla, CA

Donald B. Peck, M.S.
*Motion and Forces; Waves, Sound, and Light;
   Electricity and Magnetism*
Director of the Center for Science Education (retired)
Fairleigh Dickinson University
Madison, NJ

Javier Penalosa, Ph.D.
Diversity of Living Things, Plants
Associate Professor, Biology Department
Buffalo State College
Buffalo, NY

Raymond T. Pierrehumbert, Ph.D.
Earth's Atmosphere
Professor in Geophysical Sciences (Atmospheric Science)
The University of Chicago
Chicago, IL

Brian J. Skinner, Ph.D.
Earth's Surface
Eugene Higgins Professor of Geology and Geophysics
Yale University
New Haven, CT

Nancy E. Spaulding, M.S.
Earth's Surface, The Changing Earth, Earth's Waters
Earth Science Teacher (retired)
Elmira Free Academy
Elmira, NY

Steven S. Zumdahl, Ph.D.
Matter and Energy, Chemical Interactions
Professor Emeritus of Chemistry
University of Illinois
Urbana-Champaign, IL

Susan L. Zumdahl, M.S.
Matter and Energy, Chemical Interactions
Chemistry Education Specialist
University of Illinois
Urbana-Champaign, IL

Safety Consultant

Juliana Texley, Ph.D.
Former K–12 Science Teacher and School Superintendent
Boca Raton, FL

English Language Advisor

Judy Lewis, M.A.
Director, State and Federal Programs for reading proficiency
and high risk populations
Rancho Cordova, CA

iv

Teacher Panel Members

Carol Arbour
Tallmadge Middle School,
Tallmadge, OH

Patty Belcher
Goodrich Middle School,
Akron, OH

Gwen Broestl
Luis Munoz Marin Middle School,
Cleveland, OH

Al Brofman
Tehipite Middle School,
Fresno, CA

John Cockrell
Clinton Middle School,
Columbus, OH

Jenifer Cox
Sylvan Middle School,
Citrus Heights, CA

Linda Culpepper
Martin Middle School,
Charlotte, NC

Kathleen Ann DeMatteo
Margate Middle School,
Margate, FL

Melvin Figueroa
New River Middle School,
Ft. Lauderdale, FL

Doretha Grier
Kannapolis Middle School,
Kannapolis, NC

Robert Hood
Alexander Hamilton Middle School,
Cleveland, OH

Scott Hudson
Coverdale Elementary School,
Cincinnati, OH

Loretta Langdon
Princeton Middle School,
Princeton, NC

Carlyn Little
Glades Middle School,
Miami, FL

Ann Marie Lynn
Amelia Earhart Middle School,
Riverside, CA

James Minogue
Lowe's Grove Middle School,
Durham, NC

Joann Myers
Buchanan Middle School,
Tampa, FL

Barbara Newell
Charles Evans Hughes Middle School,
Long Beach, CA

Anita Parker
Kannapolis Middle School,
Kannapolis, NC

Greg Pirolo
Golden Valley Middle School,
San Bernardino, CA

Laura Pottmyer
Apex Middle School,
Apex, NC

Lynn Prichard
Booker T. Washington Middle Magnet
School, Tampa, FL

Jacque Quick
Walter Williams High School,
Burlington, NC

Robert Glenn Reynolds
Hillman Middle School,
Youngstown, OH

Theresa Short
Abbott Middle School,
Fayetteville, NC

Rita Slivka
Alexander Hamilton Middle School,
Cleveland, OH

Marie Sofsak
B F Stanton Middle School,
Alliance, OH

Nancy Stubbs
Sweetwater Union Unified School District,
Chula Vista, CA

Sharon Stull
Quail Hollow Middle School,
Charlotte, NC

Donna Taylor
Okeeheelee Middle School,
West Palm Beach, FL

Sandi Thompson
Harding Middle School,
Lakewood, OH

Lori Walker
Audubon Middle School & Magnet Center,
Los Angeles, CA

Teacher Lab Evaluators

Jill Brimm-Byrne
Albany Park Academy,
Chicago, IL

Gwen Broestl
Luis Munoz Marin Middle School,
Cleveland, OH

Al Brofman
Tehipite Middle School,
Fresno, CA

Michael A. Burstein
The Rashi School,
Newton, MA

Trudi Coutts
Madison Middle School,
Naperville, IL

Jenifer Cox
Sylvan Middle School,
Citrus Heights, CA

Larry Cwik
Madison Middle School,
Naperville, IL

Jennifer Donatelli
Kennedy Junior High School,
Lisle, IL

Paige Fullhart
Highland Middle School,
Libertyville, IL

Sue Hood
Glen Crest Middle School,
Glen Ellyn, IL

Ann Min
Beardsley Middle School,
Crystal Lake, IL

Aileen Mueller
Kennedy Junior High School,
Lisle, IL

Nancy Nega
Churchville Middle School,
Elmhurst, IL

Oscar Newman
Sumner Math and Science Academy,
Chicago, IL

Marina Penalver
Moore Middle School,
Portland, ME

Lynn Prichard
Booker T. Washington Middle Magnet
School, Tampa, FL

Jacque Quick
Walter Williams High School,
Burlington, NC

Seth Robey
Gwendolyn Brooks Middle School,
Oak Park, IL

Kevin Steele
Grissom Middle School,
Tinley Park, IL

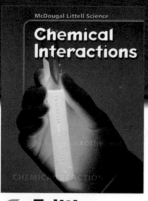

Chemical Interactions

eEdition

Chemical Interactions

Unit Features

SCIENTIFIC AMERICAN

1 Atomic Structure and the Periodic Table   6

the BIG idea

A substance's atomic structure determines its physical and chemical properties.

2 Chemical Bonds and Compounds   38

the BIG idea

The properties of compounds depend on their atoms and chemical bonds.

How do these skydivers stay together? How is this similar to the way atoms stay together? page 38

What changes are happening in this chemical reaction?
page 66

Features

Visual Highlights

Internet Resources @ ClassZone.com

INVESTIGATIONS AND ACTIVITIES

Standards and Benchmarks

Each chapter in **Chemical Interactions** covers some of the learning goals that are described in the *National Science Education Standards* (NSES) and the Project 2061 *Benchmarks for Science Literacy*. Selected content and skill standards are shown below in shortened form. The following National Science Education Standards are covered on pages xii–xxvii, in Frontiers in Science, and in Timelines in Science, as well as in chapter features and laboratory investigations: Understandings About Scientific Inquiry (A.9), Understandings About Science and Technology (E.6), Science and Technology in Society (F.5), Science as a Human Endeavor (G.1), Nature of Science (G.2), and History of Science (G.3).

Content Standards

1 Atomic Structure and the Periodic Table

National Science Education Standards

B.1.b	Substances react chemically in characteristic ways with other substances to form new substances.
B.1.c	There are more than 100 known elements that combine to produce compounds.

Project 2061 Benchmarks

4.D.1	The atoms of any element are alike but are different from atoms of other elements.
4.D.6	There are groups of elements that have similar properties, including • highly reactive metals • less-reactive metals • highly reactive nonmetals • some almost completely nonreactive gases
10.F.2	Scientists are still learning about the basic kinds of matter and how they combine.
10.G.1	The discovery that minerals containing uranium darken photographic film led to the idea of radioactivity.
10.G.2	Scientists Marie Curie and Pierre Curie isolated the elements radium and polonium.

2 Chemical Bonds and Compounds

National Science Education Standards

B.1.b	Substances react chemically in characteristic ways with other substances to form new substances.

Project 2061 Benchmarks

4.D.1	Atoms may stick together in molecules or may be packed together in large arrays.
10.F.2	Scientists are still learning about the basic kinds of matter and how they combine.
11.B.1	Models are often used to explore processes that • happen too slowly, too quickly, or are on too small a scale to observe directly • are on too great a scale to be studied experimentally • are potentially dangerous
11.C.5	Symmetry or a lack of symmetry may determine the properties of many objects.

3 Chemical Reactions

National Science Education Standards

B.1.b	In chemical reactions, the total mass is conserved.
B.3.e	In most chemical reactions, energy is transferred into or out of a system.

	Project 2061 Benchmarks
4.D.4	The temperature and acidity of a solution influence reaction rates. Many substances dissolve in water, which may make reactions between them easier.
4.D.6	An important kind of reaction between substances involves a combination of oxygen with something else—as in burning and rusting.
4.D.7	No matter how substances in a system interact with one another, the total weight of the system remains the same.
10.F.3	The work of scientist Antoine Lavoisier led to the modern science of chemistry.
10.F.4	Lavoisier tested the concept of conservation of matter by measuring the substances involved in burning.

4 Solutions

	National Science Education Standards
B.1.a	A substance has characteristic properties, such as density, a boiling point, and solubility, all of which do not depend on the amount of the sample.
B.1.c	Chemical elements do not break down during normal reactions involving • heating • electric current • acids
D.1.h	The atmosphere is a mixture of nitrogen, oxygen, and small amounts of gases such as water vapor.

	Project 2061 Benchmarks
4.D.4	The temperature and acidity of a solution influence reaction rates. Many substances dissolve in water, which may make reactions between them easier.

5 Carbon in Life and Materials

	National Science Education Standards
B.1.c	There are more than 100 known elements that combine to produce compounds.
C.1.a	Living systems demonstrate the complementary nature of structure and function.

	Project 2061 Benchmarks
4.D.6	Carbon and hydrogen are essential elements of living matter.

Process and Skill Standards

	National Science Education Standards		**Project 2061 Benchmarks**
A.1	Identify questions that can be answered through investigation.	12.B.1	Find what percentage one number is of another.
A.2	Design and conduct a scientific investigation.	12.B.2	Use, interpret, and compare numbers in several equivalent forms, such as integers, fractions, decimals, and percents.
A.3	Use appropriate tools and techniques to gather and interpret data.		
A.4	Use evidence to describe, predict, explain, and model.	12.C.3	Using appropriate units, use and read instruments that measure length, volume, weight, time, rate, and temperature.
A.5	Use critical thinking to find relationships between results and interpretations.		
A.7	Communicate procedures, results, and conclusions.	12.D.1	Use tables and graphs to organize information and identify relationships.
E.1	Identify a problem to be solved.	12.D.2	Read, interpret, and describe tables and graphs.
E.2	Design a solution or product.		
E.3	Implement the proposed solution.	12.D.4	Understand information that includes different types of charts and graphs, including circle charts, bar graphs, line graphs, data tables, diagrams, and symbols.
E.4	Evaluate the solution or design.		

Introducing Physical Science

Scientists are curious. Since ancient times, they have been asking and answering questions about the world around them. Scientists are also very suspicious of the answers they get. They carefully collect evidence and test their answers many times before accepting an idea as correct.

In this book you will see how scientific knowledge keeps growing and changing as scientists ask new questions and rethink what was known before. The following sections will help get you started.

What Is Physical Science?

In the simplest terms, physical science is the study of what things are made of and how they change. It combines the studies of both physics and chemistry. Physics is the science of matter, energy, and forces. It includes the study of topics such as motion, light, and electricity and magnetism. Chemistry is the study of the structure and properties of matter, and it especially focuses on how substances change into different substances.

The text and pictures in this book will help you learn key concepts and important facts about physical science. A variety of activities will help you investigate these concepts. As you learn, it helps to have a big picture of physical science as a framework for this new information. The four unifying principles listed below will give you this big picture. Read the next few pages to get an overview of each of these principles and a sense of why they are so important.

- **Matter is made of particles too small to see.**

- **Matter changes form and moves from place to place.**

- **Energy changes from one form to another, but it cannot be created or destroyed.**

- **Physical forces affect the movement of all matter on Earth and throughout the universe.**

the **BIG** idea

Each chapter begins with a big idea. Keep in mind that each big idea relates to one or more of the unifying principles.

Matter is made of particles too small to see.

This simple statement is the basis for explaining an amazing variety of things about the world. For example, it explains why substances can exist as solids, liquids, and gases, and why wood burns but iron does not. Like the tiles that make up this mosaic picture, the particles that make up all substances combine to make patterns and structures that can be seen. Unlike these tiles, the individual particles themselves are far too small to see.

What It Means

To understand this principle better, let's take a closer look at the two key words: *matter* and *particles.*

Matter

Objects you can see and touch are all around you. The materials that these objects are made of are called **matter.** All living things—even you—are also matter. Even though you can't see it, the air around you is matter too. Scientists often say that matter is anything that has mass and takes up space. **Mass** is a measure of the amount of matter in an object. We use the word **volume** to refer to the amount of space an object or a substance takes up.

Particles

The tiny particles that make up all matter are called **atoms.** Just how tiny are atoms? They are far too small to see, even through a powerful microscope. In fact, an atom is more than a million times smaller than the period at the end of this sentence.

There are more than 100 basic kinds of matter called **elements.** For example, iron, gold, and oxygen are three common elements. Each element has its own unique kind of atom. The atoms of any element are all alike but different from the atoms of any other element.

Many familiar materials are made of particles called molecules. In a **molecule,** two or more atoms stick together to form a larger particle. For example, a water molecule is made of two atoms of hydrogen and one atom of oxygen.

Why It's Important

Understanding atoms and molecules makes it possible to explain and predict the behavior of matter. Among other things, this knowledge allows scientists to

- explain why different materials have different characteristics
- predict how a material will change when heated or cooled
- figure out how to combine atoms and molecules to make new and useful materials

Matter changes form and moves from place to place.

You see matter change form every day. You see the ice in your glass of juice disappear without a trace. You see a black metal gate slowly develop a flaky, orange coating. Matter is constantly changing and moving.

What It Means

Remember that matter is made of tiny particles called atoms. Atoms are constantly moving and combining with one another. All changes in matter are the result of atoms moving and combining in different ways.

Matter Changes and Moves

You can look at water to see how matter changes and moves. A block of ice is hard like a rock. Leave the ice out in sunlight, however, and it changes into a puddle of water. That puddle of water can eventually change into water vapor and disappear into the air. The water vapor in the air can become raindrops, which may fall on rocks, causing them to weather and wear away. The water that flows in rivers and streams picks up tiny bits of rock and carries them from one shore to another. Understanding how the world works requires an understanding of how matter changes and moves.

Matter Is Conserved

No matter was lost in any of the changes described above. The ice turned to water because its molecules began to move more quickly as they got warmer. The bits of rock carried away by the flowing river were not gone forever. They simply ended up farther down the river. The puddles of rainwater didn't really disappear; their molecules slowly mixed with molecules in the air.

Under ordinary conditions, when matter changes form, no matter is created or destroyed. The water created by melting ice has the same mass as the ice did. If you could measure the water vapor that mixes with the air, you would find it had the same mass as the water in the puddle did.

Why It's Important

Understanding how mass is conserved when matter changes form has helped scientists to

- describe changes they see in the world
- predict what will happen when two substances are mixed
- explain where matter goes when it seems to disappear

Energy changes from one form to another, but it cannot be created or destroyed.

When you use energy to warm your food or to turn on a flashlight, you may think that you "use up" the energy. Even though the camp-stove fuel is gone and the flashlight battery no longer functions, the energy they provided has not disappeared. It has been changed into a form you can no longer use. Understanding how energy changes forms is the basis for understanding how heat, light, and motion are produced.

What It Means

Changes that you see around you depend on energy. **Energy,** in fact, means the ability to cause change. The electrical energy from an outlet changes into light and heat in a light bulb. Plants change the light energy from the Sun into chemical energy, which animals use to power their muscles.

Energy Changes Forms

Using energy means changing energy. You probably have seen electric energy changing into light, heat, sound, and mechanical energy in household appliances. Fuels like wood, coal, and oil contain chemical energy that produces heat when burned. Electric power plants make electrical energy from a variety of energy sources, including falling water, nuclear energy, and fossil fuels.

Energy Is Conserved

Energy can be converted into forms that can be used for specific purposes. During the conversion, some of the original energy is converted into unwanted forms. For instance, when a power plant converts the energy of falling water into electrical energy, some of the energy is lost to friction and sound.

Similarly, when electrical energy is used to run an appliance, some of the energy is converted into forms that are not useful. Only a small percentage of the energy used in a light bulb, for instance, produces light; most of the energy becomes heat. Nonetheless, the total amount of energy remains the same through all these conversions.

The fact that energy does not disappear is a law of physical science. The **law of conservation of energy** states that energy cannot be created or destroyed. It can only change form.

Why It's Important

Understanding that energy changes form but does not disappear has helped scientists to

- predict how energy will change form
- manage energy conversions in useful ways
- build and improve machines

Physical forces affect the movement of all matter on Earth and throughout the universe.

What makes the world go around? The answer is simple: forces. Forces allow you to walk across the room, and forces keep the stars together in galaxies. Consider the forces acting on the rafts below. The rushing water is pushing the rafts forward. The force from the people paddling helps to steer the rafts.

What It Means

A **force** is a push or a pull. Every time you push or pull an object, you're applying a force to that object, whether or not the object moves. There are several forces—several pushes and pulls—acting on you right now. All these forces are necessary for you to do the things you do, even sitting and reading.

- You are already familiar with the force of gravity. **Gravity** is the force of attraction between two objects. Right now gravity is at work pulling you to Earth and Earth to you. The Moon stays in orbit around Earth because gravity holds it close.

- A contact force occurs when one object pushes or pulls another object by touching it. If you kick a soccer ball, for instance, you apply a contact force to the ball. You apply a contact force to a shopping cart that you push down a grocery aisle or a sled that you pull up a hill.

- **Friction** is the force that resists motion between two surfaces pressed together. If you've ever tried to walk on an icy sidewalk, you know how important friction can be. If you lightly rub your finger across a smooth page in a book and then across a piece of sandpaper, you can feel how the different surfaces produce different frictional forces. Which is easier to do?

- There are other forces at work in the world too. For example, a compass needle responds to the magnetic force exerted by Earth's magnetic field, and objects made of certain metals are attracted by magnets. In addition to magnetic forces, there are electrical forces operating between particles and between objects. For example, you can demonstrate electrical forces by rubbing an inflated balloon on your hair. The balloon will then stick to your head or to a wall without additional means of support.

Why It's Important

Although some of these forces are more obvious than others, physical forces at work in the world are necessary for you to do the things you do. Understanding forces allows scientists to

- predict how objects will move
- design machines that perform complex tasks
- predict where planets and stars will be in the sky from one night to the next

The Nature of Science

You may think of science as a body of knowledge or a collection of facts. More important, however, science is an active process that involves certain ways of looking at the world.

Scientific Habits of Mind

Scientists are curious. They are always asking questions. Scientists have asked questions such as, "What is the smallest form of matter?" and "How do the smallest particles behave?" These and other important questions are being investigated by scientists around the world.

Scientists are observant. They are always looking closely at the world around them. Scientists once thought the smallest parts of atoms were protons, neutrons, and electrons. Later, protons and neutrons were found to be made of even smaller particles called quarks.

Scientists are creative. They draw on what they know to form possible explanations for a pattern, an event, or an interesting phenomenon that they have observed. Then scientists create a plan for testing their ideas.

Scientists are skeptical. Scientists don't accept an explanation or answer unless it is based on evidence and logical reasoning. They continually question their own conclusions and the conclusions suggested by other scientists. Scientists trust only evidence that is confirmed by other people or methods.

Scientists cannot always make observations with their own eyes. They have developed technology, such as this particle detector, to help them gather information about the smallest particles of matter.

Scientists ask questions about the physical world and seek answers through carefully controlled procedures. Here a researcher works with supercooled magnets.

Science Processes at Work

You can think of science as a continuous cycle of asking and seeking answers to questions about the world. Although there are many processes that scientists use, scientists typically do each of the following:

- Ask a question
- Determine what is known
- Investigate
- Interpret results
- Share results

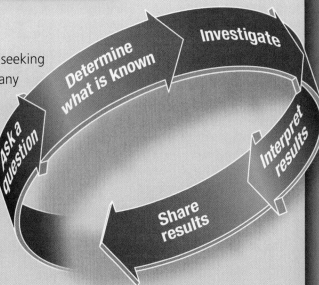

Ask a Question

It may surprise you that asking questions is an important skill. A scientific process may start when a scientist asks a question. Perhaps scientists observe an event or a process that they don't understand, or perhaps answering one question leads to another.

Determine What Is Known

When beginning an inquiry, scientists find out what is already known about a question. They study results from other scientific investigations, read journals, and talk with other scientists. A scientist working on subatomic particles is most likely a member of a large team using sophisticated equipment. Before beginning original research, the team analyzes results from previous studies.

Investigate

Investigating is the process of collecting evidence. Two important ways of investigating are observing and experimenting.

Observing is the act of noting and recording an event, a characteristic, or anything else detected with an instrument or with the senses. A researcher may study the properties of a substance by handling it, finding its mass, warming or cooling it, stretching it, and so on. For information about the behavior of subatomic particles, however, a researcher may rely on technology such as scanning tunneling microscopes, which produce images of structures that cannot be seen with the eye.

An **experiment** is an organized procedure to study something under controlled conditions. In order to study the effect of wing shape on the motion of a glider, for instance, a researcher would need to conduct controlled studies in which gliders made of the same materials and with the same masses differed only in the shape of their wings.

Scanning tunneling microscopes create images that allow scientists to observe molecular structure.

Physical chemists have found a way to observe chemical reactions at the atomic level. Using lasers, they can watch bonds breaking and new bonds forming.

Forming hypotheses and making predictions are two of the skills involved in scientific investigations. A **hypothesis** is a tentative explanation for an observation, a phenomenon, or a scientific problem that can be tested by further investigation. For example, in the mid-1800s astronomers noticed that the planet Uranus departed slightly from its expected orbit. One astronomer hypothesized that the irregularities in the planet's orbit were due to the gravitational effect of another planet—one that had not yet been detected. A **prediction** is an expectation of what will be observed or what will happen. A prediction can be used to test a hypothesis. The astronomers predicted that they would discover a new planet in the position calculated, and their prediction was confirmed with the discovery of the planet Neptune.

Interpret Results

As scientists investigate, they analyze their evidence, or data, and begin to draw conclusions. **Analyzing data** involves looking at the evidence gathered through observations or experiments and trying to identify any patterns that might exist in the data. Scientists often need to make additional observations or perform more experiments before they are sure of their conclusions. Many times scientists make new predictions or revise their hypotheses.

Often scientists use computers to help them analyze data. Computers reveal patterns that might otherwise be missed.

Scientists use computers to create models of objects or processes they are studying. This model shows carbon atoms forming a sphere.

Share Results

An important part of scientific investigation is sharing results of experiments. Scientists read and publish in journals and attend conferences to communicate with other scientists around the world. Sharing data and procedures gives them a way to test one another's results. They also share results with the public through newspapers, television, and other media.

The Nature of Technology

When you think of technology, you may think of cars, computers, and cell phones, as well as refrigerators, radios, and bicycles. Technology is not only the machines and devices that make modern lives easier, however. It is also a process in which new methods and devices are created. Technology makes use of scientific knowledge to design solutions to real-world problems.

Science and Technology

Science and technology go hand in hand. Each depends upon the other. Even designing a device as simple as a toaster requires knowledge of how heat flows and which materials are the best conductors of heat. Just as technology based on scientific knowledge makes our lives easier, some technology is used to advance scientific inquiry itself. For example, researchers use a number of specialized instruments to help them collect data. Microscopes, telescopes, spectrographs, and computers are just a few of the tools that help scientists learn more about the world. The more information these tools provide, the more devices can be developed to aid scientific research and to improve modern lives.

The Process of Technological Design

The process of technology involves many choices. For example, how does an automobile engineer design a better car? Is a better car faster? safer? cheaper? Before designing any new machine, the engineer must decide exactly what he or she wants the machine to do as well as what may be given up for the machine to do it. A faster car may get people to their destinations more quickly, but it may cost more and be less safe. As you study the technological process, think about all the choices that were made to build the technologies you use.

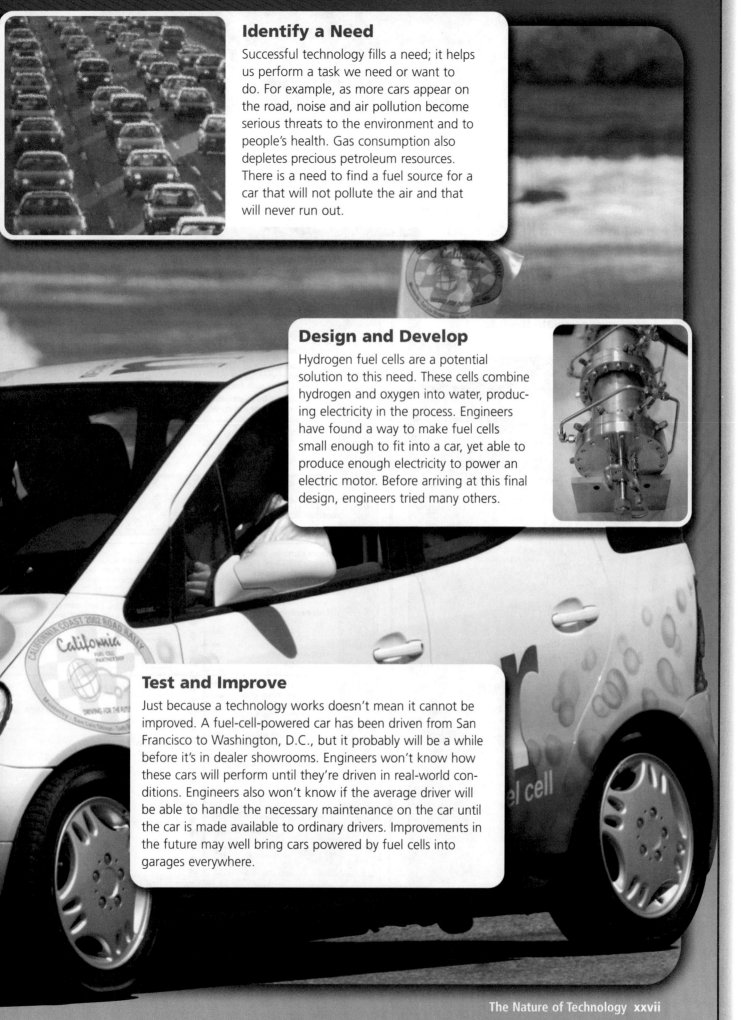

Identify a Need

Successful technology fills a need; it helps us perform a task we need or want to do. For example, as more cars appear on the road, noise and air pollution become serious threats to the environment and to people's health. Gas consumption also depletes precious petroleum resources. There is a need to find a fuel source for a car that will not pollute the air and that will never run out.

Design and Develop

Hydrogen fuel cells are a potential solution to this need. These cells combine hydrogen and oxygen into water, producing electricity in the process. Engineers have found a way to make fuel cells small enough to fit into a car, yet able to produce enough electricity to power an electric motor. Before arriving at this final design, engineers tried many others.

Test and Improve

Just because a technology works doesn't mean it cannot be improved. A fuel-cell-powered car has been driven from San Francisco to Washington, D.C., but it probably will be a while before it's in dealer showrooms. Engineers won't know how these cars will perform until they're driven in real-world conditions. Engineers also won't know if the average driver will be able to handle the necessary maintenance on the car until the car is made available to ordinary drivers. Improvements in the future may well bring cars powered by fuel cells into garages everywhere.

Using McDougal Littell Science

Reading Text and Visuals

This book is organized to help you learn. Use these boxed pointers as a path to help you learn and remember the **Big Ideas** and **Key Concepts**.

Read the Big Idea.

As you read **Key Concepts** for the chapter, relate them to **the Big Idea**.

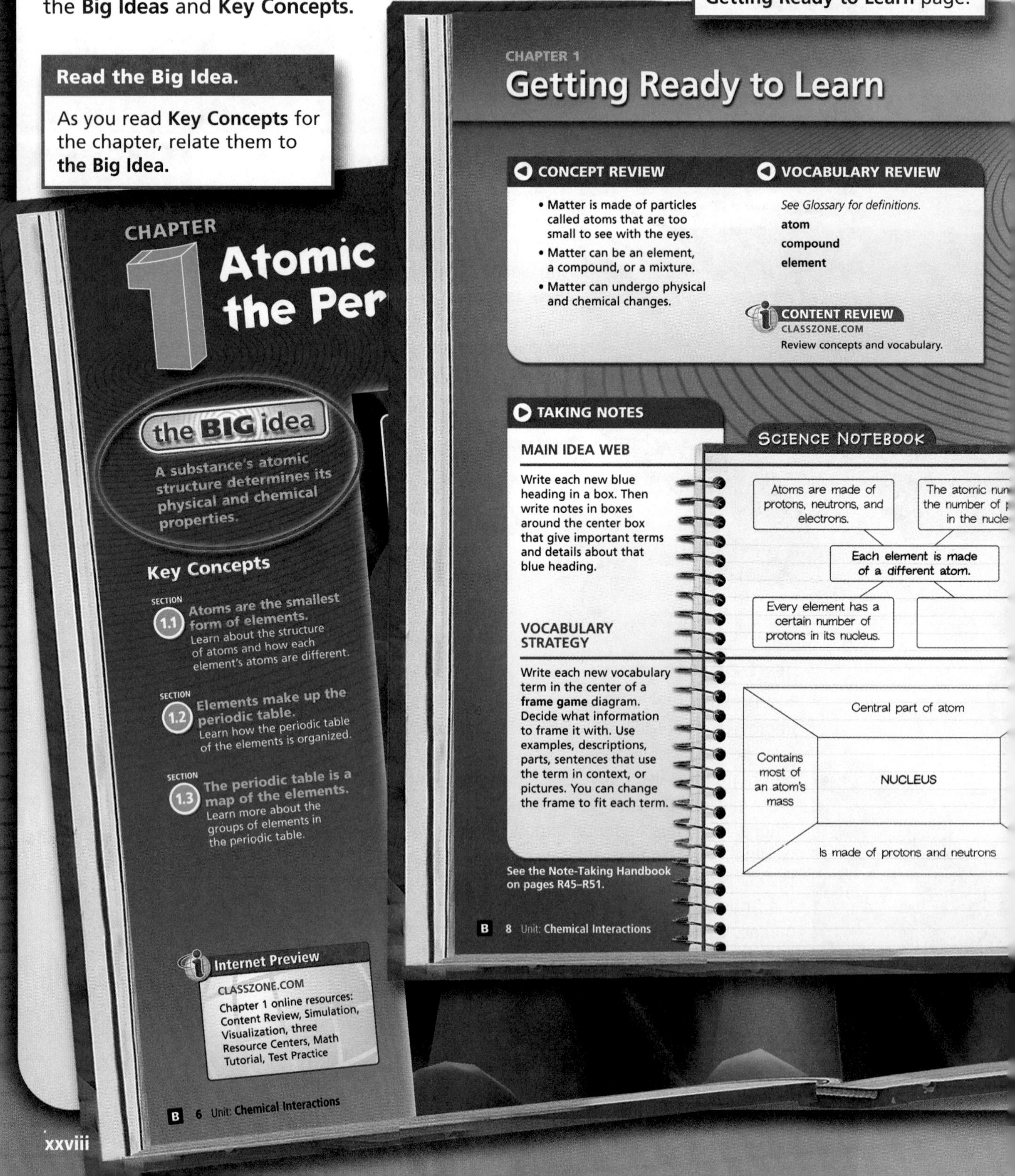

Take notes.

Use the strategies on the **Getting Ready to Learn** page.

CHAPTER

1 Atomic *the Per*

the BIG idea

A substance's atomic structure determines its physical and chemical properties.

Key Concepts

SECTION 1.1 Atoms are the smallest form of elements.
Learn about the structure of atoms and how each element's atoms are different.

SECTION 1.2 Elements make up the periodic table.
Learn how the periodic table of the elements is organized.

SECTION 1.3 The periodic table is a map of the elements.
Learn more about the groups of elements in the periodic table.

Internet Preview

CLASSZONE.COM
Chapter 1 online resources: Content Review, Simulation, Visualization, three Resource Centers, Math Tutorial, Test Practice

---

CHAPTER 1

Getting Ready to Learn

CONCEPT REVIEW

- Matter is made of particles called atoms that are too small to see with the eyes.
- Matter can be an element, a compound, or a mixture.
- Matter can undergo physical and chemical changes.

VOCABULARY REVIEW

See Glossary for definitions.

atom

compound

element

CONTENT REVIEW
CLASSZONE.COM
Review concepts and vocabulary.

TAKING NOTES

MAIN IDEA WEB

Write each new blue heading in a box. Then write notes in boxes around the center box that give important terms and details about that blue heading.

VOCABULARY STRATEGY

Write each new vocabulary term in the center of a **frame game** diagram. Decide what information to frame it with. Use examples, descriptions, parts, sentences that use the term in context, or pictures. You can change the frame to fit each term.

See the Note-Taking Handbook on pages R45–R51.

SCIENCE NOTEBOOK

Atoms are made of protons, neutrons, and electrons.

The atomic num the number of in the nucle

Each element is made of a different atom.

Every element has a certain number of protons in its nucleus.

Central part of atom

Contains most of an atom's mass

NUCLEUS

Is made of protons and neutrons

Read each heading.

See how it fits into the outline of the chapter.

KEY CONCEPT

Atoms are the smallest form of elements.

1.1

◀ **BEFORE, you learned**

- All matter is made of atoms
- Elements are the simplest substances

▶ **NOW, you will learn**

- Where atoms are found and how they are named
- About the structure of atoms
- How ions are formed from atoms

Remember what you know.

Think about concepts you learned earlier and preview what you'll learn now.

•CABULARY

•ton p. 11
•tron p. 11
•leus p. 11
•tron p. 11
•mic number p. 12
•mic mass number p. 12
•ope p. 12
• p. 14

EXPLORE The Size of Atoms

How small can you cut paper?

PROCEDURE

① Cut the strip of paper in half. Cut one of these halves in half.

② Continue cutting one piece of paper in half as many times as you can.

WHAT DO YOU THINK?

- How many cuts were you able to make?
- Do you think you could keep cutting the paper forever? Why or why not?

MATERIALS

- strip of paper about 30 centimeters long
- scissors

Try the activities.

They will introduce you to science concepts.

All matter is made of atoms.

Think of all the substances you see and touch every day. Are all of these substances the same? Obviously, the substances that make up this book you're reading are quite different from the substances in the air around you. So how many different substances can there be? This is a question people have been asking for thousands of years.

About 2400 years ago, Greek philosophers proposed that everything on Earth was made of only four basic substances—air, water, fire, and earth. Everything else contained a mixture of these four substances. As time went on, chemists came to realize that there had to be more than four basic substances. Today chemists know that about 100 basic substances, or elements, account for everything we see and touch. Sometimes these elements appear by themselves. Most often, however, these elements appear in combination with other elements to make new substances. In this section, you'll learn about the atoms of the elements that make up the world and how these atoms differ from one another.

Learn the vocabulary.

Take notes on each term.

•DING TiP

• word *element* is •ted to *elementary,* •ich means "basic."

Chapter 1: **Atomic Structure and the Periodic Table 9** **B**

Reading Text and Visuals

Study the visuals.

- Read the title.

- Read all labels and captions.

- Figure out what the picture is showing. Notice colors, arrows, and lines.

- Answer the question. **Reading Visuals** questions will help you understand the picture.

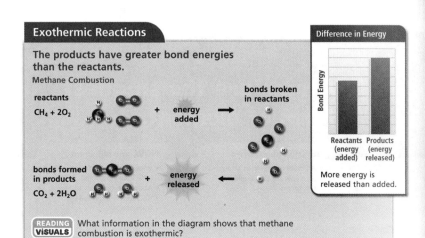

Exothermic Reactions

The products have greater bond energies than the reactants.

Methane Combustion

reactants

$CH_4 + 2O_2$ + energy added → bonds broken in reactants

bonds formed in products + energy released ← $CO_2 + 2H_2O$

Difference in Energy

Bond Energy

Reactants (energy added) Products (energy released)

More energy is released than added.

READING VISUALS What information in the diagram shows that methane combustion is exothermic?

Read one paragraph at a time.

Look for a topic sentence that explains the main idea of the paragraph. Figure out how the details relate to that idea. One paragraph might have several important ideas; you may have to reread to understand.

All common combustion reactions, such as the combustion of methane, are exothermic. To determine how energy changes in this reaction, the bond energies in the reactants—oxygen and methane—and in the products—carbon dioxide and water—can be added and compared. This process is illustrated by the diagram shown above. The difference in energy is released to the surrounding air as heat.

Some chemical reactions release excess energy as light instead of heat. For example, glow sticks work by a chemical reaction that releases energy as light. One of the reactants, a solution of hydrogen peroxide, is contained in a thin glass tube within the plastic stick. The rest of the stick is filled with a second chemical and a brightly colored dye. When you bend the stick, the glass tube inside it breaks and the two solutions mix. The result is a bright glow of light.

These jellyfish glow because of exothermic chemical reactions.

Answer the questions.

Check Your Reading questions will help you remember what you read.

Exothermic chemical reactions also occur in living things. Some of these reactions release energy as heat, and others release energy as light. Fireflies light up due to a reaction that takes place between oxygen and a chemical called luciferin. This type of exothermic reaction is not unique to fireflies. In fact, similar reactions are found in several different species of fish, squid, jellyfish, and shrimp.

CHECK YOUR READING In which ways might an exothermic reaction release energy?

B 88 Unit: **Chemical Interactions**

Doing Labs

To understand science, you have to see it in action. Doing labs helps you understand how things really work.

① Read the entire lab first.

② Form a hypothesis.

③ Follow the procedure.

④ Record the data.

CHAPTER INVESTIGATION

Chemical Bonds

OVERVIEW AND PURPOSE Chemists can identify the type of bonds in a substance by examining its properties. In this investigation you will examine the properties of different substances and use what you have learned about chemical bonds to identify the type of bond each substance contains. You will

- observe the structure of substances with a hand lens
- test the conductivity of substances
- determine the melting point of substances

▶ Problem [Write It Up]

How can you determine the type of chemical bond a substance has?

▶ Hypothesize [Write It Up]

Write three hypotheses in "If . . . , then . . . , because . . ." form to answer the problem question for each bond type—ionic, covalent, and metallic.

▶ Procedure

1. Create a data table similar to the one shown on the sample notebook page.

2. To build the conductivity tester, connect the first wire to one terminal of the battery and to one of the metal strips. Attach the second wire to the other terminal and to the lamp socket. Finally, connect the lamp socket to the third wire, and connect the other end of this wire to the second metal strip.

3. To make sure your tester works properly, touch the tips of the metal strips together. If the bulb lights, the tester is working properly. If not, check the connections carefully.

4. Get the following test compounds from your teacher: Epsom salts (MgSO₄), sugar (C₁₂H₂₂O₁₁), and iron filings (Fe). For each substance, put about 20 grams in a cup and label it.

MATERIALS
- 3 wire leads with alligator clips
- battery
- zinc and copper strips
- light bulb and socket
- test compounds
- 3 plastic cups
- distilled water
- beaker
- construction paper
- hand lens
- plastic spoon
- 3 test tubes
- test-tube rack
- candle
- wire test-tube holder

B   60 unit: Chemical Interactions

5. Test the conductivity of distilled water. Fill the beaker with 30 mL of water. Place the two metal strips into the water. Does the bulb light? Record your observations. Dry the strips completely.

6. Place dry Epsom salts on dark paper. Observe them with a hand lens. Do you see any kind of patterns in the different grains? Put the salts between the metal strips. Can you get the bulb to light by bringing the strips closer together? Record your observations.

7. Add all but a small amount of the Epsom salts to the beaker of water. Stir well. Repeat the conductivity test. What happens when you put the metal strips into the solution? Record your results.

8. Rinse and dry the beaker. Repeat steps 6–7 with other test substances. Record your results.

9. Put the remainder of each test substance into its own clean, dry test tube. Label the tubes. Light the candle. Use a test tube holder to hold each compound over the candle flame for 2 minutes. Do you notice any signs of melting? Record your observations.

▶ Observe and Analyze [Write It Up]

1. **RECORD OBSERVATIONS** Be sure you have entered all your observations in your data table.

2. **CLASSIFY** Using the periodic table, find the elements these compounds contain. How might consulting the periodic table help you determine what type of bond exists in the compound?

▶ Conclude [Write It Up]

1. **INTERPRET** Review your recorded observations. Classify the compounds as having ionic, covalent, or metallic bonds. Fill in the last column of the the data table with your conclusions.

2. **INFER** Compare your results with your hypotheses. Did your results support your hypotheses?

3. **EVALUATE** Describe possible limitations, errors, or places where errors might have occurred.

4. **APPLY** Electrocardiograms are graphs that show the electrical activity of the heart. When an electrocardiogram is made, a paste of sodium chloride is used to hold small metal discs on the patient's skin. What property of ionic compounds does this medical test make use of?

▶ INVESTIGATE Further

CHALLENGE To grow crystals, put about 60 grams of Epsom salts into a baby-food jar that is half full of hot water. Do the same using a second jar containing about 60 grams of sugar. Cover and shake the jars for a count of 60. Line two clean jar lids with dark paper. Brush or spoon a thin coating of each liquid over the paper. Let them stand in a warm place. After several days, observe the crystals that form, using a hand lens.

Chemical Bonds
Problem How can you determine the type of chemical bond a substance has?
Hypothesize
Observe and Analyze
Table 1: Properties of Bonds

Property	Epsom Salts (MgSO₄)	Sugar (C₁₂H₂₂O₁₁)	Iron Filings (Fe)	Bond Type
Crystal structure				
Conductivity of solid				
Conductivity in water				
Melting				

Conclude

Chapter 2: Chemical Bonds and Compounds 61 B

⑤ Analyze your results.

⑥ Write your lab report.

Using Technology

The Internet is a great source of information about up-to-date science. The ClassZone Web site and SciLinks have exciting sites for you to explore. Video clips and simulations can make science come alive.

Look for red banners.

Go to **classzone.com** to see simulations, visualizations, and content review.

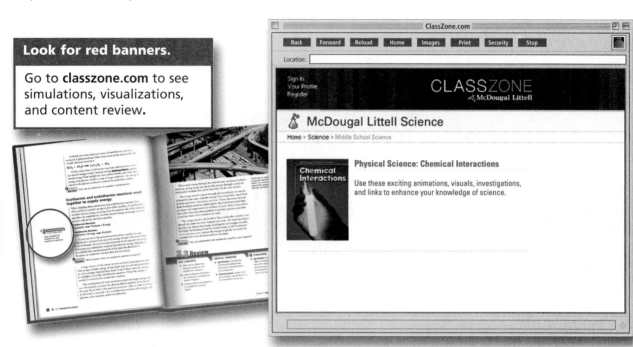

Watch the videos.

See science at work in the **Scientific American Frontiers video.**

Look up SciLinks.

Go to **scilinks.org** to explore the topic.

Forces **Code: MDL005**

Chemical Interactions
Contents Overview

VIDEO SUMMARY

SCIENTIFIC AMERICAN FRONTIERS

"Endangered Wonder Drug," a segment of the *Scientific American Frontiers* series that aired on PBS stations, traces the development of the cancer-fighting drug Taxol. The drug comes from the bark of the increasingly rare Pacific yew tree. Scientists are trying to find alternative sources in nature and in the laboratory. Taxol has been clinically tested as a treatment for women with ovarian cancer. If the drug succeeds as a cancer treatment and the demand increases, the natural supply of Taxol will soon be used up. One alternative is to grow yew trees in nurseries; another is to extract Taxol from the needles rather than the bark. Chemists are also working to make synthetic Taxol in large quantities. The ultimate solution may be a hybrid of natural and synthetic chemical compounds.

National Science Education Standards

A.9.a–d Understandings About Scientific Inquiry

E.6.a–f Understandings About Science and Technology

F.5.a–e Science and Technology in Society

G.1.a–b Science as a Human Endeavor

G.2.a Nature of Science

FRONTIERS in Science

Medicines from Nature

Where have people found medicines?

SCIENTIFIC AMERICAN FRONTIERS

View the "Endangered Wonder Drug" segment of your Scientific American Frontiers video to see how chemicals found in nature can improve the health of people.

B 2 Unit: Chemical Interactions

ADDITIONAL RESOURCES

Technology Resources

 Scientific American Frontiers Video: *Endangered Wonder Drug:* 11-minute video segment that introduces the unit.

 ClassZone.com
CAREER LINK, Chemist

Guide student viewing and comprehension of the video:

 Video Teaching Guide, pp. 1–2; Video Viewing Guide, p. 3; Video Wrap-Up, p. 4

Scientific American Frontiers Video Guide, pp. 43–46

Unit projects procedures and rubrics:

 Unit Projects, pp. 5–10

In Brazil, extracts from plants are used to treat everything from Parkinson's Disease to arthritis.

Finding Natural Remedies

In the 1960s, people were searching desperately for new cancer-fighting agents. Scientists tested over 35,000 compounds, some of which came from the bark of the Pacific yew tree, long known to have strong effects on the body. The tests indicated that something in the bark stopped the growth of cancerous tumors. Scientists eventually derived the drug Taxol from the compounds found in the yew tree. **?** **A**

Natural medicines are much more than simple folk cures, like the sapi karta leaves that the Kuna people of Panama believe increase creativity. Many powerful medicines are based on compounds found in nature. But even though these natural compounds may be very effective at treating diseases, they can be limited in supply and can have harmful side effects. Organic chemists must find ways to make these compounds safer and produce them in greater amounts. **?** **B**

Modeling the Molecule

To make a compound, a chemist must know what its molecule looks like, atom by atom. Many useful drugs have structures that contain many atoms arranged in complicated ways. The chemist must know exactly how many atoms of each kind are in the molecule and how they are arranged. One atom in the wrong place might mean that the drug won't work the way it should.

Frontiers in Science **3** **B**

FOCUS

◉ Set Learning Goals

Students will

- Learn where people find medicines.
- Compare the benefits and costs of using natural medicines.
- Analyze the process of bringing a drug to market.
- Examine molecular structure and uses of plant-derived medicines.

Some folk cures have turned out to contain drugs that perform well in scientific tests. Point out that the active ingredients in green tea, which has been used for centuries, have only been scientifically tested for efficacy during the last 40 years. Tell students that frontiers may include looking at old ways in a new light.

INSTRUCT

Scientific Process

Out of 110,000 tested compounds, very few have led to cancer-fighting drugs. Ask: What do you think are important personal qualities for a chemist who is looking for new medicines? *Sample answer: optimism and determination*

Technology Design

Point out that some natural remedies are produced in tiny quantities by plants or animals. Ask: What is the first step to finding out whether enough of a useful chemical can be synthesized? *analyze its composition and structure*

DIFFERENTIATE INSTRUCTION

◗ More Reading Support

A What disease can be treated with extracts from yew bark? *cancer*

B How do chemists solve the problem of the small quantities of natural medicines? *manufacture more*

Advanced Have students research natural remedies that have been used for so long that most people forget they were once extracted from plants, such as the compound in aspirin, a drug that originally came from willow trees.

Scientific Process

Natural sources of drugs can be unreliable because climate variation and differences in growing conditions, harvesting, and storage affect the amount of the compound produced. This variability makes it hard to test a drug or assign a dosage. Ask students to brainstorm some of the issues involved in producing reliable results under these circumstances.

Technology Design

Discuss the problem presented in "Assembling the Puzzle." Ask:

- Suppose you know the types and numbers of atoms in a molecule. What else do you need to find out to synthesize the molecule? *its structure and how to assemble it*

- How can you learn about molecular structure? *using spectroscopy and modeling*

History of Science

The drug-development process is rarely straightforward. Sometimes, a drug developed for one purpose is found to treat a completely unrelated medical condition. For example, many years ago a drug developed to treat tuberculosis was administered to patients suffering from tuberculosis in a psychiatric hospital. The drug was found to be effective, but it was also observed to have an unexpected effect. The patients given the drug were observed to become cheerful and more optimistic. By chance, a new antidepressant drug had been developed.

Updates in Science

Remind students that science changes very quickly. For example, despite the complexity of the molecule (as shown in the video segment), Taxol has now been synthesized in the laboratory. Taxol has been a boon to many cancer patients, although it is not a cure. However, some patients develop a resistance to it over time, and some types of tumors do not respond to Taxol at all.

To study the structures of molecules, chemists use a method called spectroscopy. Spectroscopy is a process that shows how the molecules of a compound respond to certain forms of radiation. Three important types of spectroscopy are

- NMR (nuclear magnetic resonance) spectroscopy, which allows chemists to identify small groups of atoms within larger molecules

- IR (infrared) spectroscopy, which shows the presence of certain types of bonds in molecules

- X-ray studies, which show details such as how much space there is between atoms and what the overall physical shapes of molecules are

Chemists put all this information together to determine the structure of a molecule. They might even build a model of the molecule.

Assembling the Puzzle

Once chemists know the structure of the molecule, they must figure out the starting reactants and the specific sequence of chemical reactions that will produce that molecule as a final product. It is a lot like doing a jigsaw puzzle when you know what the final picture looks like but still have to fit together all the pieces. Only in this case, the chemists may not even be sure what the little pieces look like.

Organic chemists often prefer to complete the process backward. They look at a model of the complete molecule and then figure out how they might build one just like it. How do chemists know what kinds of reactions might produce a certain molecule? Chemists have classified chemical reactions into different types. They determine how combinations of reactions will put the various kinds of atoms into their correct places in the molecule. Chemists may need to combine dozens of reactions to get the desired molecule.

Testing the Medicine

Once chemists have produced the desired drug molecule, the synthetic compound must be carefully tested to make sure it works like the natural substance does. The sequence of reactions must also be tested to make sure they produce the same compound when larger amounts of chemicals are used.

SCIENTIFIC AMERICAN FRONTIERS

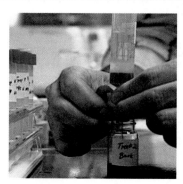

View the "Endangered Wonder Drug" segment of your *Scientific American Frontiers* video to see how modern medicines can be developed from chemical compounds found in nature.

IN THIS SCENE FROM THE VIDEO ▶

A researcher works with a substance found in bark.

SAVING LIVES THROUGH NATURE AND CHEMISTRY
Medicines from plants and other natural sources have been used by different cultures around the world for thousands of years. The ephedra plant contains the raw material for many decongestants, which help shrink swollen nasal passages. It was used by the Chinese more than 5000 years ago. Today, the bark of the Pacific yew tree is being used as the source of the anticancer drug Taxol. A large amount of bark from the tree, however, is needed to make just one dose of the drug, and very few Pacific yew trees are available. Chemists, therefore, are trying to make this medicine in the laboratory.

DIFFERENTIATE INSTRUCTION

More Reading Support

C What process uses radiation to show the shape of molecules in a compound? *spectroscopy*

Below Level Ask students to make a chart showing the steps in synthesizing a molecule. Have a group of students write down each task described in the text, then organize them into a flow chart of the sequence of events.

Once a potential new drug is found in nature, it may take several years, or even decades, to figure out how to produce the drug synthetically and test it for safety. Only a small percentage of drugs tested ever goes to market, because the drugs must undergo several stages of testing on both animals and humans. Today, chemists routinely search the seas and the forests for marine organisms and rare plants that might have the power to fight cancer, heart disease, or viruses.

Chemists often use computers to make models of drug molecules. Computers allow the chemists to see how the drug molecules will interact with other molecules.

UNIT PROJECTS

As you study this unit, work alone or with a group on one of these projects.

Medicines Around You

Present a report about a plant in your region that has medicinal properties.

- Collect samples of a plant that has medicinal properties.
- Bring your plant samples into your classroom. Prepare and present a report about the plant and the way it is used in medicine.

Model Medicine

Build a scale model of a molecule that is used to treat a certain illness.

- Using the Internet or an encyclopedia, determine the structure of a compound that interests you.
- Using foam balls, toothpicks, water colors, string, and other materials, construct a model of the molecule. Describe your model to the class.

Remedies

Write a news report about a popular herbal remedy, such as Saint John's Wort.

- To learn more about the herbal remedy, try interviewing a personal fitness trainer or an employee of a health-food store.
- Deliver a news report to the class telling of the advantages of the remedy and warning of its potential dangers.

 **CAREER CENTER**
CLASSZONE.COM

Learn more about careers in chemistry.

Have students read the questions and think of some of their own. Remind them that scientists usually end up with more questions—that inquiry is the driving force of science.

- With the class, generate on the board a list of new questions.
- Students can add to the list after they watch the Scientific American Frontiers Video.
- Students can use the list as a springboard for choosing their Unit Projects.

UNIT PROJECTS

Encourage students to pick the project that most appeals to them. Point out that each is long-term and will take several weeks to complete. You might group or pair students to work on projects and in some cases guide student choice. Some of the projects have student choice built into them. Each project has two worksheet pages, including a rubric. Use the pages to guide students through criteria, process, and schedule.

 Unit Projects, pp. 5–10

REVISIT concepts introduced in this article:

Chapter 1
- Atoms and elements, pp. 9–15
- The periodic table, pp. 17–23
- Map of the elements, pp. 26–32

Chapter 2
- Elements form compounds, pp. 41–45
- Chemical bonds, pp. 47–54
- Substances' properties depend on bonds, pp. 56–59

Chapter 3
- Chemical reactions alter arrangements, pp. 69–76
- Masses of reactants and products, pp. 78–84
- Life and industry, pp. 94–99

Chapter 4
- Solutes, pp. 117–123

Chapter 5
- Carbon-based molecules, pp. 147–152
- Carbon-based molecules are life's building blocks, pp. 154–161

DIFFERENTIATE INSTRUCTION

? More Reading Support

D Where are chemists looking for natural drug sources? *the sea and forest*

Atomic Structure and the Periodic Table

Physical Science
UNIFYING PRINCIPLES

PRINCIPLE 1

Matter is made of particles too small to see.

PRINCIPLE 2

Matter changes form and moves from place to place.

PRINCIPLE 3

Energy changes from one form to another, but it cannot be created or destroyed.

PRINCIPLE 4

Physical forces affect the movement of all matter on Earth and throughout the universe.

Unit: Chemical Interactions
BIG IDEAS

CHAPTER 1
Atomic Structure and the Periodic Table
A substance's atomic structure determines its physical and chemical properties.

CHAPTER 2
Chemical Bonds and Compounds
The properties of compounds depend on their atoms and chemical bonds.

CHAPTER 3
Chemical Reactions
Chemical reactions form new substances by breaking and making chemical bonds.

CHAPTER 4
Solutions
When substances dissolve to form a solution, the properties of a mixture change.

CHAPTER 5
Carbon in Life and Materials
Carbon is essential to living things and to modern materials.

CHAPTER 1
KEY CONCEPTS

SECTION 1.1

Atoms are the smallest form of elements.

1. All matter is made of atoms.

2. Each element is made of a different atom.

3. Atoms form ions.

SECTION 1.2

Elements make up the periodic table.

1. Elements can be organized by similarities.

2. The periodic table organizes the atoms of the elements by properties and atomic number.

SECTION 1.3

The periodic table is a map of the elements.

1. The periodic table has distinct regions.

2. Most elements are metals.

3. Nonmetals and metalloids have a wide range of properties.

4. Some atoms can change their identity.

The Big Idea Flow Chart is available on p. T1 in the **UNIT TRANSPARENCY BOOK.**

Previewing Content

1.1 Atoms are the smallest form of elements. pp. 9–16

1. All matter is made of atoms.
All matter is made up of atoms of about 100 elements, or basic substances. Hydrogen is the most abundant element in the universe; oxygen is the most abundant element in Earth's crust. Every element has a unique name and symbol. Names and symbols of the elements come from many sources.

2. Each element is made of a different atom.
Atoms are made of three smaller particles: protons, neutrons, electrons.
- At the center of an atom is the **nucleus**, which contains almost all of the atom's mass.
- The nucleus contains protons and neutrons. Protons have a positive charge and neutrons have no charge. Protons and neutrons have approximately the same mass.
- In a cloud around the nucleus are **electrons.** Electrons are 2000 times smaller than protons or neutrons. The figure below shows the position of the electron cloud and the nucleus in an atom.

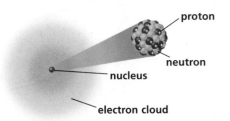

The **atomic number** is the number of protons in an atom. **Atomic mass number** is the number of protons plus the number of neutrons in the nucleus. Electrons have negligible mass. **Isotopes** are atoms of the same element with different numbers of neutrons. Since isotopes occur in various amounts in nature, the atomic mass number of an atom is the average mass of all its isotopes.

3. Atoms form ions.
Atoms form **ions** when they gain or lose electrons. Gaining electrons results in negative ions. Losing electrons results in positive ions. Atoms normally lose and gain electrons in pairs.

1.2 Elements make up the periodic table. pp. 17–25

1. Elements can be organized by similarities.
Many scientists thought the elements could be organized by their properties. Dmitri Mendeleev made the first periodic table. Mendeleev used atomic mass to order the elements and placed elements with similar properties in the same rows.

2. The periodic table organizes the atoms of the elements by properties and atomic number.
The modern periodic table is organized by atomic number. The periodic table gives the following information about each element: atomic number, chemical symbol, name, average atomic mass. It also indicates state at room temperature.

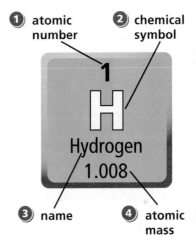

- A **group** is a column of elements. The elements in a group have similar properties.
- A **period** is a row of elements. These elements have chemical properties that tend to change the same way across the table. Properties like atomic size, density, and likelihood to form ions vary in regular ways up, down, and across the periodic table.

Common Misconceptions

NATURE OF ATOMS Students may hold the misconception that atoms are solid bits that vary in size and shape and have no space between them. Atoms of the same substance are identical in size and shape and are mostly empty space, not solid.

 This misconception is addressed on p. 10.

MISCONCEPTION DATABASE
CLASSZONE.COM Background on student misconceptions

SIZE OF ATOMS Students may hold the misconception that atoms can be viewed under a regular optical microscope. Atoms are unimaginably small, many magnitudes smaller than microscopic objects such as cells of organisms.

 This misconception is addressed on p. 12.

Previewing Content

1.3 The periodic table is a map of the elements. pp. 26–33

1. The periodic table has distinct regions.

Position in the periodic table reveals something about how **reactive** an element is. Elements in Groups 1 and 17 are especially reactive. Elements in Group 18 are the least reactive.

2. Most elements are metals.

Metals are usually shiny, often conduct electricity and heat well, and can be easily shaped and drawn into a wire.

- Alkali metals and alkaline earth metals are at the left of the periodic table and are very reactive.
- Transition metals are near the center of the periodic table and include copper, gold, silver, and iron.
- Rare earth metals are in the top row of the two rows of metals shown outside the main body of the periodic table.
- The two bottom rows are separated from the table to save space.

3. Nonmetals and metalloids have a wide range of properties.

Nonmetals appear on the right side of the periodic table. They include elements with a wide range of properties. Carbon, nitrogen, oxygen, and sulfur are nonmetals, as are the extremely reactive halogens, such as chlorine, and noble, or inert, gases, such as neon.

Metalloids lie between metals and nonmetals in the periodic table. They have characteristics of both. An important use of metalloids is in the making of semiconductors for electronic devices.

4. Some atoms can change their identity.

The atomic nucleus is held together by forces. Sometimes there can be too many or too few neutrons in a nucleus, and so the forces holding it together cannot hold it together properly. To regain its stability, the nucleus will produce particles and eject them. This process is called radioactivity. The identity of radioactive atoms changes when the number of protons change. This is called radioactive decay. Radioactive decay occurs at a steady rate that is characteristic of the particular isotope. The amount of time that it takes for one-half of the atoms in a particular sample to decay is called the half-life of the isotope. The chart shown below illustrates the progress of radioactive decay.

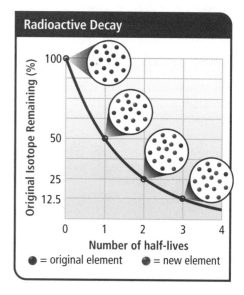

Radioactive Decay

Common Misconceptions

PROPERTIES OF ELEMENTS Students may hold the misconception that a single atom of an element will exhibit the same properties as the element in bulk. Samples of elements exhibit particular, identifiable chemical properties. At the atomic level, properties such as color and texture are meaningless.

 This misconception is addressed on p. 27.

MISCONCEPTION DATABASE
CLASSZONE.COM Background on student misconceptions

Previewing Labs

Lab Generator CD-ROM
Edit these Pupil Edition and generate alternative.

EXPLORE the BIG idea

That's Far! p. 7 Students pace out the relative distance between an electron and a nucleus, demonstrating that most of the volume of an atom is empty space.	**TIME** 10 minutes **MATERIALS** baseball, dime
Element Safari, p. 7 Students look at labels on food packages to learn that elements make up common materials.	**TIME** 10 minutes **MATERIALS** periodic table; packages of (or labels from) baking soda, vinegar, cereal flakes, antacid tablets
Internet Activity: Periodic Table, p. 7 Students explore the organization of the periodic table.	**TIME** 20 minutes **MATERIALS** computer with Internet access

SECTION 1.1

EXPLORE The Size of Atoms, p. 9 Students model the particulate nature of matter by cutting paper into smaller and smaller pieces.	**TIME** 10 minutes **MATERIALS** 30 cm strip of paper, pair of scissors
INVESTIGATE Masses of Atomic Particles, p. 13 Students use common objects to model the relative masses of atomic particles.	**TIME** 20 minutes **MATERIALS** balance; large paper clip; items available in bulk, such as sand, clay, or water

SECTION 1.2

EXPLORE Similarities and Differences of Objects, p. 17 Students classify an assortment of buttons to see how objects can be organized using different criteria.	**TIME** 15 minutes **MATERIALS** assorted buttons like those found on shirts, etc.
CHAPTER INVESTIGATION Modeling Atomic Masses, pp. 24–25 Students make model atoms from film cans and washers to determine the atom's relative masses.	**TIME** 40 minutes **MATERIALS** 5 film cans, 21 washers, balance

SECTION 1.3

INVESTIGATE Radioactivity, p. 31 Students use pennies to model the half-life of radioactive elements.	**TIME** 30 minutes **MATERIALS** 50 pennies, bag, graph paper

Additional INVESTIGATION, Investigating the Unseen, A, B, & C, pp. 59–67; Teacher Instructions, pp. 329–330

	INTEGRATED TECHNOLOGY	LABS AND ACTIVITIES

CHAPTER 1
Atomic Structure and the Periodic Table

 CLASSZONE.COM
- eEdition Plus
- EasyPlanner
- Misconception Database
- Content Review
- Test Practice
- Simulation
- Visualization
- Resource Centers
- Internet Activity: Periodic Table
- Math Tutorial

 CD-ROMS
- eEdition
- EasyPlanner
- Power Presentations
- Content Review
- Lab Generator
- Test Generator

 AUDIO CDS
- Audio Readings
- Audio Readings in Spanish

SCILINKS.ORG
 SCI LINKS

 EXPLORE the Big Idea, p. 7
- That's Far!
- Element Safari
- Internet Activity: Periodic Table

 UNIT RESOURCE BOOK
- Family Letter, p. ix
- Spanish Family Letter, p. x
- Unit Projects, pp. 5–10

 Lab Generator CD-ROM
Generate customized labs.

SECTION
1.1 Atoms are the smallest form of elements.
pp. 9–16

Time: 2 periods (1 block)
 Lesson Plan, pp. 11–12

- **RESOURCE CENTER,** The Atom, Elements Important to Life
- **SIMULATION,** Build an Atom

 UNIT TRANSPARENCY BOOK
- Big Idea Flow Chart, p. T1
- Daily Vocabulary Scaffolding, p. T2
- Note-Taking Model, p. T3
- 3-Minute Warm-Up, p. T4
- "The Atomic Model" Visual, "Isotopes" Visual, p. T6

- EXPLORE The Size of Atoms, p. 9
- INVESTIGATE Masses of Atomic Particles, p. 13
- Connecting Sciences, p. 16

 UNIT RESOURCE BOOK
- Datasheet, Masses of Atomic Particles, p. 20
- Additional INVESTIGATION, Investigating the Unseen, A, B, & C, pp. 59–67

SECTION
1.2 Elements make up the periodic table.
pp. 17–25

Time: 3 periods (1.5 blocks)
 Lesson Plan, pp. 22–23

 UNIT TRANSPARENCY BOOK
- Daily Vocabulary Scaffolding, p. T2
- 3-Minute Warm-Up, p. T4

- EXPLORE Similarities and Differences of Objects, p. 17
- CHAPTER INVESTIGATION, Modeling Atomic Masses, pp. 24–25

 UNIT RESOURCE BOOK
CHAPTER INVESTIGATION, Modeling Atomic Masses, A, B, & C, pp. 50–58

SECTION
1.3 The periodic table is a map of the elements.
pp. 26–33

Time: 3 periods (1.5 block)
 Lesson Plan, pp. 32–33

- **VISUALIZATION,** Radioactive Decay
- **MATH TUTORIAL**

 UNIT TRANSPARENCY BOOK
- Big Idea Flow Chart, p. T1
- Daily Vocabulary Scaffolding, p. T2
- 3-Minute Warm-Up, p. T5
- Chapter Outline, pp. T7–T8

- INVESTIGATE Radioactivity, p. 31
- Math in Science, p. 33

 UNIT RESOURCE BOOK
- Datasheet, Radioactivity, p. 41
- Math Support, p. 48
- Math Practice, p. 49

READING AND REINFORCEMENT

ASSESSMENT

STANDARDS

- Frame Game, B26–27
- Main Idea Web, C38–39
- Daily Vocabulary Scaffolding, H1–8

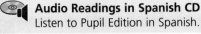 **UNIT RESOURCE BOOK**
- Vocabulary Practice, pp. 45–46
- Decoding Support, p. 47
- Summarizing the Chapter, pp. 68–69

 • Chapter Review, pp. 35–36
- Standardized Test Practice, p. 37

 UNIT ASSESSMENT BOOK
- Diagnostic Test, pp. 1–2
- Chapter Test, A, B, & C, pp. 6–17
- Alternative Assessment, pp. 18–19

 Spanish Chapter Test, pp. 233–236

National Standards
A.2–8, A.9.a–c, A.9.e–f, B.1.b, B.1.c, G.1.b, G.2.a–b

See p. 6 for the standards.

Audio Readings CD
Listen to Pupil Edition.

Audio Readings in Spanish CD
Listen to Pupil Edition in Spanish.

Test Generator CD-ROM
Generate customized tests.

Lab Generator CD-ROM
Rubrics for Labs

 UNIT RESOURCE BOOK
- Reading Study Guide, A & B, pp. 13–16
- Spanish Reading Study Guide, pp. 17–18
- Challenge and Extension, p. 19
- Reinforcing Key Concepts, p. 21

 Ongoing Assessment, pp. 10–12, 14

 Section 1.1 Review, p. 15

 UNIT ASSESSMENT BOOK
Section 1.1 Quiz, p. 3

National Standards
A.2–8, A.9.a–b, A.9.e–f, B.1.c

UNIT RESOURCE BOOK
- Reading Study Guide, A & B, pp. 24–27
- Spanish Reading Study Guide, pp. 28–29
- Challenge and Extension, p. 30
- Reinforcing Key Concepts, p. 31

 Ongoing Assessment, pp. 18, 22–23

 Section 1.2 Review, p. 23

 UNIT ASSESSMENT BOOK
Section 1.2 Quiz, p. 4

National Standards
A.2–8, A.9.a–c, A.9.e–f, B.1.b, G.2.a–b

 UNIT RESOURCE BOOK
- Reading Study Guide, A & B, pp. 34–37
- Spanish Reading Study Guide, pp. 38–39
- Challenge and Extension, p. 40
- Reinforcing Key Concepts, p. 42
- Challenge Reading, pp. 43–44

 Ongoing Assessment, pp. 26–27, 29–30

 Section 1.3 Review, p. 32

 UNIT ASSESSMENT BOOK
Section 1.3 Quiz, p. 5

National Standards
A.2–8, A-9.a–c, A.9.e–f, B.1.b

Previewing Resources for Differentiated Instruction

CHAPTER INVESTIGATION

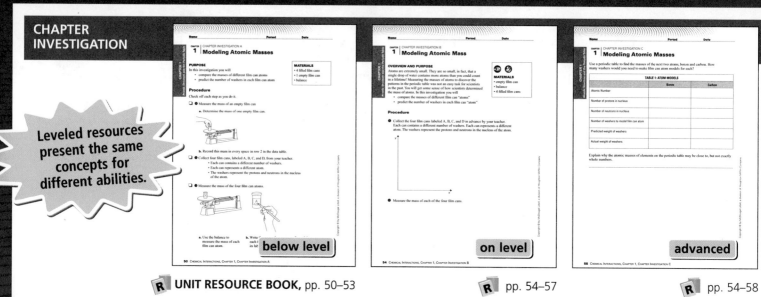

Leveled resources present the same concepts for different abilities.

below level

on level

advanced

R **UNIT RESOURCE BOOK,** pp. 50–53   R pp. 54–57   R pp. 54–58

READING STUDY GUIDE

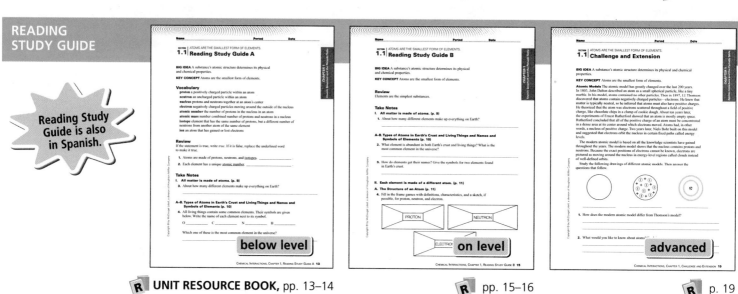

Reading Study Guide is also in Spanish.

below level

on level

advanced

R **UNIT RESOURCE BOOK,** pp. 13–14   R pp. 15–16   R p. 19

CHAPTER TEST

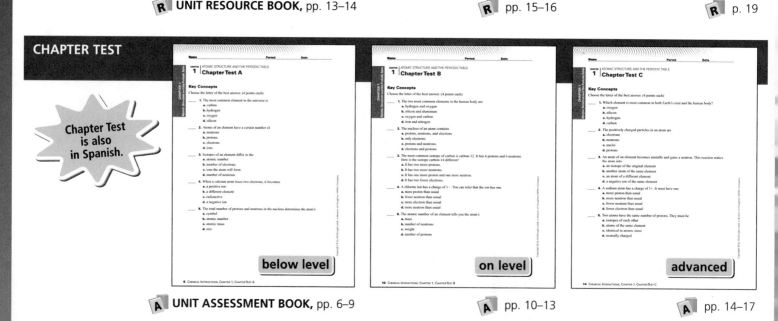

Chapter Test is also in Spanish.

below level

on level

advanced

A **UNIT ASSESSMENT BOOK,** pp. 6–9   A pp. 10–13   A pp. 14–17

TECHNOLOGY

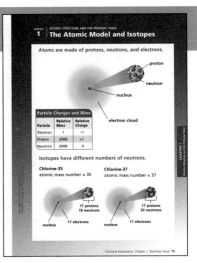

There are three Resource Centers for this chapter.

CLASSZONE.COM　　　**CD/CD-ROMS**　　　**CLASSZONE.COM**

VISUAL CONTENT

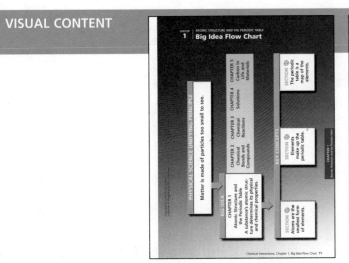

UNIT TRANSPARENCY BOOK, p. T1　　　**T** p. T3　　　**T** p. T6

MORE SUPPORT

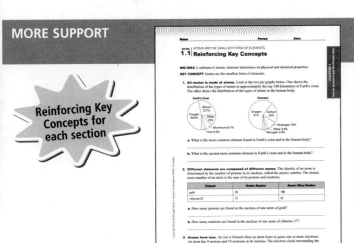

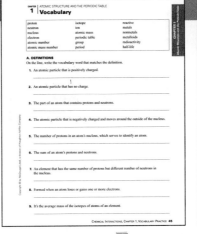

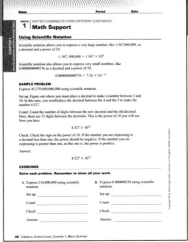

Reinforcing Key Concepts for each section

UNIT RESOURCE BOOK, p. 21　　　**R** pp. 45–46　　　**R** p. 48

INTRODUCE

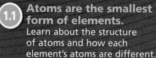

Have students look at the image of nickel atoms and discuss how the question in the box links to the Big Idea:

- How do you know the picture of this object is not the way you would normally view the object?

- Could you reproduce this view by holding an ordinary magnifying glass over a nickel-plated cup? Explain why or why not.

- What might account for the differences in the way nickel looks in the photo and on a nickel-plated cup?

National Science Education Standards

Content

B.1.b Substances often are placed in categories or groups if they react in similar ways; metals is such a group.

B.1.c Chemical elements do not break down during normal laboratory reactions involving such treatments as heating, exposure to electric current, or reaction with acids. There are more than 100 known elements that combine in a multitude of ways to produce compounds, which account for the living and non-living substances that we encounter.

Process

A.2–8 Design and conduct an investigation; use tools to gather and interpret data; use evidence; think critically between evidence and explanation; recognize different explanations and predictions; communicate scientific procedures and explanations; use mathematics.

A.9.a–c, A.9.e–f Understand scientific inquiry by using different investigations, methods, mathematics, and explanations based on logic, evidence, and skepticism.

G.1.b Science requires different abilities.

G.2.a–b Nature of Science

CHAPTER

Atomic Structure and the Periodic Table

the BIG idea

A substance's atomic structure determines its physical and chemical properties.

You can't zoom in any closer than this! The picture is an extremely close-up view of nickel. How do things look different the closer you get to them?

Key Concepts

SECTION
 Atoms are the smallest form of elements.
Learn about the structure of atoms and how each element's atoms are different.

SECTION
 Elements make up the periodic table.
Learn how the periodic table of the elements is organized.

SECTION
 The periodic table is a map of the elements.
Learn more about the groups of elements in the periodic table.

Internet Preview

CLASSZONE.COM

Chapter 1 online resources: Content Review, Simulation, Visualization, three Resource Centers, Math Tutorial, Test Practice

B 6 Unit: **Chemical Interactions**

INTERNET PREVIEW

CLASSZONE.COM For student use with the following pages:

Review and Practice
- Content Review, pp. 8, 34
- Math Tutorial: Scientific Notation, p. 33
- Test Practice, p. 37

Activities and Resources
- Internet Activity, p. 7
- Resource Center: The Atom, p. 11, Elements Important to Life, p. 16
- Simulation: Build an Atom, p. 12
- Visualization: Radioactive Decay, p. 32

NSTA
scilinks.org

SCILINKS

Atomic Theory
Code: MDL022

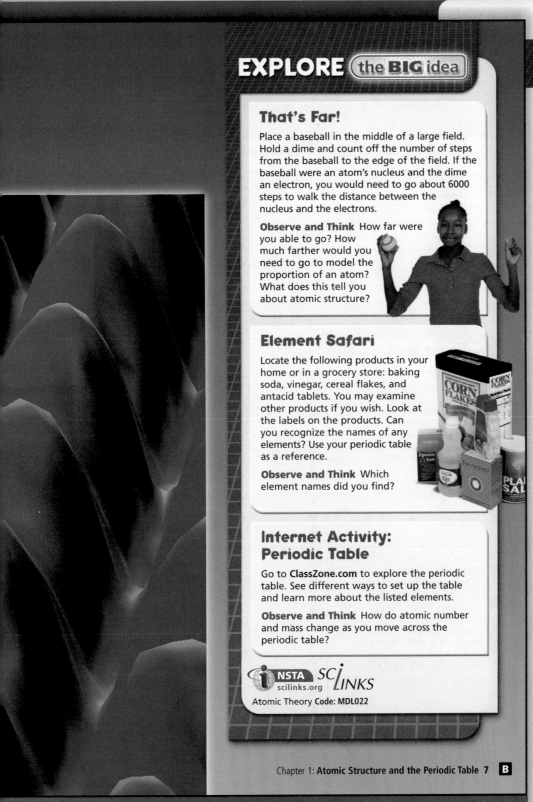

That's Far!

Place a baseball in the middle of a large field. Hold a dime and count off the number of steps from the baseball to the edge of the field. If the baseball were an atom's nucleus and the dime an electron, you would need to go about 6000 steps to walk the distance between the nucleus and the electrons.

Observe and Think How far were you able to go? How much farther would you need to go to model the proportion of an atom? What does this tell you about atomic structure?

Element Safari

Locate the following products in your home or in a grocery store: baking soda, vinegar, cereal flakes, and antacid tablets. You may examine other products if you wish. Look at the labels on the products. Can you recognize the names of any elements? Use your periodic table as a reference.

Observe and Think Which element names did you find?

Internet Activity: Periodic Table

Go to **ClassZone.com** to explore the periodic table. See different ways to set up the table and learn more about the listed elements.

Observe and Think How do atomic number and mass change as you move across the periodic table?

NSTA scilinks.org SC*L*INKS

Atomic Theory Code: MDL022

Chapter 1: **Atomic Structure and the Periodic Table 7** **B**

EXPLORE (the **BIG** idea)

These inquiry-based activities are appropriate for use at home or as a supplement to classroom instruction.

That's Far!

PURPOSE To introduce the concept that most of the volume of an atom is empty space between the nucleus and electrons. Students demonstrate this by pacing out the distance between an electron and a nucleus.

TIP *10 min.* Emphasize that although the scale isn't exactly correct, the idea is the great amount of space.

Answer: probably 50 to 100 steps; if students took 50 steps, they'd need to go more than 100 times that distance to model the proportion of an atom. An atom has a lot of empty space.

REVISIT after p. 12.

Element Safari

PURPOSE To introduce the concept that common materials are made up of elements. Students look for names on product packages.

TIP *10 min.* Encourage students to work with their parents on this activity. Inform students that the names of elements in compound names may be spelled differently than they are in the periodic table.

Answer: Students will probably find sodium (Na), hydrogen (H), and oxygen (O), and perhaps chlorine (Cl).

REVISIT after p. 21.

Internet Activity: Periodic Table

PURPOSE To introduce the logical arrangement of the periodic table.

TIP *20 min.* Encourage students to have a parent assist them with this activity.

Answer: They increase.

REVISIT after p. 30.

TEACHING WITH TECHNOLOGY

Graphing Software If you have graphing software, have students make a pie chart using figures from the "Mass of Elements in 50 kg Human" table on p. 16. Students can also use software or a graphing calculator to create graphs for "Modeling Atomic Masses," pp. 24–25, or "Investigate Radioactivity" on p. 31.

Scanning Electron Microscope Tell students that the photograph on pp. 6–7 is an image from a scanning electron microscope (SEM). Search the Internet for other images from SEM or STM (scanning tunneling microscope) technology.

PREPARE

CONCEPT REVIEW

Activate Prior Knowledge

- Place a penny, an iron nail, and a piece of graphite from a broken pencil point on a white sheet of paper.

- Ask students if they think that each of these samples is a basic substance that contains only one kind of atom.

- Remind students that even the tiniest sample of a simple substance contains millions and billions of the same kind of atom, because atoms are so tiny.

- Point out that if any of the three substances contains more than one kind of atom, it can be broken down chemically into two or more different substances. Substances are not simple because they look simple; substances are simple because they contain only one kind of atom.

TAKING NOTES

Main Idea Web

Students should make clear which box contains the main idea by centering the box and highlighting it in some way.

Vocabulary Strategy

The frame game is a good way for students to make associations by grouping information around a vocabulary word. The framing data might include examples, a definition, details, and characteristics.

Vocabulary and Note-Taking Resources

- Vocabulary Practice, pp. 45–46
- Decoding Support, p. 47

- Daily Vocabulary Scaffolding, p. T2
- Note-Taking Model, p. T3

- Frame Game, B26–27
- Main Idea Web, C38–39
- Daily Vocabulary Scaffolding, H1–8

CONCEPT REVIEW

- Matter is made of particles called atoms that are too small to see with the eyes.
- Matter can be an element, a compound, or a mixture.
- Matter can undergo physical and chemical changes.

VOCABULARY REVIEW

See Glossary for definitions.

atom

compound

element

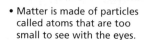
CONTENT REVIEW
CLASSZONE.COM
Review concepts and vocabulary.

TAKING NOTES

MAIN IDEA WEB

Write each new blue heading in a box. Then write notes in boxes around the center box that give important terms and details about that blue heading.

VOCABULARY STRATEGY

Write each new vocabulary term in the center of a **frame game** diagram. Decide what information to frame it with. Use examples, descriptions, parts, sentences that use the term in context, or pictures. You can change the frame to fit each term.

See the Note-Taking Handbook on pages R45–R51.

SCIENCE NOTEBOOK

Atoms are made of protons, neutrons, and electrons.

The atomic number is the number of protons in the nucleus.

Each element is made of a different atom.

Every element has a certain number of protons in its nucleus.

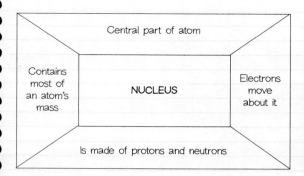

Central part of atom

Contains most of an atom's mass

NUCLEUS

Electrons move about it

Is made of protons and neutrons

CHECK READINESS

Administer the Diagnostic Test to determine students' readiness for new science content and their mastery of requisite math skills.

Diagnostic Test, pp. 1–2

Technology Resources

Students needing content and math skills should visit **ClassZone.com**.

- **CONTENT REVIEW**
- **MATH TUTORIAL**

CONTENT REVIEW CD-ROM

KEY CONCEPT

1.1 Atoms are the smallest form of elements.

◀ **BEFORE, you learned**
- All matter is made of atoms
- Elements are the simplest substances

▶ **NOW, you will learn**
- Where atoms are found and how they are named
- About the structure of atoms
- How ions are formed from atoms

VOCABULARY

proton p. 11
neutron p. 11
nucleus p. 11
electron p. 11
atomic number p. 12
atomic mass number p. 12
isotope p. 12
ion p. 14

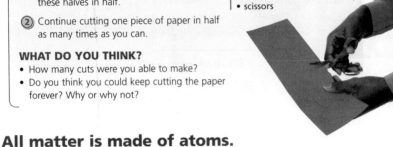

EXPLORE The Size of Atoms

How small can you cut paper?

PROCEDURE

1. Cut the strip of paper in half. Cut one of these halves in half.

2. Continue cutting one piece of paper in half as many times as you can.

WHAT DO YOU THINK?
- How many cuts were you able to make?
- Do you think you could keep cutting the paper forever? Why or why not?

MATERIALS
- strip of paper about 30 centimeters long
- scissors

All matter is made of atoms.

Think of all the substances you see and touch every day. Are all of these substances the same? Obviously, the substances that make up this book you're reading are quite different from the substances in the air around you. So how many different substances can there be? This is a question people have been asking for thousands of years.

About 2400 years ago, Greek philosophers proposed that everything on Earth was made of only four basic substances—air, water, fire, and earth. Everything else contained a mixture of these four substances. As time went on, chemists came to realize that there had to be more than four basic substances. Today chemists know that about 100 basic substances, or elements, account for everything we see and touch. Sometimes these elements appear by themselves. Most often, however, these elements appear in combination with other elements to make new substances. In this section, you'll learn about the atoms of the elements that make up the world and how these atoms differ from one another.

READING TiP

The word *element* is related to *elementary*, which means "basic."

◐ Set Learning Goals

Students will

- Recognize where atoms of some common elements are found and how they are named.

- Describe atomic structure and how that structure determines an element's identity.

- Explain how ions form from atoms.

- Model the relative masses of atomic particles in an experiment.

◔ 3-Minute Warm-Up

Display Transparency 4 or copy this exercise on the board:

Match the definitions with the terms.

Definitions

1. Measure of how much matter an object contains. *c*

2. Smallest particle of an element that has the chemical properties of that element. *a*

3. Amount of space matter takes up. *b*

Terms

a. atom

b. volume

c. mass

 3-Minute Warm-Up, p. T4

1.1 MOTIVATE

EXPLORE The Size of Atoms

PURPOSE To introduce the particulate nature of matter by modeling the tiny size of atoms

TIP *10 min.* Have students predict how many times they will cut the paper in half.

WHAT DO YOU THINK? *Students will make about five to seven cuts. Every substance has a smallest possible piece. Tell students that it would take about 31 cuts to get to the size of an average atom.*

RESOURCES FOR DIFFERENTIATED INSTRUCTION

Below Level

UNIT RESOURCE BOOK
- Reading Study Guide A, pp. 13–14
- Decoding Support, p. 47

AUDIO CDS

R Additional INVESTIGATION,
Investigating the Unseen, A, B, & C, pp. 59–67;
Teacher Instructions, pp. 329–330

Advanced

UNIT RESOURCE BOOK
Challenge and Extension, p. 19

English Learners

UNIT RESOURCE BOOK
Spanish Reading Study Guide, pp. 17–18

AUDIO CDS

- Audio Readings in Spanish
- Audio Readings (English)

Mathematics Connection

Explain to students that not all pie charts (or other charts that display gathered information) are completely secure in their data. Deviations might occur.

Address Misconceptions

IDENTIFY Ask: If you had a microscope powerful enough to see atoms in pure gold, what would the atoms look like? Ask students to sketch them. If students draw atoms of various shapes and sizes with no space between them, they hold the misconception that atoms of the same element are solid bits that vary in size and shape and are jammed together.

CORRECT Have students imagine a wall covered with evenly spaced black dots. Then have them imagine backing away from the wall: the dots would appear smaller and closer together as students get farther away. Eventually, the wall would look solid black. Say that the black dots are like the atoms in matter, which merge into the form of the matter—solid, liquid, or gas.

REASSESS Ask: If you could see the atoms in a grain of sand, what would they look like? *dots with space around them*

Technology Resources

Visit **ClassZone.com** for background on common student misconceptions.

 MISCONCEPTION DATABASE

Ongoing Assessment

Recognize where atoms of common elements are found and how they are named.

Ask: In terms of percentages, which has more oxygen, Earth's crust or a human body? *a human body*

 Answer: hydrogen
CHECK YOUR READING

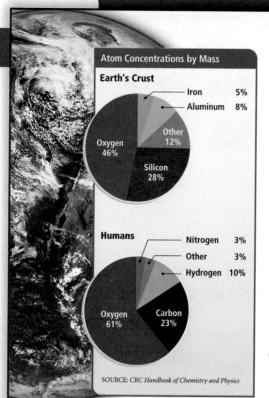

Atom Concentrations by Mass

Earth's Crust

- Iron 5%
- Aluminum 8%
- Other 12%
- Oxygen 46%
- Silicon 28%

Humans

- Nitrogen 3%
- Other 3%
- Hydrogen 10%
- Oxygen 61%
- Carbon 23%

SOURCE: *CRC Handbook of Chemistry and Physics*

Types of Atoms in Earth's Crust and Living Things

Atoms of the element hydrogen account for about 90 percent of the total mass of the universe. Hydrogen atoms make up only about 1 percent of Earth's crust, however, and most of those hydrogen atoms are combined with oxygen atoms in the form of water. The graph on the left shows the types of atoms in approximately the top 100 kilometers of Earth's crust.

The distribution of the atoms of the elements in living things is very different from what it is in Earth's crust. Living things contain at least 25 types of atoms. Although the amounts of these atoms vary somewhat, all living things—animals, plants, and bacteria—are composed primarily of atoms of oxygen, carbon, hydrogen, and nitrogen. As you can see in the lower graph on the left, oxygen atoms account for more than half your body's mass.

 CHECK YOUR READING What is the most common element in the universe?

Names and Symbols of Elements

Elements get their names in many different ways. Magnesium, for example, was named for the region in Greece known as Magnesia. Lithium comes from the Greek word *lithos,* which means "stone." Neptunium was named after the planet Neptune. The elements einsteinium and fermium were named after scientists Albert Einstein and Enrico Fermi.

Each element has its own unique symbol. For some elements, the symbol is simply the first letter of its name.

hydrogen (H)   sulfur (S)   carbon (C)

The symbols for other elements use the first letter plus one other letter of the element's name. Notice that the first letter is capitalized but the second letter is not.

aluminum (Al)   platinum (Pt)   cadmium (Cd)   zinc (Zn)

The origins of some symbols, however, are less obvious. The symbol for gold (Au), for example, doesn't seem to have anything to do with the element's name. The symbol refers instead to gold's name in Latin, *aurum.* Lead (Pb), iron (Fe), and copper (Cu) are a few other elements whose symbols come from Latin names.

B 10 Unit: **Chemical Interactions**

DIFFERENTIATE INSTRUCTION

? More Reading Support

A What element makes up more than half of your body's mass? *oxygen*

B From where do elements get their names? *from people, places, Greek words*

English Learners Some English learners may not be familiar with how pie charts display information. Use a simpler example than the "Atoms Concentrations by Mass" charts. For example, create a pie chart displaying the numbers of different hair colors among students in the class.

Inclusion Help students with hearing impairments identify the names of elements presented on this page. Write each element's name on the board; then pronounce it distinctly for the students.

Each element is made of a different atom.

In the early 1800s British scientist John Dalton proposed that each element is made of tiny particles called atoms. Dalton stated that all of the atoms of a particular element are identical but are different from atoms of all other elements. Every atom of silver, for example, is similar to every other atom of silver but different from an atom of iron.

Dalton's theory also assumed that atoms could not be divided into anything simpler. Scientists later discovered that this was not exactly true. They found that atoms are made of even smaller particles.

RESOURCE CENTER
CLASSZONE.COM
Learn more about the atom.

The Structure of an Atom

A key discovery leading to the current model of the atom was that atoms contain charged particles. The charge on a particle can be either positive or negative. Particles with the same type of charge repel each other—they are pushed apart. Particles with different charges attract each other—they are drawn toward each other.

Atoms are composed of three types of particles—electrons, protons, and neutrons. A **proton** is a positively charged particle, and a **neutron** is an uncharged particle. The neutron has approximately the same mass as a proton. The protons and neutrons of an atom are grouped together in the atom's center. This combination of protons and neutrons is called the **nucleus** of the atom. Because it contains protons, the nucleus has a positive charge. **Electrons** are negatively charged particles that move around outside the nucleus.

VOCABULARY
Remember to make a frame for *neutron, proton,* and *electron* and for other vocabulary terms.

The Atomic Model

Atoms are made of protons, neutrons, and electrons.

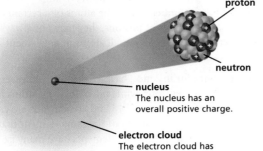

proton

neutron

nucleus
The nucleus has an overall positive charge.

electron cloud
The electron cloud has a negative charge.

Particle Charges and Mass		
Particle	Relative Mass	Relative Charge
Electron	1	−1
Proton	2000	+1
Neutron	2000	0

READING VISUALS Which part of the atom has a negative charge?

History of Science

Around 450 B.C., the Greek philosopher Democritus described matter as being made of small, indestructible atoms. Although some other Greeks contributed to this theory, most others rejected it, and it was discredited until the 18th century. In 1704, Isaac Newton described matter as being made of indivisible particles. In 1803, John Dalton proposed the modern atomic theory, which stated that each element had its own atom.

Real World Example

Beams of electrons have many practical uses. In a television or computer, rapidly moving electron beams create the images on the screen. In a scanning electron microscope, an electron beam produces images of very small objects.

Teach from Visuals

To help students interpret the diagram "The Atomic Model," ask:

- What part of the atom is pulled out and enlarged to show more detail? *the nucleus, in order to show protons and neutrons*

- Why would it be impossible to show the exact dimensions of an atom on paper? *because the nucleus is so small compared with total atom space, and because the distance between the nucleus and electrons is so great*

 This visual is also available as T6 in the Unit Transparency Book.

Ongoing Assessment

Describe atomic structure.

Ask: What are the major parts of an atom? *protons and neutrons in the nucleus, electrons in a cloud around the nucleus*

READING VISUALS *Answer: the electron*

DIFFERENTIATE INSTRUCTION

 More Reading Support

C What particles are grouped in the atom's nucleus? *protons and neutrons*

D What particles move around outside the nucleus? *electrons*

Advanced Challenge students to find out about quarks—smaller, more fundamental particles of atoms. Since quarks' existence is still hypothetical, have students report on the evidence used to support their existence.

R Challenge and Extension, p. 19

English Learners Have English learners write each vocabulary term and its definition on separate index cards, or put the terms on the Science Word Wall so students have quick and easy reminders to refer to during the lesson.

Teach Difficult Concepts

Students may conceive of the electron cloud as a mostly filled-in space, such as cotton candy or plastic foam. Point out that electrons take up almost no space—they have a tiny mass. The electron cloud is drawn to show where electrons are likely to be around the atom. The electron cloud is not a physical thing.

EXPLORE (the BIG idea)

Revisit "That's Far!" on p. 7. Have students reinterpret their observations.

Address Misconceptions

IDENTIFY Ask: If you put a grain of sand under a microscope, would you be able to see the atoms in the sand, or would the atoms be too small? If students answer that they would be able to see the atoms, they hold a misconception of the size of atoms.

CORRECT Point out that even a grain of sand has room for millions of atoms. Nothing in the classroom could possibly magnify an atom enough to make it visible.

REASSESS Ask: How many times would you have to enlarge the period at the end of a sentence in order to view any atoms? *at least a million times*

Technology Resources

Visit **ClassZone.com** for background on common student misconceptions.

MISCONCEPTION DATABASE

Ongoing Assessment

CHECK YOUR READING *Answer: Atomic mass number includes the protons and neutrons in the nucleus. Atomic number is the number of protons in the nucleus.*

SIMULATION
CLASSZONE.COM

Build a model of an atom.

Gold has 79 protons and 79 electrons.

READING TiP
The *iso-* in *isotope* is from the Greek language, and it means "equal."

Atoms are extremely small, about 10^{-10} meters in diameter. This means that you could fit millions of atoms in the period at the end of this sentence. The diagram on page 11, picturing the basic structure of the atom, is not drawn to scale. In an atom the electron cloud is about 10,000 times the diameter of the nucleus.

Electrons are much smaller than protons or neutrons—about 2000 times smaller. Electrons also move about the nucleus very quickly. Scientists have found that it is not possible to determine their exact positions with any certainty. This is why we picture the electrons as being in a cloud around the nucleus.

The negative electrons remain associated with the nucleus because they are attracted to the positively charged protons. Also, because electrical charges that are alike (such as two negative charges) repel each other, electrons remain spread out in the electron cloud. Neutral atoms have no overall electrical charge because they have an equal number of protons and electrons.

Atomic Numbers

If all atoms are composed of the same particles, how can there be more than 100 different elements? The identity of an atom is determined by the number of protons in its nucleus, called the **atomic number.** Every hydrogen atom—atomic number 1—has exactly one proton in its nucleus. Every gold atom has 79 protons, which means the atomic number of gold is 79.

? E

Atomic Mass Numbers

? F

The total number of protons and neutrons in an atom's nucleus is called its **atomic mass number.** While the atoms of a certain element always have the same number of protons, they may not always have the same number of neutrons, so not all atoms of an element have the same atomic mass number.

All chlorine atoms, for instance, have 17 protons. However, some chlorine atoms have 18 neutrons, while other chlorine atoms have 20 neutrons. Atoms of chlorine with 18 and 20 neutrons are called chlorine isotopes. **Isotopes** are atoms of the same element that have a different number of neutrons. Some elements have many isotopes, while other elements have just a few.

CHECK YOUR READING How is atomic mass number different from atomic number?

Atom Size

Millions of atoms could fit in a space the size of this dot. It would take you 500 years to count the number of atoms in a grain of salt.

DIFFERENTIATE INSTRUCTION

? **More Reading Support**

E What particles are counted to determine atomic number? *protons*

F What particles are counted for atomic mass number? *protons and neutrons*

Below Level Use this analogy to illustrate the distinction between atomic number and atomic mass number: Suppose you have a bag of hard candies. There are 10 "hot red" candies and 10 "cool white" candies in the bag. Imagine that this bag stands for an atom, and that the hot red candies are protons and the cool white candies are neutrons. How would you figure out the atomic number? *You'd count the hot red candies—10.* How would you figure out the atomic mass number? *You'd add up all the candies—20.*

Isotopes

Isotopes have different numbers of neutrons.

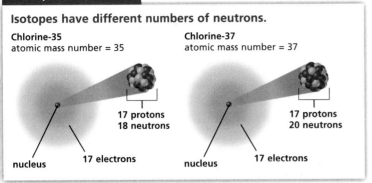

Chlorine-35
atomic mass number = 35

17 protons
18 neutrons

nucleus   17 electrons

Chlorine-37
atomic mass number = 37

17 protons
20 neutrons

nucleus   17 electrons

A particular isotope is designated by the name of the element and the total number of its protons and neutrons. You can find the number of neutrons in a particular isotope by subtracting the atomic number from the atomic mass number. For example, chlorine-35 indicates the isotope of chlorine that has 18 neutrons. Chlorine-37 has 20 neutrons. Every atom of a given element always has the same atomic number because it has the same number of protons. However, the atomic mass number varies depending on the number of neutrons.

INVESTIGATE Masses of Atomic Particles

How can you model the relative masses of atomic particles?

PROCEDURE

1. Use a paper clip to represent the mass of an electron. Determine its mass.
2. Find a substance in the classroom (sand, clay, water) from which you could make a model representing the mass of a proton or neutron. The mass of a proton or neutron is about 2000 times the mass of an electron.
3. Measure out the substance until you have enough of it to make your model.

WHAT DO YOU THINK?
- What substance did you use to make your model?
- What was the model's mass?
- What do you conclude about the masses of atomic particles?

CHALLENGE The diameter of an electron is approximately 1/2000 that of a proton. What two objects could represent each of these to scale?

SKILL FOCUS
Modeling

MATERIALS
- balance
- large paper clip
- other items

TIME
20 minutes

Chapter 1: **Atomic Structure and the Periodic Table** 13 **B**

DIFFERENTIATE INSTRUCTION

More Reading Support

G Do isotopes have different numbers of protons or of neutrons? *neutrons*

Additional Investigation To reinforce Section 1.1 learning goals, use the following full-period investigation:

Additional INVESTIGATION, Investigating the Unseen, A, B, & C, pp. 59–67, 329–330
(Advanced students should complete Levels B and C.)

Teach from Visuals

To help students interpret the "Isotopes" visual, ask:
- How are the two chlorine atoms similar? *Both have 17 electrons, and a nucleus with 17 protons.*
- How are they different? *The first has 18 neutrons, and the second has 20 neutrons.*

T This visual is also available as T6 in the Unit Transparency Book.

INVESTIGATE Masses of Atomic Particles

PURPOSE To model the relative masses of atomic particles

TIPS *20 min.*
- At the beginning of the activity, ask students to state what the paper clip stands for (electron) and what the other object stands for (proton or neutron).
- During the activity, have students restate what the objects with which they are working represent.

WHAT DO YOU THINK? *Answers will vary. Students should conclude that protons (or neutrons) are much more massive than electrons.*

CHALLENGE *The diameter of a proton is 1×10^{-15} meter and of the electron, about 1×10^{-18} meter. Emphasize to students that as yet it is impossible for scientists to determine exactly the relative sizes of protons and electrons.*

R Datasheet, Masses of Atomic Particles, p. 20

Technology Resources

Customize this student lab as needed or look for an alternative. Print rubrics to assess student lab reports.

Lab Generator CD-ROM

Teach from Visuals

To help students interpret the diagrams of the sodium atom and ion, ask:

- In each of the two pictures, what does the dark clump in the center stand for? *the nucleus*

- What does the surrounding cloud stand for? *the electrons*

- What charge does each part have? *The nucleus has a positive charge. The electron cloud has a negative charge.*

- What creates the difference in size between the atom and the ion? *the number of electrons; the positive ion has fewer electrons*

Teach Difficult Concepts

Students may have difficulty understanding why atoms sometimes lose or gain electrons to form ions. Explain that atoms often form ions because having a certain number of electrons in the electron cloud is more electrically stable. Also emphasize that ions form in pairs: when one atom loses an electron, another atom gains an electron.

Ongoing Assessment

Explain how ions form from atoms.

Ask: If an atom gains an electron, what does it become? *a negative ion*

CHECK YOUR READING *Answer: An atom must lose an electron.*

MAIN IDEA WEB
Make a main idea web to organize what you know about ions.

H

Atoms form ions.

An atom has an equal number of electrons and protons. Since each electron has one negative charge and each proton has one positive charge, atoms have no overall electrical charge. An **ion** is formed when an atom loses or gains one or more electrons. Because the number of electrons in an ion is different from the number of protons, an ion does have an overall electric charge.

Formation of Positive Ions

?
I

Consider how a positive ion can be formed from an atom. The left side of the illustration below represents a sodium (Na) atom. Its nucleus contains 11 protons and some neutrons. Because the electron cloud surrounding the nucleus consists of 11 electrons, there is no overall charge on the atom. If the atom loses one electron, however, the charges are no longer balanced. There is now one more proton than there are electrons. The ion formed, therefore, has a positive charge.

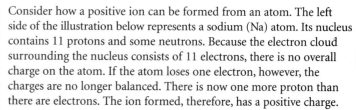
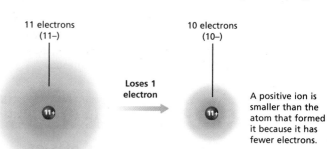

11 electrons (11−)

10 electrons (10−)

Loses 1 electron

A positive ion is smaller than the atom that formed it because it has fewer electrons.

Sodium Atom (Na)

Sodium Ion (Na⁺)

Notice the size of the positive ion. Because there are fewer electrons, there is less of a repulsion among the remaining electrons. Therefore, the positive ion is smaller than the neutral atom.

Positive ions are represented by the symbol for the element with a raised plus sign to indicate the positive charge. In the above example, the sodium ion is represented as Na^+.

Some atoms form positive ions by losing more than one electron. In those cases, the symbol for the ion also indicates the number of positive charges on the ion. For example, calcium loses two electrons to form an ion Ca^{2+}, and aluminum loses three electrons to form Al^{3+}.

CHECK YOUR READING What must happen to form a positive ion?

DIFFERENTIATE INSTRUCTION

? **More Reading Support**

H What kind of charge does an electron have? *negative*

I After an atom has lost an electron, what kind of charge does it have? *positive*

Below Level To help students understand ions, draw a diagram of an atom on the board. Draw the nucleus and the surrounding electron cloud. In the nucleus, draw three plus marks; in the cloud, draw three minus marks. Have students identify the marks as protons and electrons. Emphasize that the protons' positive charges balance the electrons' negative charges. Erase one minus sign in the electron cloud. Have students discuss how this change will affect the atom.

Formation of Negative Ions

The illustration below shows how a negative ion is formed. In this case the atom is chlorine (Cl). The nucleus of a chlorine atom contains 17 protons and some neutrons. The electron cloud has 17 electrons, so the atom has no overall charge. When an electron is added to the chlorine atom, a negatively charged ion is formed. Notice that a negative ion is larger than the neutral atom that formed it. The extra electron increases the repulsion within the cloud, causing it to expand.

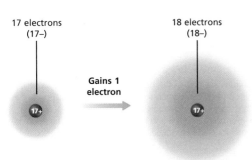

17 electrons
(17–)

18 electrons
(18–)

Gains 1 electron

A negative ion is larger than the atom that formed it because it has more electrons.

Chlorine Atom (Cl)

Chloride Ion (Cl⁻)

Negative ions are represented by placing a minus sign to the right and slightly above the element's symbol. The negative chloride ion in the example, therefore, would be written as Cl^-. If an ion has gained more than one electron, the number of added electrons is indicated by a number in front of the minus sign. Oxygen (O), for example, gains two electrons when it forms an ion. Its symbol is O^{2-}.

1.1 Review

KEY CONCEPTS

1. Which two atoms are most common in Earth's crust? in the human body?

2. What are the particles that make up an atom?

3. What happens when an atom forms an ion?

CRITICAL THINKING

4. **Infer** Magnesium and sodium atoms are about the same size. How does the size of a magnesium ion with a 2+ charge compare with that of a sodium ion with a single + charge?

5. **Compare** The atomic number of potassium is 19. How does potassium-39 differ from potassium-41?

◇ CHALLENGE

6. **Analyze** When determining the mass of an atom, the electrons are not considered. Why can scientists disregard the electrons?

ANSWERS

. oxygen and silicon; oxygen nd carbon

. protons, neutrons, and ·lectrons

. It gains or loses an electron.

4. The magnesium ion is smaller.

5. Potassium-41 has two more neutrons.

6. Scientists can disregard electrons because they have such a small mass.

Develop Critical Thinking

COMPARE AND CONTRAST Have students compare and contrast negative and positive ions. To compare, students might say that both types of ions have unbalanced charges (or unequal numbers of protons and electrons). To contrast, they might say that a negative ion has gained an electron, acquiring an overall negative charge, while a positive ion has lost an electron, acquiring an overall positive charge.

Reinforce (the BIG idea)

Have students relate the section to the Big Idea.

 R Reinforcing Key Concepts, p. 21

1.1 ASSESS & RETEACH

Assess

 A Section 1.1 Quiz, p. 3

Reteach

Draw a three-column chart on the board. Label the first column "Particle," the second "Charge," and the third "Location." Ask students to list the three atomic particles (*proton, neutron, electron*), the charge each particle has (*positive, neutral, negative*), and the location of each in an atom (*nucleus, nucleus, electron cloud around nucleus*). Encourage students to copy the chart into their notebooks for later reference.

Technology Resources

Have students visit **ClassZone.com** for reteaching of Key Concepts.

 CONTENT REVIEW

 CONTENT REVIEW CD-ROM

Set Learning Goal

To identify chemical elements in the human body and the roles they play

Present the Science

Fluoride ions protect teeth by replacing other ions in the tooth's enamel, or hard covering. The new compound is much more resistant to the acids that can form in the mouth and so helps prevent the spread of tooth decay.

Discussion Questions

- Ask: Where do you find iron in the human body? *in the hemoglobin in red blood cells*

- Ask: What are two important functions of sodium and potassium in the body? *to regulate the amount and location of water and to make up sweat to control temperature*

- Ask: In what parts of the body is most of the calcium found? *bones and teeth*

- Ask: What element makes up part of the hard coating on teeth? *fluorine*

Teaching with Technology

Have students use graphics software to construct a pie graph of the chart data. They can convert each amount into a percentage by dividing by 50 kilograms.

Close

Tell students that any element, no matter how important for life, can be harmful if taken in too large a dose. Ask: Why do you think this is so? *A human's body chemistry will be upset if elements are not balanced as they need to be.*

Technology Resources

Have students visit **ClassZone.com** to find more about the elements important to life.

 **RESOURCE CENTER**

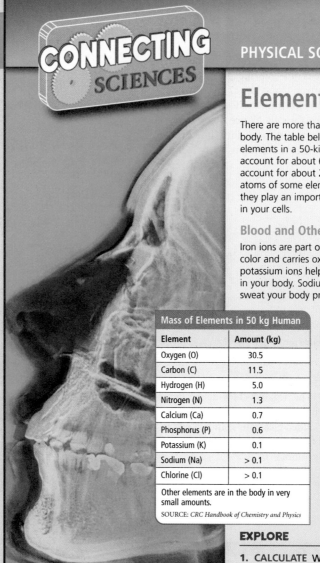

CONNECTING SCIENCES

Elements of Life

There are more than 25 different types of atoms in the cells of your body. The table below shows the amount of atoms of some of the elements in a 50-kilogram human. Atoms of the element oxygen account for about 61 percent of a person's mass. Atoms of carbon account for about 23 percent of a person's mass. Although the atoms of some elements are present only in very small amounts, they play an important role in the chemical processes that occur in your cells.

Blood and Other Fluids

Iron ions are part of the hemoglobin that gives your blood its red color and carries oxygen to cells throughout your body. Sodium and potassium ions help regulate the amount and location of the water in your body. Sodium and potassium ions also make up part of the sweat your body produces to regulate temperature.

Mass of Elements in 50 kg Human	
Element	**Amount (kg)**
Oxygen (O)	30.5
Carbon (C)	11.5
Hydrogen (H)	5.0
Nitrogen (N)	1.3
Calcium (Ca)	0.7
Phosphorus (P)	0.6
Potassium (K)	0.1
Sodium (Na)	> 0.1
Chlorine (Cl)	> 0.1

Other elements are in the body in very small amounts.

SOURCE: *CRC Handbook of Chemistry and Physics*

Bones and Teeth

The sturdier structures of your body get their strength from the calcium, magnesium, and phosphorus. You have less than a kilogram of calcium in your body, almost all of which is in your bones and teeth. Fluoride ions make up part of the hard coating on your teeth. This is why you'll often find fluoride ions added to toothpaste.

Elements to Avoid

In some way, the atoms of every element in the periodic table play a role in human lives. Many of them, however, can be hazardous if handled improperly. For example, arsenic and mercury are poisonous.

EXPLORE

1. **CALCULATE** What percentage of your body is made up of oxygen, carbon, hydrogen, and nitrogen?

2. **CHALLENGE** Salt, made of sodium ions and chloride ions, is an essential part of your diet. However, too much salt can cause health problems. Use the Internet to find out about the problems caused by too much or too little salt in your diet.

RESOURCE CENTER Find out more about the
CLASSZONE.COM elements important to life.

This photo shows a false color X-ray of the human skull. X-rays show the bones in the human body. Bones contain calcium.

B 16 Unit: **Chemical Interactions**

EXPLORE

1. *CALCULATE* oxygen: 30.5 kg/50 kg = 61.0%;
   carbon: 11.5 kg/50 kg = 23.0%;
   hydrogen: 5 kg/50 kg = 10.0%;
   nitrogen: 1.3 kg/50 kg = 2.6%

2. *CHALLENGE* Too much salt can lead to too much sodium, which can cause high blood pressure. Sodium is an essential nutrient, however, in maintaining the electrolyte balance in the body.

KEY CONCEPT

1.2 Elements make up the periodic table.

◀ **BEFORE, you learned**
- Atoms have a structure
- Every element is made from a different type of atom

▶ **NOW, you will learn**
- How the periodic table is organized
- How properties of elements are shown by the periodic table

VOCABULARY
atomic mass p. 17
periodic table p. 18
group p. 22
period p. 22

EXPLORE Similarities and Differences of Objects

How can different objects be organized?

PROCEDURE

① With several classmates, organize the buttons into three or more groups.

② Compare your team's organization of the buttons with another team's organization.

WHAT DO YOU THINK?
- What characteristics did you use to organize the buttons?
- In what other ways could you have organized the buttons?

MATERIALS
buttons

Elements can be organized by similarities.

One way of organizing elements is by the masses of their atoms. Finding the masses of atoms was a difficult task for the chemists of the past. They could not place an atom on a pan balance. All they could do was find the mass of a very large number of atoms of a certain element and then infer the mass of a single one of them.

Remember that not all the atoms of an element have the same atomic mass number. Elements have isotopes. When chemists attempt to measure the mass of an atom, therefore, they are actually finding the average mass of all its isotopes. The **atomic mass** of the atoms of an element is the average mass of all the element's isotopes. Even before chemists knew how the atoms of different elements could be different, they knew atoms had different atomic masses.

Chapter 1: **Atomic Structure and the Periodic Table 17** **B**

1.2 FOCUS

● Set Learning Goals
Students will
- Describe how the periodic table is organized.
- Identify how properties of elements are shown by the periodic table.

◐ 3-Minute Warm-Up
Display Transparency 4 or copy the following exercise on the board:

Correct any statements that are not true.

1. The atomic number of an atom is the number of protons and neutrons in the nucleus. *atomic mass number*

2. Isotopes are atoms of the same element, but with a different number of neutrons. *true*

3. When an atom gives up an electron, it becomes an ion with a negative charge. *positive*

　3-Minute Warm-Up, p. T4

1.2 MOTIVATE

EXPLORE Similarities and Differences of Objects

PURPOSE To classify logically a variety of objects

TIP *15 min.* Even if students do not physically sort the buttons, have them suggest ways they can organize them.

WHAT DO YOU THINK? *Size, shape, color, number of holes, and texture are likely criteria for organizing the buttons.*

RESOURCES FOR DIFFERENTIATED INSTRUCTION

Below Level
UNIT RESOURCE BOOK
Reading Study Guide A, pp. 24–25
Decoding Support, p. 47

🔊 **AUDIO CDS**

Advanced
UNIT RESOURCE BOOK
Challenge and Extension, p. 30

English Learners
UNIT RESOURCE BOOK
Spanish Reading Study Guide, pp. 28–29

💿 **AUDIO CDS**
- Audio Readings in Spanish
- Audio Readings (English)

Teach from Visuals

To help students interpret the visual of Mendeleev's periodic table, ask: What do the question marks in the table represent? *missing elements that Mendeleev thought should be there, based on the properties of other elements in the column and the changes in atomic mass.*

Integrate the Sciences

The concept of periodic patterns—patterns that repeat consistently at regular intervals—appears in many areas of science. In astronomy, the planets move in periodic patterns: they travel in elliptical paths around the Sun, passing the same points in space repeatedly. This particular periodic movement is the basis of a year as a measure of time.

Social Studies Connection

In his periodic table, Mendeleev used the Latin alphabet for element symbols and the Cyrillic alphabet for the text. Cyrillic letters are based on the Greek alphabet. Slavic languages that use Cyrillic include Russian, Serbian, Bulgarian, and Ukrainian.

Ongoing Assessment

increasing atomic mass and similar chemical properties

Mendeleev's Periodic Table

In the early 1800s several scientists proposed systems to organize the elements based on their properties. None of these suggested methods worked very well until a Russian chemist named Dmitri Mendeleev (MENH-duh-LAY-uhf) decided to work on the problem.

In the 1860s, Mendeleev began thinking about how he could organize the elements based on their physical and chemical properties. He made a set of element cards. Each card contained the atomic mass of an atom of an element as well as any information about the element's properties. Mendeleev spent hours arranging the cards in various ways, looking for a relationship between properties and atomic mass.

The exercise led Mendeleev to think of listing the elements in a chart. In the rows of the chart, he placed those elements showing similar chemical properties. He arranged the rows so the atomic masses increased as one moved down each vertical column. It took Mendeleev quite a bit of thinking and rethinking to get all the relationships correct, but in 1869 he produced the first **periodic table** of the elements. We call it the periodic table because it shows a periodic, or repeating, pattern of properties of the elements. In the reproduction of Mendeleev's first table shown below, notice how he placed carbon (C) and silicon (Si), two elements known for their similarities, in the same row.

?A **?B**

CHECK YOUR READING What organizing method did Mendeleev use?

Dmitri Mendeleev (1834–1907) first published a periodic table of the elements in 1869.

— 70 —

ъ ней, мнѣ кажется, уже ясно выражается примѣнимость вы ллемаго мною. начала ко всей совокупности элементовъ, пай ыхъ извѣстенъ съ достовѣрностію. На этотъ разъ я и желалъ ущественно найти общую систему элементовъ. Вотъ этотъ

	Ti=50	Zr=90	?=180.
	V=51	Nb=94	Ta=182.
	Cr=52	Mo=96	W=186.
	Mn=55	Rh=104,4	Pt=197,4
	Fe=56	Ru=104,4	Ir=198.
	Ni=Co=59	Pl=106₆,	Os=199.
H=1		Cu=63,4	Ag=108 Hg=200.
Be=9,4 Mg=24	Zn=65,2	Cd=112	
B=11 Al=27,4	?=68	Ur=116 Au=197?	
C=12 Si=28	?=70	Sn=118	
N=14 P=31	As=75	Sb=122 Bi=210	
O=16 S=32	Se=79,4	Te=128?	

DIFFERENTIATE INSTRUCTION

? More Reading Support

A Who produced the first periodic table of elements? *Dmitri Mendeleev*

B What pattern repeats in the periodic table? *the properties of the elements*

English Learners Within this section are sentences with a variety of introductory clauses and phrases. Give students the following examples and have them identify the subject of the sentence: "In the rows of the chart, *he* placed those elements showing similar chemical properties." (p. 18)

"At the start, many *chemists* found it hard to accept Mendeleev's predictions of unknown elements." (p. 19)

Encourage students to use introductory clauses and phrases in their own writing.

Predicting New Elements

When Mendeleev constructed his table, he left some empty spaces where no known elements fit the pattern. He predicted that new elements that would complete the chart would eventually be discovered. He even described some of the properties of these unknown elements.

At the start, many chemists found it hard to accept Mendeleev's predictions of unknown elements. Only six years after he published the table, however, the first of these elements—represented by the question mark after aluminum (Al) on his table—was discovered. This element was given the name gallium, after the country France (Gaul) where it was discovered. In the next 20 years, two other elements Mendeleev predicted would be discovered.

The periodic table organizes the atoms of the elements by properties and atomic number.

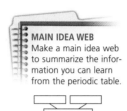

MAIN IDEA WEB
Make a main idea web to summarize the information you can learn from the periodic table.

The modern periodic table on pages 20 and 21 differs from Mendeleev's table in several ways. For one thing, elements with similar properties are found in columns, not rows. More important, the elements are not arranged by atomic mass but by atomic number.

Reading the Periodic Table

Each square of the periodic table gives particular information about the atoms of an element.

❶ The number at the top of the square is the atomic number, which is the number of protons in the nucleus of an atom of that element.

❷ The chemical symbol is an abbreviation for the element's name. It contains one or two letters. Some elements that have not yet been named are designated by temporary three-letter symbols.

❸ The name of the element is written below the symbol.

❹ The number below the name indicates the average atomic mass of all the isotopes of the element.

The color of the element's symbol indicates the physical state of the element at room temperature. White letters—such as the *H* for hydrogen in the box to the right—indicate a gas. Blue letters indicate a liquid, and black letters indicate a solid. The background colors of the squares indicate whether the element is a metal, nonmetal, or metalloid. These terms will be explained in the next section.

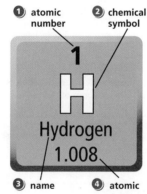

❶ atomic number ❷ chemical symbol

1

H

Hydrogen

1.008

❸ name ❹ atomic mass

Mathematics Connection

Atomic masses that are not whole numbers may baffle students. Remind students that different isotopes of an element have different atomic mass numbers. The atomic mass number represents the average atomic mass of all the isotopes of the elements. If there is only 30 percent of isotope A and 70 percent of isotope B, the average mass will be closer to the mass of isotope B, but not that exact mass.

History of Science

In 1913, the British chemist Henry Moseley began examining the x-ray radiation produced by some of the elements in the periodic table. Moseley noticed that the wavelengths produced by the elements changed in a predictable way that coincided with the element's position in the periodic table. Moseley concluded that there must be some other fundamental quantity that determined an element's position in the periodic table. This was how scientists came to reorder the periodic table by atomic number instead of atomic mass.

Social Studies Connection

Many of today's chemical symbols are based on old names for elements that were in use hundreds or thousands of years ago. Tell students that old alchemical symbols exist for elements such as iron, mercury, gold, and silver. Students may want to find pictures of alchemical symbols on the Internet.

DIFFERENTIATE INSTRUCTION

More Reading Support

C In the periodic table, where does the atomic number appear? *at the top of each square*

Below Level Help students interpret the details of the element square. Ask: Can different elements have the same atomic number? Why? *No; the atomic number is the number of protons, which is used to identify elements. Every element has a different atomic number.* What parts of the square besides atomic number are unique? *symbol, name*

Discuss whether the atomic mass is unique for each element, and why it could be the same.

Teach from Visuals

To help students interpret the periodic table, ask:

- What color are most of the elements in the table? *yellow*
- According to the key at the bottom, what are those elements? *metals*
- How many nonmetals are in the table? *17*
- Why do you think the lanthanide and actinide series are listed below the others? *to save space*

Teacher Demo

You can demonstrate a flame test of elements if you have the following equipment: Bunsen burner; flame-test wires; test solutions of 0.5M barium chloride, calcium chloride, copper sulfate, potassium nitrate, and sodium chloride; 5M hydrochloric acid. Clean the metal loop by dipping it in the hydrochloric acid and putting it in the flame: the flame should not change color. Then dip the loop in a test solution (use the NaCl last, since it can be hard to clean from the wire). Put the loop in the flame and ask students to identify the color change. Tell students what element the solution contains. (Repeat this process for each solution.) Barium produces a light green flame; calcium, reddish yellow; copper, blue green; potassium, lilac; and sodium, yellow.

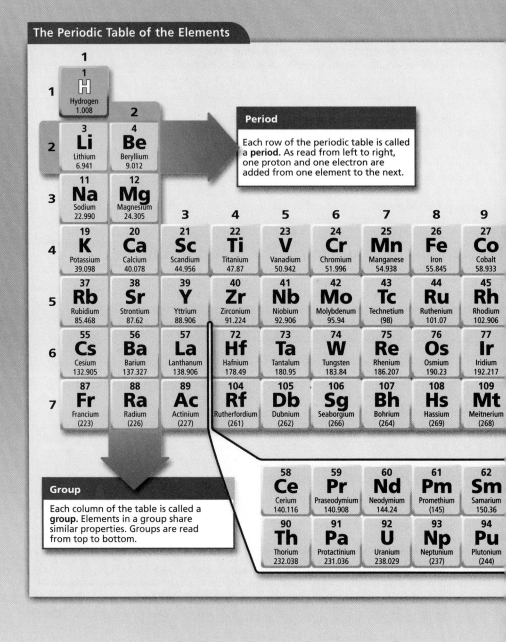

The Periodic Table of the Elements

Period
Each row of the periodic table is called a **period**. As read from left to right, one proton and one electron are added from one element to the next.

Group
Each column of the table is called a **group**. Elements in a group share similar properties. Groups are read from top to bottom.

 Metal   Metalloid Nonmetal **Fe** Solid **Hg** Liquid (O) G

DIFFERENTIATE INSTRUCTION

Inclusion Distribute enlarged copies of the periodic table so students with visual impairment can understand the table's layout. Tactile periodic tables are available from the American Printing House for the Blind.

Advanced Have students identify elements that are named after scientists, and then have them research one of the scientists. Point out that the higher-numbered elements are the most likely to have such names.

 Challenge and Extension, p. 30

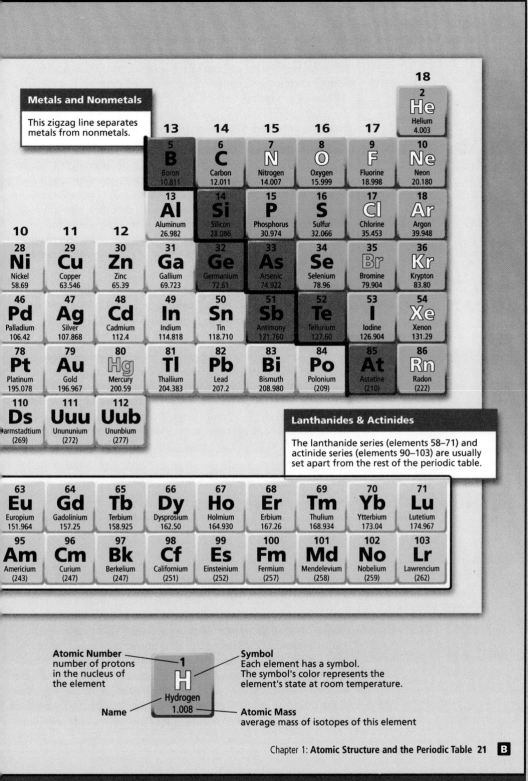

Metals and Nonmetals

This zigzag line separates metals from nonmetals.

Lanthanides & Actinides

The lanthanide series (elements 58–71) and actinide series (elements 90–103) are usually set apart from the rest of the periodic table.

Atomic Number — number of protons in the nucleus of the element

Symbol — Each element has a symbol. The symbol's color represents the element's state at room temperature.

Name

1
H
Hydrogen
1.008

Atomic Mass — average mass of isotopes of this element

Chapter 1: **Atomic Structure and the Periodic Table** 21 **B**

Develop Critical Thinking

SEQUENCE Have students relate sequence to periodicity in the periodic table. Ask:

- What role does sequence of numbers play in the periodic table? *The atomic number increases by one as you go from left to right and then jump to the start of the next row (period).*
- Why are the rows broken where they are? *to align groups that have similar properties*

EXPLORE (the **BIG** idea)

Revisit "Element Safari" on p. 7. Have students locate on the periodic table each element they found in the safari, and then have them find the symbol, atomic number, and atomic mass for each element.

Language Arts Connection

Students have learned that some symbols for elements are based on Latin words. One interesting example is Hg, for mercury. The symbol comes from the Latin *hydrargyrum,* meaning "liquid silver." Students may have seen a drop of mercury, which does indeed look like liquid silver.

DIFFERENTIATE INSTRUCTION

Advanced Have students investigate an element and determine when it was discovered and how it was named. Helium, for example, was found first in the Sun (Greek *hēlios*) later on Earth.

Teach from Visuals

To help students connect the visuals of a row and a column to the larger periodic table, ask:

- From what part of the periodic table is the column labeled "Group 17" taken? *the far right*
- From what part is the row labeled "Period 3" taken? *the upper part*

Metacognitive Strategy

Discuss mnemonic devices and other strategies that students can use to remember the difference between group and period. For example, periods run horizontally, like a sentence, and a sentence ends with a period.

Health Connection

People who for health reasons cannot have table salt (NaCl) can instead use potassium chloride (KCl). The taste is somewhat bitter. Potassium and sodium are in the same group and so have similar properties.

Ongoing Assessment

Describe how the periodic table is organized.

Ask: Why is a group sometimes called a family of elements? *A group is sometimes called a family of elements because the elements are related in terms of properties and traits they share.*

Groups and Periods

The elements in Group 17, the halogens, show many similarities.

Elements in a vertical column of the periodic table show similarities in their chemical and physical properties. The elements in a column are known as a **group,** and they are labeled by a number at the top of the column. Sometimes a group is called a family of elements, because these elements seem to be related.

The illustration at the left shows Group 17, commonly referred to as the halogen group. Halogens tend to combine easily with many other elements and compounds, especially with the elements in Groups 1 and 2. Although the halogens have some similarities to one another, you can see from the periodic table that their physical properties are not the same. Fluorine and chlorine are gases, bromine is a liquid, and iodine and astatine are solids at room temperature. Remember that the members of a family of elements are related but not identical.

Metals like copper can be used to make containers for water. Some metals—such as lithium, sodium, and potassium—however, react violently if they come in contact with water. They are all in the same group, the vertical column labeled 1 on the table.

Each horizontal row in the periodic table is called a **period.** Properties of elements change in a predictable way from one end of a period to the other. In the illustration below, which shows Period 3, the elements on the far left are metals and the ones on the far right are nonmetals. The chemical properties of the elements show a progression; similar progressions appear in the periods above and below this one.

Period 3 contains elements with a wide range of properties. Aluminum (Al) is used to make drink cans, while argon (Ar) is a gas used in light bulbs.

Trends in the Periodic Table

Because the periodic table organizes elements by properties, an element's position in the table can give information about the element. Remember that atoms form ions by gaining or losing electrons. Atoms of elements on the left side of the table form positive ions easily. For example, Group 1 atoms lose an electron to form ions with one positive charge (1+). Atoms of the elements in Group 2, likewise, can lose two electrons to form ions with a charge of 2+. At the other side of the table, the atoms of elements in Group 18 normally do not form ions at all. Atoms of elements in Group 17, however, often gain one

DIFFERENTIATE INSTRUCTION

?) More Reading Support

D What do you call a column of the periodic table? *a group*

E What do you call a row of the periodic table? *a period*

English Learners The use of dashes in writing may be confusing to English learners. The dash can function like a comma or parentheses, and introduce an appositive phrase. For example, "Some metals—such as lithium, sodium, and potassium—however, react violently if they come in contact with water."

electron to form a negative ion (1–). Similarly, the atoms of elements in Group 16 can gain two electrons to form a 2– ion. The atoms of the elements in Groups 3 to 12 all form positive ions, but the charge can vary.

Other information about atoms can be determined by their position in the table. The illustration to the right shows how the sizes of atoms vary across periods and within groups. An atom's size is important because it affects how the atom will react with another atom.

The densities of elements also follow a pattern. Density generally increases from the top of a group to the bottom. Within a period, however, the elements at the left and right sides of the table are the least dense, and the elements in the middle are the most dense. The element osmium (Os) has the highest known density, and it is located at the center of the table.

Chemists cannot predict the exact size or density of an atom of one element based on that of another. These trends, nonetheless, are a valuable tool in predicting the properties of different substances. The fact that the trends appeared after the periodic table was organized by atomic number was a victory for all of the scientists like Mendeleev who went looking for them all those years before.

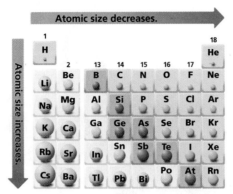

Atomic size decreases.

Atomic size increases.

Atomic size is one property that changes in a predictable way across, up, and down the periodic table.

 CHECK YOUR READING What are some properties that can be related to position on the periodic table?

1.2 Review

KEY CONCEPTS

1. How is the modern periodic table organized?

2. What information about an atom's properties can you read from the periodic table?

3. How are the relationships of elements in a group different from the relationships of elements in a period?

CRITICAL THINKING

4. **Infer** Would you expect strontium (Sr) to be more like potassium (K) or bromine (Br)? Why?

5. **Predict** Barium (Ba) is in Group 2. Recall that atoms in Group 1 lose one electron to form ions with a 1+ charge. What type of ion does barium form?

CHALLENGE

6. **Analyze** Explain how chemists can state with certainty that no one will discover an element between sulfur (S) and chlorine (Cl).

 Chapter 1: **Atomic Structure and the Periodic Table** 23 **B**

ANSWERS

1. by atomic number

2. the number of protons in an atom's nucleus, the average mass of one atom of that element, the element's symbol, and the element's name

3. The elements in a group have similar properties. The elements in a period have varying properties.

4. potassium, because it is closer on the periodic table

5. positive ion with 2+ charge

6. Chlorine has an atomic number one greater than sulfur. This means it has one more proton in its nucleus. Each element is defined by its atomic number, which is an integer.

Ongoing Assessment

Identify how properties of elements are shown by the periodic table.

Ask: As you go right on the table, how are the atoms of the elements changing? *Their number of protons (and therefore their atomic masses) are increasing.*

CHECK YOUR READING *Answer: atomic size and density*

Teach from Visuals

To help students interpret the chart of atomic sizes, direct their attention to the two arrows. Ask:

• Which side of the chart has the largest atoms? *the left side*

• Where would the largest atom on the chart be? *at bottom left*

Reinforce (the **BIG** idea)

Have students relate the section to the Big Idea.

R Reinforcing Key Concepts, p. 31

1.2 ASSESS & RETEACH

Assess

A Section 1.2 Quiz, p. 4

Reteach

Point out any square on the table and ask students what the numbers and letters mean. *Sample: 40 = atomic number (number of protons); Zr = atomic symbol; Zirconium = name; 91.224 = atomic mass*

Technology Resources

Have students visit **ClassZone.com** for reaching of Key Concepts.

 CONTENT REVIEW

 CONTENT REVIEW CD-ROM

CHAPTER INVESTIGATION

Focus

PURPOSE Students model atoms and determine the atoms' relative masses.

OVERVIEW Students will use washers in film cans to represent protons and neutrons in atomic nuclei. They will designate one can to be the baseline with an atomic mass of 1 and will compare the masses of the remaining cans with the baseline. Students will find that you can use relative weights to determine atomic mass number.

Lab Preparation

- Ahead of time fill four film cans for each group of students as follows: 1 washer in A, 4 washers in B, 7 washers in C, and 9 washers in D. One empty can is also used in the investigation.
- You could use pennies or quarters instead of washers.
- Prior to the investigation, have students read through the investigation and prepare their data tables. Or you may wish to copy and distribute datasheets and rubrics.

 UNIT RESOURCE BOOK, pp. 50–58

 SCIENCE TOOLKIT, F12

Lab Management

- Encourage multiple measurements of the masses to ensure accuracy.
- Students should use the same mass of the empty can for all the samples.
- Make sure students understand that the data in rows 1–4 of the table represent actual measurements that may not be all whole numbers. The data in row 5, however, are their estimations of the numbers of washers in each film can and so must be a whole number.

INCLUSION Have students with visual impairments work the balance so they can feel when the mass is equal, and have a partner relate the measured mass.

CHAPTER INVESTIGATION

Modeling Atomic Masses

OVERVIEW AND PURPOSE Atoms are extremely small. They are so small, in fact, that a single drop of water contains more atoms than you could count in a lifetime! Measuring the masses of atoms to discover the patterns in the periodic table was not an easy task for scientists in the past. This investigation will give you some sense of how scientists determined the mass of atoms. You will

- compare the masses of different film can "atoms"
- predict the number of washers in each film can "atom"

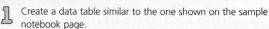
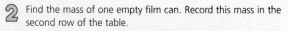

▶ Procedure

MATERIALS
- empty film can
- balance
- 4 filled film cans

1. Create a data table similar to the one shown on the sample notebook page.

2. Find the mass of one empty film can. Record this mass in the second row of the table.

3. Collect the four film cans labeled A, B, C, and D in advance by your teacher. Each can contains a different number of washers and represents a different atom. The washers represent the protons and neutrons in an atom's nucleus.

4. Measure the mass of each of the four film cans.

5. Record the masses of the film can atoms in the first row of your data table.

6. Subtract the mass of an empty film can from the mass of each film can atom. Record the differences in the correct spaces in your data table. These masses represent the masses of the washers in your film can atoms. Think of this mass as the mass of the nucleus.

7. Subtract the mass of the washers in film can A from the mass of the washers in film can B. Record this value in the fourth row in your data table.

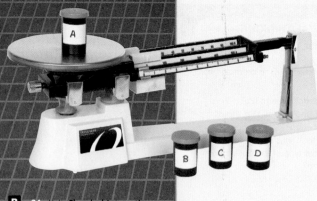

INVESTIGATION RESOURCES

 CHAPTER INVESTIGATION, Modeling Atomic Masses
- Level A, pp. 50–53
- Level B, pp. 54–57
- Level C, p. 58

Advanced students should complete Levels B & C.

 Writing a Lab Report, D12–13

Technology Resources

Customize this student lab as needed or look for an alternative. Print rubrics to assess student lab reports.

 Lab Generator CD-ROM

Subtract the mass of the washers in film can B from the mass of the washers in film can C. Record this value in the fourth row of column B in your data table.

⑨ Subtract the mass of the washers in film can C from the mass of the washers in film can D. Record this mass in the fourth row of column C in your data table.

⑩ Subtract the mass of the washers in film can A from the mass of the washers in film can D. Record this mass in the fourth row of column D in your data table.

▶ Observe and Analyze [Write It Up]

1. **RECORD OBSERVATIONS** Be sure your data table and calculations are complete. Double-check your arithmetic.

2. **ANALYZE DATA** Examine your data table. Do you notice any patterns in how the masses increase? Given that all the washers in the film can atoms have identical masses, what might the ratio of the mass of the washers to the smallest mass tell you?

3. **PREDICT** Assume that the lightest film can contains only one washer. Predict the numbers of washers in the other film cans. Write your predictions in your data table. (**Hint:** You might try dividing the masses of the washers in film can atoms B, C, and D by the mass of the washer in film can atom A. Round your results to the nearest whole number.)

4. **GRAPH DATA** On a sheet of graph paper, plot the masses (in grams) of the washers in the film can atoms on the *y*-axis and the number of washers in each can on the *x*-axis. Connect the points on the graph.

5. **INTERPRET DATA** Compare the masses of your film can atoms with the masses of the first four atoms on the periodic table. Which represents which?

▶ Conclude [Write It Up]

1. **IDENTIFY LIMITS** What can't this activity tell you about the identity of your film can atoms? (**Hint:** Protons and neutrons in real atoms have about the same mass.)

2. **INFER** Hydrogen has only a single proton in its nucleus. If your film can atoms represent the first four elements in the periodic table, what are the numbers of protons and neutrons in each atom?

3. **APPLY** Single atoms are far too small to place on a balance. How do you think scientists determine the masses of real atoms?

▶ INVESTIGATE Further

CHALLENGE Use a periodic table to find the masses of the next two atoms (boron and carbon). How many washers would you need to make film can atom models for each?

Modeling Atomic Masses
Observe and Analyze
Table 1. Masses of Film Can Atoms

	A	B	C	D
Mass of film can atom (g)				
Mass of empty film can (g)				
Mass of washers (g)				
Mass of washers compared with nearest neighbor (g)				
Number of washers in each can				

Atomic Structure and the Periodic Table 25 **B**

▶ Observe and Analyze [Write It Up]

1. *SAMPLE DATA Empty film can, 7.3 g. Can A, 11.8 g; washers, 4.5 g. Can B, 24.4 g; washers, 17.1 g. Can C, 39.4 g; washers, 32.1 g. Can D, 47.9 g; washers, 42.6 g.*

2. *Students may notice that the mass of B is about four times the mass of A. The ratio of the mass of the washers to the smallest mass might tell you that the smallest mass might be the mass of one washer.*

3. *1, 4, 7, 9*

4. *See students' graphs.*

5. *A represents hydrogen; B represents helium; C represents lithium; D represents beryllium*

▶ Conclude [Write It Up]

1. *The activity cannot tell you how the mass divides between protons and neutrons.*

2. *A = 1 proton, 0 neutrons; B = 2 protons and 2 neutrons; C = 3 protons and 4 neutrons; D = 4 protons and 5 neutrons*

3. *They measure the masses of many atoms, then divide by the number of atoms to find the mass of a single atom.*

▶ INVESTIGATE Further

CHALLENGE 11 washers for boron and 12 washers for carbon

Post-Lab Discussion

Ask: Why did you subtract the mass of the empty film can from the mass of each of the filled cans? *You want to know the mass of the "atoms" in the can; adding the mass of the container would skew the results.* Also ask if students have an appreciation for what Mendeleev and others went through to determine atomic masses.

○ Set Learning Goals

Students will

- Classify elements as metals, non-metals, and metalloids.
- Identify different groups of elements.
- Describe radioactive elements.
- Model half-life in an experiment.

◁ 3-Minute Warm-Up

Display Transparency 5 or copy the following exercise on the board:

Look for the elements below in the periodic table on pp. 20–21. Write how each pair of elements are related.

1. calcium and barium *in the same group*
2. lithium and carbon *in the same period*
3. uranium and curium *in the same period and in the actinide series*

 3-Minute Warm-Up, p. T5

1.3 MOTIVATE

THINK ABOUT

PURPOSE To reinforce the concept that elements with similar properties are in the same part of the periodic table

DISCUSS Help students locate these elements in the periodic table. Discuss the other elements that are near them. *Fluorine, krypton, neon, chlorine, and bromine are near argon. Boron, carbon, silicon, gallium, and germanium are near aluminum. Nickel, palladium, silver, zinc, and cadmium are near copper.*

Ongoing Assessment

CHECK YOUR READING *Answer: Elements with similar properties are near each other in the periodic table.*

KEY CONCEPT

1.3 The periodic table is a map of the elements.

◁ BEFORE, you learned	▷ NOW, you will learn
• The periodic table is organized into groups of elements with similar characteristics	• How elements are classified as metals, nonmetals, and metalloids
• The periodic table organizes elements according to their properties	• About different groups of elements
	• About radioactive elements

VOCABULARY

reactive p. 26
metal p. 27
nonmetal p. 29
metalloid p. 30
radioactivity p. 30
half-life p. 32

THINK ABOUT

How are elements different?

The photograph shows common uses of the elements copper, aluminum, and argon: copper in a penny, aluminum in a pie plate, and argon in a light bulb. The atoms of each element is located in a different part of the periodic table, and each has a very different use. Find these elements on the periodic table. What other elements are near these?

■ metal ■ metalloid ■ nonmetal

The periodic table has distinct regions.

The periodic table is a kind of map of the elements. Just as a country's location on the globe gives you information about its climate, an atom's position on the periodic table indicates the properties of its element. The periodic table has three main regions—metals on the left, nonmetals (except hydrogen) on the right, and metalloids in between. The periodic table on pages 20 and 21 indicates these regions with different colors. A yellow box indicates a metal; green, a nonmetal; and purple, a metalloid.

Its position in the table also indicates how reactive an element is. The term **reactive** indicates how likely an element is to undergo a chemical change. Most elements are somewhat reactive and combine with other materials. The atoms of the elements in Groups 1 and 17 are the most reactive. The elements of Group 18 are the least reactive of all the elements.

CHECK YOUR READING How does the periodic table resemble a map?

RESOURCES FOR DIFFERENTIATED INSTRUCTION

Below Level	**Advanced**	**English Learners**
UNIT RESOURCE BOOK	**UNIT RESOURCE BOOK**	**UNIT RESOURCE BOOK**
• Reading Study Guide A, pp. 34–35	• Challenge and Extension, p. 40	Spanish Reading Study Guide, pp. 38–39
• Decoding Support, p. 47	• Challenge Reading, pp. 43–44	
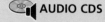 **AUDIO CDS**		**AUDIO CDS**
		• Audio Readings in Spanish
		• Audio Readings (English)

Most elements are metals.

When you look at the periodic table, it is obvious from the color that most of the elements are metals. In general, **metals** are elements that conduct electricity and heat well and have a shiny appearance. Metals can be shaped easily by pounding, bending, or being drawn into a long wire. Except for mercury, which is a liquid, metals are solids at room temperature.

Sodium is a metal that is so soft it can be cut with a knife at room temperature.

You probably can name many uses for the metal **copper**.

Aluminum is often used for devices that must be strong and light.

Reactive Metals

The metals in Group 1 of the periodic table, the alkali metals, are very reactive. Sodium and potassium are often stored in oil to keep them away from air. When exposed to air, these elements react rapidly with oxygen and water vapor. The ions of these metals, Na^+ and K^+, are important for life, and play an essential role in the functioning of living cells.

The metals in Group 2, the alkaline earth metals, are less reactive than the alkali metals. They are still more reactive than most other metals, however. Calcium ions are an essential part of your diet. Your bones and teeth contain calcium ions. Magnesium is a light, inexpensive metal that is often combined with other metals when a lightweight material is needed, such as for airplane frames.

Reactive Metals

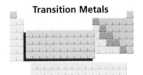

Transition Metals

The elements in Groups 3–12 are called the transition metals. Among these metals are some of the earliest known elements, such as copper, gold, silver, and iron. Transition metals are generally less reactive than most other metals. Because gold and silver are easily shaped and do not react easily, they have been used for thousands of years to make jewelry and coins. Ancient artifacts made from transition metals can be found in many museums and remain relatively unchanged since the time they were made. Today, dimes and quarters are made of copper and nickel, and pennies are made of zinc with a coating of copper. Transition metal ions even are found in the foods you eat.

Transition Metals

Address Misconceptions

IDENTIFY Ask: If you could see a tiny nickel atom, would it be silver and shiny like a coin, or have a different appearance? If students answer that the nickel atom would look like a nickel coin, they hold the misconception that atoms of an element have the same properties as the element in bulk.

CORRECT Point out that atoms are too tiny to resemble anything we can see, touch, or smell. It is the atoms interacting in a mass that determines what we sense.

REASSESS Ask: Describe the picture of nickel atoms on pp. 6–7. Do the atoms resemble the metal in a nickel coin, or look different? *The atoms look like crowded mountain peaks in the picture, while the metal in a nickel is shiny and smooth. The atoms do not resemble the appearance of a nickel.*

Technology Resources

Visit **ClassZone.com** for background on common student misconceptions.

MISCONCEPTION DATABASE

Ongoing Assessment

Classify elements as metals, nonmetals, and metalloids.

Ask: In what regions of the periodic table are the nonmetals and the metals? *The nonmetals are on the far right; most of the elements are metals, which stretch from the left through the center.*

On the table, what colors are metals? nonmetals? *yellow; green*

DIFFERENTIATE INSTRUCTION

More Reading Support

A What elements conduct heat and electricity well? *metals*

B Why are sodium and potassium stored in oil? *to protect them from reacting with air*

English Learners Tell English learners that "although" (p. 28) is a conjunction that indicates an exception or something that is not necessarily obvious. Students may not be familiar with the phrase "tend to" (p. 29); tell them it means "usually."

Inclusion To help students with visual impairments, translate visually oriented physical properties, such as shininess of metals, into tactile properties: a penny, for example, would feel smooth, hard, and cold at room temperature.

Integrate the Sciences

The human body needs a number of transition metals, in small amounts, to function properly. One of the most important transition metals for humans is iron, which is a key component of hemoglobin in red blood cells. Other important transition metals include zinc, molybdenum, copper, chromium, iron, and manganese.

Teacher Demo

Bring a box of iron-rich breakfast cereal and a vial of iron filings to class. Crush a small sample of cereal. Put it in a beaker of hot distilled water and stir with a magnetic stirrer. After about 15 minutes of stirring, remove the stirrer and show the dark slivers of iron on the magnet. Have students inspect the iron from the cereal and compare it with the iron filings. Point out that humans need iron in their diet for several important functions, including the carrying of oxygen in the red blood cells.

Real World Example

Although many transition metals are widespread in Earth's crust, their distribution in economically useful concentrations (ores) is uneven. The United States and Canada have many of these metals in abundance, but often choose to acquire them from countries that have higher-quality ores or lower costs for the labor to recover them.

The properties of the transition metals make them particularly important to industry. Iron is the main part of steel, a material used for bridges and buildings. Most electric wires and many other electrical devices are made of copper. Copper is also used to make water pipes. Indeed, it would be hard to think of an industry that doesn't make use of transition metals.

Although other transition metals may be less familiar, many of them are important for modern technology. The tiny coil of wire inside incandescent light bulbs is made of tungsten. Platinum is in the catalytic converters that reduce pollution from automobile engines.

For many applications, two or more metals are combined to form an alloy. Alloys can be stronger, less likely to corrode, or easier to shape than pure metals. Steel, which is stronger than the pure iron it contains, often includes other transition metals, such as nickel, chromium, or manganese. Brass, an alloy of copper and zinc, is stronger than either metal alone. Jewelry is often made of an alloy of silver and copper, which is stronger than pure silver.

Rare Earth Elements

Rare Earth Elements

The rare earth elements are the elements in the top row of the two rows of metals that are usually shown outside the main body of the periodic table. Taking these elements out of the main body of the table makes the table more compact. The rare earth elements are often referred to as lanthanides because they follow the element lanthanum (La) on the table. They are called rare earth elements because scientists once thought that these elements were available only in tiny amounts in Earth's crust. As mining methods improved, scientists learned that the rare earths were actually not so rare at all—only hard to isolate in pure form.

More and more uses are being found for the rare earth elements. Europium (Eu), for example, is used as a coating for some television tubes. Praseodymium (Pr) provides a protective coating against harmful radiation in the welder's helmet in the photograph on the right.

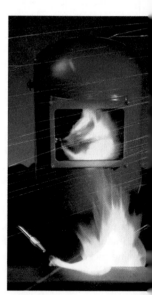

DIFFERENTIATE INSTRUCTION

 More Reading Support

C What is the combination of two or more metals called? *alloy*

D Where are the rare earth metals in the table? *in the top row of the rows below the table*

Advanced Have students use the Internet to research and report on the discovery of an element. You might suggest Marie Curie's discovery of polonium and radium or the contributions of Karl Scheele, Sir Humphry Davy, and Bernard Courtois in the discovery of chlorine and iodine.

R
• Challenge and Extension, p. 40
• Challenge Reading, pp. 43–44

Nonmetals and metalloids have a wide range of properties.

The elements to the right side of the periodic table are called **nonmetals.** As the name implies, the properties of nonmetals tend to be the opposite of those of metals. The properties of nonmetals also tend to vary more from element to element than the properties of the metals do. Many of them are gases at room temperature, and one—bromine—is a liquid. The solid nonmetals often have dull surfaces and cannot be shaped by hammering or drawing into wires. Nonmetals are generally poor conductors of heat and electric current.

The main components of the air that you breathe are the nonmetal elements nitrogen and oxygen. Nitrogen is a fairly unreactive element, but oxygen reacts easily to form compounds with many other elements. Burning and rusting are two familiar types of reactions involving oxygen. Compounds containing carbon are essential to living things. Two forms of the element carbon are graphite, which is a soft, slippery black material, and diamond, a hard crystal. Sulfur is a bright yellow powder that can be mined from deposits of the pure element.

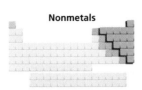
Nonmetals

Halogens

The elements in Group 17 are commonly known as halogens, from Greek words meaning "forming salts." Halogens are very reactive non-metals that easily form compounds called salts with many metals. Because they are so reactive, halogens are often used to kill harmful microorganisms. For example, the halogen chlorine is used to clean drinking water and to prevent the growth of algae in swimming pools. Solutions containing iodine are often used in hospitals and doctors' offices to kill germs on skin.

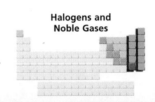
Halogens and Noble Gases

Noble Gases

Group 18 elements are called the noble, or inert, gases because they almost never react with other elements. Argon gas makes up about one percent of the atmosphere. The other noble gases are found in the atmosphere in smaller amounts. Colorful lights, such as those in the photograph on the right, are made by passing an electric current through tubes filled with neon, krypton, xenon, or argon gas. Argon gas also is placed in tungsten filament light bulbs, because it will not react with the hot filament.

 CHECK YOUR READING Where on Earth can you find noble gases?

Noble gases produce the light for many signs.

Chapter 1: **Atomic Structure and the Periodic Table** 29 **B**

Develop Critical Thinking

CLASSIFY Have students classify each of the following elements as a metal or a nonmetal. They should be able to explain their classification.

calcium	*metal*
sulfur	*nonmetal*
tin	*metal*
neon	*nonmetal*
nitrogen	*nonmetal*
carbon	*nonmetal*
mercury	*metal*

Art Connection

Have students look for interesting photographs of "neon" signs. Point out that each noble gas produces a particular color; for example, neon glows reddish orange. A greater variety of colors can be produced by having a glowing gas interact with a coating on the inside of a tube.

Ongoing Assessment

Identify different groups of elements.

Ask: Where in the periodic table are the gases that almost never react with other elements? *Group 18 (noble gases) on the right side* Where in the periodic table are the very reactive nonmetals that easily form salt compounds with metals? *Group 17 (halogens) on the right side*

CHECK YOUR READING *Answer: in the colorful lights in many signs*

DIFFERENTIATE INSTRUCTION

 More Reading Support

E What type of elements are on the right of the table? *nonmetals*

F What is true of group 18 elements? *They almost never react with other elements.*

Alternative Assessment Have students create a two-column chart to list the differences between noble gases and halogens. *Halogens: group 17 in periodic table, very reactive nonmetals that form salt compounds with metals, used to kill harmful microorganisms because they are so reactive. Noble gases: group 16 in periodic table, almost never react with other elements, used in lighting materials because they are not reactive.*

Chapter 1 **29** **B**

Revisit "Internet Activity: Periodic Table" on p. 7. Have students review the definition of atomic number, the definition of atomic mass, and the relationship between these measures.

History of Science

Marie Curie (1867–1934) blazed a path for women scientists by making huge contributions to physics and chemistry. Curie was the first woman to win a Nobel Prize in physics (1903, shared) and in chemistry (1911, alone). In 1906, she became the first woman professor at the Paris Sorbonne, one of France's great universities.

Ongoing Assessment

Describe radioactive elements.

Ask: What makes a radioactive isotope unstable? *If the nucleus has too many or too few neutrons, it may become unstable and emit particles and energy.*

CHECK YOUR READING *Answer: radioactivity*

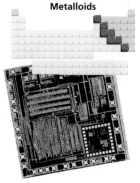

Metalloids

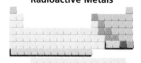

The metalloid silicon is found in sand and in computer microchips.

Radioactive Metals

Metalloids

Metalloids are elements that have properties of both metals and nonmetals. In the periodic table, they lie on either side of a zigzag line separating metals from nonmetals. The most common metalloid is silicon. Silicon atoms are the second most common atoms in Earth's crust.

Metalloids often make up the semiconductors found in electronic devices. Semiconductors are special materials that conduct electricity under some conditions and not under others. Silicon, gallium, and germanium are three semiconductors used in computer chips.

Some atoms can change their identity.

The identity of an element is determined by the number of protons in its nucleus. Chemical changes do not affect the nucleus, so chemical changes don't change one type of atom into another. There are, however, conditions under which the number of protons in a nucleus can change and so change the identity of an atom.

Recall that the nucleus of an atom contains protons and neutrons. Attractive forces between protons and neutrons hold the nucleus together even though protons repel one another. We say an atomic nucleus is stable when these attractive forces keep it together.

Each element has isotopes with different numbers of neutrons. The stability of a nucleus depends on the right balance of protons and neutrons. If there are too few or too many neutrons, the nucleus may become unstable. When this happens, particles are produced from the nucleus of the atom to restore the balance. This change is accompanied by a release of energy.

If the production of particles changes the number of protons, the atom is transformed into an atom of a different element. In the early 1900s, the Polish physicist Marie Curie named the process by which atoms produce energy and particles **radioactivity.** Curie was the first person to isolate polonium and radium, two radioactive elements.

An isotope is radioactive if the nucleus has too many or too few neutrons. Most elements have radioactive isotopes, although these isotopes are rare for small atoms. For the heaviest of elements—those beyond bismuth (Bi)—all of the isotopes are radioactive.

Scientists study radioactivity with a device called a Geiger counter. The Geiger counter detects the particles from the breakup of the atomic nucleus with audible clicks. More clicks indicate that more particles are being produced.

? **H**

CHECK YOUR READING How can an atom of one element change into an atom of a different element?

DIFFERENTIATE INSTRUCTION

? **More Reading Support**

G What are metalloids? *elements that have properties of both metals and nonmetals*

H What do radioactive atoms produce? *particles and energy*

Below Level Use wheel-and-spoke diagrams to clarify relationships among element groupings. At the center of one wheel will be metals; spokes will be reactive metals, transition metals, and rare earth elements. At the center of a second wheel will be nonmetals; spokes will be halogens, nonhalogens, and noble gases. The two wheels can be connected by a dotted line labeled *metalloids,* signifying that metalloids bridge the metals and nonmetals. Ask students to list the symbol of each metalloid on the line.

Uses of Radioactivity in Medicine

The radiation produced from unstable nuclei is used in hospitals to diagnose and treat patients. Some forms of radiation from nuclei are used to destroy harmful tumors inside a person's body without performing an operation. Another medical use of radiation is to monitor the activity of certain organs in the body. A patient is injected with a solution containing a radioactive isotope. Isotopes of a given atom move through the body in the same way whether or not they are radioactive. Doctors detect the particles produced by the radioactive isotopes to determine where and how the body is using the substance.

Although radiation has its benefits, in large doses it is harmful to living things and should be avoided. Radiation can damage or kill cells, and the energy from its particles can burn the skin. Prolonged exposure to radiation has been linked to cancer and other health problems.

A technician prepares a patient for radiotherapy. Radiation can be used to kill cancerous tissue, such as a brain tumor.

INVESTIGATE Radioactivity

How quickly can atoms change?

PROCEDURE

1. Put 50 pennies in a bag. The pennies represent 50 atoms.
2. Pour out the pennies.
3. Count the number of pennies that landed head side up. These represent atoms whose nuclei changed.
4. Refill the bag with only the pennies that landed tail side up.
5. Repeat steps 2–4 until all of the pennies have landed head side up. Each time you pour out the pennies counts as one turn.
6. Construct a graph with the number of atoms that changed on the *y*-axis and the number of turns on the *x*-axis.

WHAT DO YOU THINK?
- After one turn, how many atoms had changed? had not changed?
- In how many turns did all the atoms change?
- From looking at your graph, what can you conclude about the rate of radioactive change?

CHALLENGE If you used a different number of pennies, would your results be different? In what way?

SKILL FOCUS
Modeling

MATERIALS
- 50 pennies
- bag
- graph paper

TIME
30 minutes

Chapter 1: **Atomic Structure and the Periodic Table** **31** **B**

INVESTIGATE Radioactivity

PURPOSE To model how quickly atoms of radioactive elements can change

TIP *30 min.* Any common coin will work; all coins should be the same denomination.

WHAT DO YOU THINK? *About one-half of the atoms did not change. Two tries will change three-fourths of the atoms. It should take up to seven times for all the atoms to turn. Each time, about half the atoms change.*

CHALLENGE *Yes; it would take longer to get down to no atoms of the original. The curve of the graph would be the same.*

 Datasheet, Radioactivity, p. 41

Technology Resources

Customize this student lab as needed or look for an alternative. Print rubrics to assess student lab reports.

 Lab Generator CD-ROM

Teaching with Technology

Students can use graphing software or calculators to construct their graphs. Ask them to compare their graphs to the half-life graph pictured on p. 32.

Real World Example

One radioactive element that people often encounter in the natural environment is radon, a gas with a short half-life. Uranium-rich rock and soil emit radon, which can enter homes through the foundations. High levels of radon appear to contribute to lung cancer. What level of radon is safe for humans is a topic of debate and study.

DIFFERENTIATE INSTRUCTION

? More Reading Support

I Why should we avoid large doses of radiation? *It harms living things.*

Below Level Point out that because atoms are so small, they cannot be counted like heads and tails on a coin. Tell students that scientists have instruments, such as the Geiger counter, that can intercept and count the radiation particles from an unstable element. These instruments are analogous to a baseball backstop that counts the number of balls that hit it; many balls hit it during a highly active "inning," and few during a stable "inning."

Teach from Visuals

To help students interpret the graph of radioactive decay, ask: How are the graph and the circles connected? *The circles show pictorially the amount of radioisotope (y-axis) that is left at each time (x-axis).*

Integrate the Sciences

Living things have the same concentration of carbon-14 when they are alive. When they die, organisms stop taking in carbon-14. Knowing how fast carbon-14 decays, scientists can calculate how long ago an organism lived.

Reinforce (the **BIG** idea)

Have students relate the section to the Big Idea.

 Reinforcing Key Concepts, p. 42

1.3 ASSESS & RETEACH

Assess

 Section 1.3 Quiz, p. 5

Reteach

Distribute copies of the periodic table to groups of students. Suggest they flip through the text in this section to help them identify the following: reactive metals *(Groups 1 and 2)*, transition metals *(Groups 3–12)*, rare earth elements *(top row of the two rows of metals outside main body of periodic table)*, nonmetals *(upper right corner of table)*, halogens *(Group 17)*, noble gases *(Group 18)*, and metalloids *(upper right corner between metals and nonmetals; B, Si, Ge, As, Sb, Te, At)*.

Technology Resources

Have students visit **ClassZone.com** for reteaching of Key Concepts.

 CONTENT REVIEW

 CONTENT REVIEW CD-ROM

Radioactive Decay

VISUALIZATION
CLASSZONE.COM
Watch how a radioactive element decays over time.

Radioactive atoms produce energy and particles from their nuclei. The identity of these atoms changes because the number of protons changes. This process is known as radioactive decay. Over time, all of the atoms of a radioactive isotope will change into atoms of another element.

Radioactive decay occurs at a steady rate that is characteristic of the particular isotope. The amount of time that it takes for one-half of the atoms in a particular sample to decay is called the **half-life** of the isotope. For example, if you had 1000 atoms of a radioactive isotope with a half-life of 1 year, 500 of the atoms would change into another element over the course of a year. In the next year, 250 more atoms would decay. The illustration to the right shows how the amount of the original isotope would decrease over time.

The half-life is a characteristic of each isotope and is independent of the amount of material. A half-life is also not affected by conditions such as temperature or pressure. Half-lives of isotopes can range from a small fraction of a second to many billions of years.

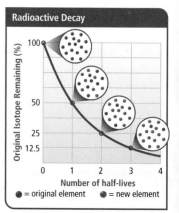

Radioactive Decay

● = original element   ● = new element

Half-Lives of Selected Elements

Isotope	Half-Life
Uranium-238	4,510,000,000 years
Carbon-14	5,730 years
Radon-222	3.82 days
Lead-214	27 minutes
Polonium-214	.00016 seconds

1.3 Review

KEY CONCEPTS

1. What are the three main classes of elements in the periodic table?
2. What are the major characteristics of metals?
3. How can an atom of one element change to an atom of another element?

CRITICAL THINKING

4. **Compare** Use the periodic table to determine whether a carbon or a fluorine atom would be more reactive.
5. **Calculate** What fraction of a radioactive sample remains after three half-lives?

● CHALLENGE

6. **Analyze** Why do you think the noble gases were among the last of the naturally occurring elements to be discovered?

ANSWERS

1. metals, metalloids, and nonmetals

2. Metals usually are shiny, easily bent or drawn into a wire, and good conductors of electricity and heat.

3. Radioactive decay causes the number of protons in the nucleus to change.

4. fluorine

5. one-eighth

6. The noble gases were hard to notice, because they don't often combine or react with other elements.

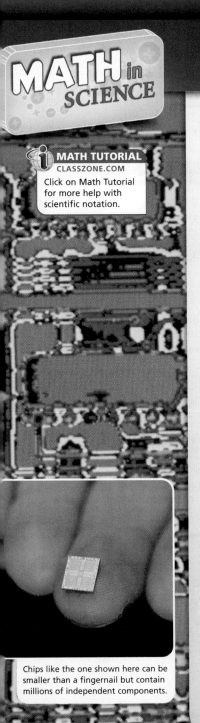

MATH TUTORIAL
CLASSZONE.COM

Click on Math Tutorial for more help with scientific notation.

Numbers with Many Zeros

Semiconductor devices are at the heart of the modern personal computer. Today tiny chips can contain more than 42,000,000 connections and perform about 3,000,000,000 calculations per second. Computers have little problem working with such large numbers. Scientists, however, use a scientific notation as a short-hand way to write large numbers. Scientific notation expresses a very large or very small number as the product of a number between 1 and 10 and a power of 10.

Example

Large Number How would you express the number 6,400,000,000—the approximate population of the world—in scientific notation?

(1) Look at the number and count how many spaces you would need to move the decimal point to get a number between 1 and 10.

(2) Place the decimal point in the space and multiply the number by the appropriate power of 10. The power of 10 will be equivalent to the number of spaces you moved the decimal point.

ANSWER 6.4×10^9

Small Number How would you express 0.0000023 in scientific notation?

(1) Count the number of places you need to move the decimal point to get a number between 1 and 10. This time you move the decimal point to the right, not the left.

(2) The power of 10 you need to multiply this number by is still equal to the number of places you moved the decimal point. Place a negative sign in front of it to indicate that you moved the decimal point to the right.

ANSWER 2.3×10^{-6}

Answer the following questions.

1. Express the following numbers in scientific notation:
   (a) 75,000   (b) 54,000,000,000   (c) 0.0000064

2. Express these numbers in decimal form:
   (a) 6.0×10^{24} (b) 7.4×10^{22} (c) 5.7×10^{-10}

CHALLENGE What is 2.2×10^{22} subtracted from 4.6×10^{22}?

Chips like the one shown here can be smaller than a fingernail but contain millions of independent components.

ANSWERS

1. 7.5×10^4; 5.4×10^{10}; 6.4×10^6

2. 6,000,000,000,000,000,000,000,000; 74,000,000,000,000,000,000,000;
   0.00000000057

CHALLENGE 2.4×10^{22}

MATH IN SCIENCE
Math Skills Practice for Science

Set Learning Goal

To express very large or very small numbers using scientific notation

Present the Science

Scientists often work with extreme numbers. A good example is the number indicating the size of an atom—or conversely, the number of atoms in a sample of matter. Writing out all the zeros would be time-consuming and very prone to error. Scientific notation avoids confusion.

Develop Number Sense

• Make sure students distinguish between the procedures for handling large numbers and small numbers. In large numbers, the zeros are strung out to the right, so the decimal is moved to the left. In small numbers, the procedure is just the reverse.

• For the Challenge question, prompt students to compare the exponents of 10 in the two scientific notations.

DIFFERENTIATION TIP Some students might benefit from counting the decimal places themselves. Write the number 6,400,000,000 or 0.0000023 on the board, leaving space between the numbers. Students can use colored chalk (or a colored dry erase marker) to count off the decimal places.

Close

Have students contrast the procedures for expressing a large number and a small decimal number in scientific notation. *For the large number, you move the decimal to the left and use a positive exponent; for the small number, you move the decimal to the right and use a negative exponent.*

R • Math Support, p. 48
   • Math Practice, p. 49

Technology Resources

Students can visit **ClassZone.com** for practice in scientific notation.

 MATH TUTORIAL

BACK TO

the BIG idea

Have students look back at the atoms photograph of nickel on pp. 6–7. Ask what an even more highly magnified photograph might show. *individual protons and neutrons, but probably not electrons*

○ KEY CONCEPTS SUMMARY

SECTION 1.1

Ask: How is the mass of an atom distributed? *most is concentrated in nucleus*
Ask: Why is an electron cloud pictured, rather than individual electrons? *Electrons are about 2000 times smaller than protons, and move very quickly around the nucleus. Since it is not possible to determine their exact positions, the electron cloud represents them.*

SECTION 1.2

Ask: Suppose an element is brittle and a poor conductor of electricity. To find another element with similar properties, would you look up and down or side to side in the periodic chart? *up and down*

SECTION 1.3

Ask: Would an element near the upper-right corner of the periodic table be shiny and easily formed into wire? Why? *No; those are properties of metals, and the elements near the upper-right corner of the periodic table are nonmetals.*

Review Concepts

- Big Idea Flow Chart, p. T1
- Chapter Outline, pp. T7–T8

---

1 Chapter Review

the BIG idea

A substance's atomic structure determines its physical and chemical properties.

 CONTENT REVIEW
CLASSZONE.COM

◐ KEY CONCEPTS SUMMARY

1.1 Atoms are the smallest form of elements.

- All matter is made of the atoms of approximately 100 elements.
- Atoms are made of protons, neutrons, and electrons.
- Different elements are made of different atoms.
- Atoms form ions by gaining or losing electrons.

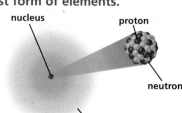

nucleus
proton
neutron
electron cloud

VOCABULARY
proton p. 11
neutron p. 11
nucleus p. 11
electron p. 11
atomic number p. 12
atomic mass number p. 12
isotope p. 12
ion p. 14

1.2 Elements make up the periodic table.

- Elements can be organized by similarities.
- The periodic table organizes the atoms of the elements by properties and atomic number.

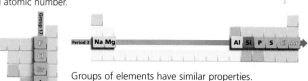

Groups of elements have similar properties.
Elements in a period have varying properties.

VOCABULARY
atomic mass p. 17
periodic table p. 18
group p. 22
period p. 22

1.3 The periodic table is a map of the elements.

- The periodic table has distinct regions.
- Most elements are metals.
- Nonmetals and metalloids have a wide range of properties.
- Some atoms can change their identity through radioactive decay.

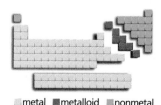

■ metal ■ metalloid ■ nonmetal

VOCABULARY
reactive p. 26
metal p. 27
nonmetal p. 29
metalloid p. 30
radioactivity p. 30
half-life p. 32

Technology Resources

Have students visit **ClassZone.com** or use the CD-ROM for a cumulative review of concepts.

 CONTENT REVIEW

 CONTENT REVIEW CD-ROM

Engage students in a whole-class interactive review of Key Concepts. Edit content as you wish.

 POWER PRESENTATIONS

Reviewing Vocabulary

Describe how the vocabulary terms in the following pairs are related to each other. Explain the relationship in a one- or two-sentence answer. Underline each vocabulary term in your answer.

1. isotope, nucleus

2. atomic mass, atomic number

3. electron, proton

4. atomic number, atomic mass number

5. group, period

6. metals, nonmetals

7. radioactivity, half-life

Reviewing Key Concepts

Multiple Choice Choose the letter of the best answer.

8. The central part of an atom is called the
   a. electron   c. proton
   b. nucleus   d. neutron

9. The electric charge on a proton is
   a. positive   c. neutral
   b. negative   d. changing

10. The number of protons in the nucleus is the
    a. atomic mass   c. atomic number
    b. isotope   d. half-life

11. Nitrogen has atomic number 7. An isotope of nitrogen containing seven neutrons would be
    a. nitrogen-13   c. nitrogen-15
    b. nitrogen-14   d. nitrogen-16

12. How does the size of a negative ion compare to the size of the atom that formed it?
    a. It's smaller.
    b. It's larger.
    c. It's the same size.
    d. It varies.

13. The modern periodic table is organized by
    a. size of atom
    b. atomic mass
    c. number of neutrons
    d. atomic number

14. Elements in a group have
    a. a wide range of chemical properties
    b. the same atomic radius
    c. similar chemical properties
    d. the same number of protons

15. Elements in a period have
    a. a wide range of chemical properties
    b. the same atomic radius
    c. similar chemical properties
    d. the same number of protons

16. From left to right in a period, the size of atoms
    a. increases   c. remains the same
    b. decreases   d. shows no pattern

17. The elements in Group 1 of the periodic table are commonly called the
    a. alkali metals   c. alkaline earth metals
    b. transition metals d. rare earth metals

18. The isotope nitrogen-13 has a half-life of 10 minutes. If you start with 40 grams of this isotope, how many grams will you have left after 20 minutes?
    a. 10   c. 20
    b. 15   d. 30

Short Answer Write a short answer to each question. You may need to consult a periodic table.

19. Rubidium forms the positive ion Rb^+. Is this ion larger or smaller than the neutral atom? Explain.

20. How can you find the number of neutrons in the isotope nitrogen-16?

21. Explain how density varies across and up and down the periodic table.

22. Place these elements in order from least reactive to most reactive: nickel (Ni), xenon (Xe), lithium (Li). How did you determine the order?

Reviewing Vocabulary

1. An isotope of an element has a different number of neutrons in its nucleus than another atom of that element.

2. The atomic mass of an element is basically the mass of the protons and neutrons in the nucleus, whereas the atomic number is the number of protons in the nucleus.

3. An electron is a particle with a negative charge, and a proton is a particle with a positive charge.

4. Atomic number is the number of protons in a nucleus, whereas atomic mass number is the number of protons and neutrons in the nucleus.

5. A group is a vertical columns in the periodic table that contains elements with similar properties, whereas a period is a horizontal row of elements with varying properties.

6. Metals are usually solids at room temperature, while nonmetals are usually gas or liquid at room temperature.

7. Radioactivity is measured in half-life, the amount of time it takes for one-half of a substance's atoms to decay.

Reviewing Key Concepts

8. b   14. c
9. a   15. a
10. c   16. b
11. b   17. a
12. b   18. a
13. d

19. smaller because the atom lost an electron

20. Subtract its atomic number from 16.

21. Density generally increases as one moves down a group. Elements at the ends of a period are least dense, and elements in the middle are most dense.

22. Xenon, nickel, lithium. Xenon is a noble gas and thus very nonreactive. Nickel is a transition metal and so not very reactive. Lithium is an alkali metal and thus very reactive.

ASSESSMENT RESOURCES

UNIT ASSESSMENT BOOK
- Chapter Test A, pp. 6–9
- Chapter Test B, pp. 10–13
- Chapter Test C, pp. 14–17
- Alternative Assessment, pp. 18–19

SPANISH ASSESSMENT BOOK
Spanish Chapter Test, pp. 233–236

Technology Resources

Edit test items and answer choices.

 Test Generator CD-ROM

Visit **ClassZone.com** to extend test practice.

 Test Practice

Thinking Critically

23. B, C, and F are metals; they look shiny and conduct electricity. E is a metalloid; it is a semiconductor. A and D are nonmetals; they do not have any of the properties of metals.

24. a cube of A

25. Element E; it is a semiconductor.

26. The decreased occurrence of thyroid disease is linked to increased ingestion of iodine.

27. It is larger.

28. because halogens are the most reactive nonmetals

29. 120 neutrons and 80 protons

30. larger

31. 12.5 g

32. thorium

the **BIG** idea

33. Nickel is a smooth, shiny metal. When magnified enough to see individual atoms, however, we see that it can appear bumpy and rough.

34. Remove three protons from the nucleus of a lead atom to make it into a gold atom (82 − 3 = 79). You might need to remove some neutrons to stabilize the new atom.

35. Atoms are arranged in the periodic table by their atomic number, which is the number of protons in an atom's nucleus. The structure of an atom's nucleus determines its place in the periodic table.

UNIT PROJECTS

Give students the appropriate Unit Project worksheets from the URB for their projects. Both directions and rubrics can be used as a guide.

 Unit Projects, pp. 5–10

Thinking Critically

The table below lists some properties of six elements. Use the information and your knowledge of the properties of elements to answer the next three questions.

Element	Appearance	Density (g/cm³)	Conducts Electricity
A	dark purple crystals	4.93	no
B	shiny silvery solid	0.97	yes
C	shiny silvery solid	22.65	yes
D	yellow powder	2.07	no
E	shiny gray solid	5.32	semiconductor
F	shiny bluish solid	8.91	yes

23. ANALYZE Based on the listed properties, identify each of the elements as a metal, nonmetal, or metalloid.

24. APPLY Which would weigh more: a cube of element A or a same-sized cube of element D?

25. HYPOTHESIZE Which element(s) do you think you might find in electronic devices? Why?

26. HYPOTHESIZE The thyroid gland, located in your throat, secretes hormones. In 1924 iodine was added to table salt. As more and more Americans used iodized salt, the number of cases of thyroid diseases decreased. Write a hypothesis that explains the observed decrease in thyroid-related diseases.

27. INFER How does the size of a beryllium (Be) atom compare with the size of an oxygen (O) atom?

28. PREDICT Although noble gases do not naturally react with other elements, xenon and krypton have been made to react with halogens such as chlorine in laboratories. Why are the halogens most likely to react with the noble gases?

Below is an element square from the periodic table. Use it to answer the next two questions.

80
Hg
Mercury
200.59

29. CALCULATE One of the more common isotopes of mercury is mercury-200. How many protons and neutrons are in the nucleus of mercury-200?

30. INFER Cadmium occupies the square directly above mercury on the periodic table. Is a cadium atom larger or smaller than a mercury atom?

31. CALCULATE An isotope has a half-life of 40 minutes. How much of a 100-gram sample would remain unchanged after two hours?

32. APPLY When a uranium atom with 92 protons and 146 neutrons undergoes radioactive decay, it produces a particle that consists of two protons and two neutrons from its nucleus. Into which element is the uranium atom transformed?

the **BIG** idea

33. ANALYZE Look again at the photograph on pages 6–7. Answer the question again, using what you have learned in the chapter.

34. DRAW CONCLUSIONS Suppose you've been given the ability to take apart and assemble atoms. How could you turn lead into gold?

35. ANALYZE Explain how the structure of an atom determines its place in the periodic table.

UNIT PROJECTS

If you are doing a unit project, make a folder for your project. Include in your folder a list of the resources you will need, the date on which the project is due, and a schedule to track your progress. Begin gathering data.

MONITOR AND RETEACH

If students still have difficulty with atomic mass number and atomic mass, create a simple atomic diagram clearly showing all the protons and neutrons in the nucleus. Have students find atomic number by counting protons, and find atomic mass by counting neutrons and protons.

To review isotopes, draw two more atomic diagrams, in which only the number of neutrons varies from the original diagram. Then have students average the atomic masses of the isotopes.

Students may benefit from summarizing sections of the chapter.

 Summarizing the Chapter, pp. 68–69

Standardized Test Practice

For practice on your state test, go to . . .
TEST PRACTICE
CLASSZONE.COM

Interpreting Tables

The table below shows part of the periodic table of elements.

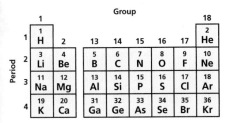

Answer the questions based on the information given in the table.

1. What does the number above the symbol for each element represent?

   a. Its number of isotopes

   b. Its atomic number

   c. Its number of neutrons

   d. Its atomic mass

2. The atom of what element is in Period 4, Group 13?

   a. Na   **c.** Al

   b. Ga   **d.** K

3. What do the elements on the far right of the table (He, Ne, Ar, and Kr) have in common?

   a. They do not generally react with other elements.

   b. They are in liquids under normal conditions.

   c. They are metals that rust easily.

   d. They are very reactive gases.

4. How many electrons does a neutral chlorine (Cl) atom contain?

   a. 16   **c.** 18

   b. 17   **d.** 19

5. If a sodium (Na) atom loses one electron to form a positive ion, how many electrons would lithium (Li) lose to form a positive ion?

   a. 0   **c.** 2

   b. 1   **d.** 3

6. If a fluorine (F) atom gains one electron to form a negative ion, how many electrons would bromine (Br) gain to form a negative ion?

   a. 0   **c.** 2

   b. 1   **d.** 3

Extended Response

Answer the following two questions in detail. Include some of the terms shown in the word box at right. Underline each term you use in your answer.

electron	nucleus	proton
isotope	neutron	radioactivity

7. Democritus was an ancient Greek philosopher who claimed that all matter was made of tiny particles he called atoms. Democritus said that all atoms were made of the same material. The objects of the world differed because each was made of atoms of different sizes and shapes. How does the modern view of atoms differ from this ancient view? How is it similar?

8. Half-life is a measure of the time it takes half of the radioactive atoms in a substance to decay into other atoms. If you know how much radioactive material an object had to begin with, how could you use half-life to determine its age now?

Interpreting Tables

Extended Response

7. RUBRIC

4 points for a response that correctly answers the question and uses the following terms accurately:

- protons
- neutrons
- electrons

Sample: Democritus' atomic theory is like the atomic theory of today because we now know that all atoms are made of the same stuff—protons, neutrons, and electrons—and that atoms are different sizes. However, we no longer think, as Democritus did, that atoms are shaped differently.

3 points correctly answers the question and uses two terms accurately

2 points correctly answers the question and uses one term accurately

1 point correctly answers the question or uses one term accurately

8. RUBRIC

4 points for a response that correctly answers the question and uses the following terms accurately:

- radioactive
- isotope

Sample: One way to figure it out is by seeing how much of the new element is there and how much of the radioactive isotope is left, and then using the ratio to find how many years since the organism died.

3 points correctly answers the question and uses one term accurately

2 points partly answers the question and uses one term accurately

1 point partly answers the question or uses one term accurately

METACOGNITIVE ACTIVITY

Have students answer the following questions in their **Science Notebook:**

1. What questions do you still have about atomic structure?

2. Which topics in this chapter would you like to learn more about?

3. What have you learned in this chapter that can be applied to your Unit Project?

CHAPTER

2 Chemical Bonds and Compounds

Physical Science
UNIFYING PRINCIPLES

PRINCIPLE 1

Matter is made of particles too small to see.

PRINCIPLE 2

Matter changes form and moves from place to place.

PRINCIPLE 3

Energy changes from one form to another, but it cannot be created or destroyed.

PRINCIPLE 4

Physical forces affect the movement of all matter on Earth and throughout the universe.

Unit: Chemical Interactions
BIG IDEAS

CHAPTER 1
Atomic Structure and the Periodic Table
A substance's atomic structure determines its physical and chemical properties.

CHAPTER 2
Chemical Bonds and Compounds
The properties of compounds depend on their atoms and chemical bonds.

CHAPTER 3
Chemical Reactions
Chemical reactions form new substances by breaking and making chemical bonds.

CHAPTER 4
Solutions
When substances dissolve to form a solution, the properties of the mixture change.

CHAPTER 5
Carbon in Life and Materials
Carbon is essential to living things and to modern materials.

CHAPTER 2
KEY CONCEPTS

SECTION 2.1

Elements combine to form compounds.

1. Compounds have different properties from the elements that make them.

2. Atoms combine in predictable numbers.

SECTION 2.2

Chemical bonds hold compounds together.

1. Chemical bonds between atoms involve electrons.

2. Atoms can transfer electrons.

3. Atoms can share electrons.

4. Chemical bonds give all materials their structures.

SECTION 2.3

Substances' properties depend on their bonds.

1. Metals have unique bonds.

2. Ionic and covalent bonds give compounds certain properties.

3. Bonds can make the same element look different.

T The Big Idea Flow Chart is available on p. T9 in the **UNIT TRANSPARENCY BOOK.**

Previewing Content

SECTION

2.1 Elements combine to form compounds. pp. 41–46

1. Compounds have different properties from the elements that make them.

A compound is a combination of two or more elements. What makes a compound different from a mixture is that atoms of the elements in a compound are held together by chemical bonds. The properties of a compound are often quite different from the properties of the elements that make it.

2. Atoms combine in predictable numbers.

Compounds have a definite composition. Each compound contains a specific ratio of atoms held together by chemical bonds. The compound formed when one nitrogen atom combines with one oxygen atom is different from the compound formed when two nitrogen atoms combine with one oxygen atom. Compounds are not like simple mixtures, which have no definite combination.

A chemical compound is represented by a **chemical formula.** Chemical formulas use the symbols for the elements to show the different elements that make up a compound. To show the ratios of the atoms of those elements in the compound, subscripts are used. A **subscript** is a number written to the right of the chemical symbol and slightly below the line. Carbon dioxide, for example, has two oxygen atoms for each carbon atom. The chemical formula for carbon dioxide is therefore CO_2. In the table below, notice the ratios of the atoms in each compound.

Compound Name	Atomic Ratio	Chemical Formula
Hydrogen chloride	1:1	HCl
Water	2:1	H_2O
Ammonia	1:3	NH_3
Methane	1:4	CH_4
Propane	3:8	C_3H_8

Different compounds, such as H_2O and H_2O_2, can be composed of the same elements. Because they are different compounds, they have different properties.

SECTION

2.2 Chemical bonds hold compounds together. pp. 47–55

1. Chemical bonds between atoms involve electrons.

Chemical bonds are the glue holding the atoms in compounds together. Chemical bonds are the result of interactions between the electron clouds of two or more atoms.

2. Atoms can transfer electrons.

Ions are formed when atoms gain or lose electrons. When one atom loses an electron, another atom picks up that electron, forming a negative and positive ion pair. A positive ion is attracted to a negative ion. This attraction is called an **ionic bond.**

Ionic bonds produce large crystal networks of atoms, because the attraction between the positive and negative ions acts in all directions. In an ionic compound, positive ions are attracted to all negative ions, and negative ions are attracted to all positive ions.

Ionic compounds bear the name of the positive ion followed by the name of the negative ion, with the suffix –*ide*. The ionic compound made of lithium and chlorine is lithium chloride.

3. Atoms can share electrons.

Atoms that share a pair of electrons have what is called a **covalent bond.** Atoms form covalent bonds because they are more energetically stable when they have a certain number of electrons around their nuclei. Covalent bonds can form between two atoms of the same element or two atoms of different elements. A **molecule** is a group of atoms held together by covalent bonds. It has no electrical charge.

In some cases, two atoms can form as many as four covalent bonds. Most often, the electrons shared in a covalent bond spend more time closer to one of the nuclei than the other. When electrons stay much closer to one nucleus than the other, this is a **polar covalent bond.**

4. Chemical bonds give all materials their structures.

The shape of the crystal formed by an ionic compound depends on the ratio, shapes, and sizes of the ions. Covalent compounds do not form crystals; they form individual molecules.

Molecules have characteristic shapes, or molecular structures. Molecular structure affects many properties of the compounds.

Common Misconceptions

PHYSICAL VERSUS CHEMICAL COMBINATIONS Students might think that chemical combinations are the same as physical combinations. For chemical compounds to be created, chemical bonds must be formed.

 This misconception is addressed on p. 48.

 MISCONCEPTION DATABASE
CLASSZONE.COM Background on student misconceptions

Previewing Content

SECTION

2.3 Substances' properties depend on their bonds. pp. 56–61

1. Metals have unique bonds.

Metal atoms share electrons in all directions with other metal atoms in a type of bond called a **metallic bond.** The figure shows how the electrons in a metal are shared by many metal atoms.

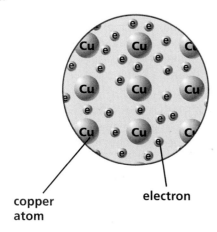

copper
atom

electron

Properties of metals are determined by the mobility of the electrons in a metallic bond. These properties include conductivity, ductility, and malleability.

2. Ionic and covalent bonds give compounds certain properties.

Ions are tightly locked into place in the structure of a crystal, so ionic bonds are difficult to break. Ionic compounds generally
• have high melting and boiling points
• are hard and brittle and do not conduct electricity when a solid
• break up into negative and positive ions when dissolved and will conduct an electric current in solution

The molecules of covalent compounds are not held together as tightly.
• Boiling and melting points of covalent compounds are relatively low.
• Molecules stay intact when dissolved in water.
• Molecule size and shape affect properties also.

3. Bonds can make the same element look different.

Different forms of the same element, called allotropes, can result from different covalent bonds. Carbon forms three different allotropes, all with different properties: diamond, graphite, and fullerene.

Previewing Labs

EXPLORE the BIG idea

Mixing It Up, p. 39 Students observe that the properties of a combination differ from the properties of the components.	**TIME** 10 minutes **MATERIALS** red and yellow modeling compound
The Shape of Things, p. 39 Students observe the crystal structure of an ionic compound, salt.	**TIME** 10 minutes **MATERIALS** table salt, dark paper, hand lens
Internet Activity: Bonding, p. 39 Students are introduced to differences in ionic and covalent bonding.	**TIME** 20 minutes **MATERIALS** computer with Internet access

SECTION 2.1

EXPLORE Compounds, p. 41 Students compare the properties of a compound with those of the elements that compose it.	**TIME** 10 minutes **MATERIALS** carbon, beaker of water, sugar, test tube, test-tube holder, candle, matches
INVESTIGATE Element Ratios, p. 43 Students model a compound and determine the ratio of its "elements." R Modeling Compounds Datasheet	**TIME** 20 minutes **MATERIALS** nuts and bolts, Modeling Compounds Datasheet

SECTION 2.2

INVESTIGATE Crystals, p. 53 Students observe how a crystal grows over a period of time.	**TIME** 30 minutes **MATERIALS** crystal-growing substance (sugar, table salt, alum, or copper sulfate), 2 glass beakers, hot tap water, stirring stick, cotton string, paper clip, pencil, hand lens

SECTION 2.3

EXPLORE Bonds in Metals, p. 56 Students examine conductivity of metals.	**TIME** 10 minutes **MATERIALS** masking tape, 3 6-inch pieces of copper wire, D-cell battery, light bulb and holder, objects to test
CHAPTER INVESTIGATION **Chemical Bonds,** pp. 60–61 Students investigate properties of substances with different types of bonds.	**TIME** 40 minutes **MATERIALS** 3 wire leads with alligator clips; battery; zinc strip; copper strip; light bulb and socket; 20 grams each of Epsom salts, sugar, and iron filings; 3 plastic cups; distilled water; beaker; constuction paper; hand lens; plastic spoon; 3 test tubes; test-tube rack; candle; matches; wire test-tube holder

R **Additional INVESTIGATION,** Weird Water, A, B, & C, pp. 119–127; Teacher Instructions, pp. 329–330

Previewing Chapter Resources

	INTEGRATED TECHNOLOGY	LABS AND ACTIVITIES

Chapter 2
Chemical Bonds and Compounds

 CLASSZONE.COM
- eEdition Plus
- EasyPlanner
- Misconception Database
- Content Review
- Test Practice
- Visualization
- Resource Centers
- Internet Activity: Bonding
- Math Tutorial

 SCILINKS.ORG
SCI LINKS

 CD-ROMS
- eEdition
- EasyPlanner
- Power Presentations
- Content Review
- Lab Generator
- Test Generator

 AUDIO CDS
- Audio Readings
- Audio Readings in Spanish

PE EXPLORE the Big Idea, p. 39
- Mixing It Up
- The Shape of Things
- Internet Activity: Bonding

R **UNIT RESOURCE BOOK**
Unit Projects, pp. 5–10

 Lab Generator CD-ROM
Generate customized labs.

SECTION
2.1 Elements combine to form compounds.
pp. 41–46

Time: 2 periods (1 block)
R Lesson Plan, pp. 70–71

 • **RESOURCE CENTER,** Chemical Formulas
• **MATH TUTORIAL**

 UNIT TRANSPARENCY BOOK
- Big Idea Flow Chart, p. T9
- Daily Vocabulary Scaffolding, p. T10
- Note-Taking Model, p. T11
- 3-Minute Warm-Up, p. T12

PE • EXPLORE Compounds, p. 41
• INVESTIGATE Element Ratios, p. 43
• Math in Science, p. 46

R **UNIT RESOURCE BOOK**
- Modeling Compounds Datasheet, p. 79
- Datasheet, Element Ratios, p. 80
- Math Support, p. 108
- Math Practice, p. 109

SECTION
2.2 Chemical bonds hold compounds together.
pp. 47–55

Time: 2 periods (1 block)
R Lesson Plan, pp. 82–83

 • **VISUALIZATION,** Polar Electron Clouds

 UNIT TRANSPARENCY BOOK
- Daily Vocabulary Scaffolding, p. T10
- 3-Minute Warm-Up, p. T12
- "Comparing Bonds" Visual, p. T14

PE • INVESTIGATE Crystals, p. 53
• Think Science, p. 55

R **UNIT RESOURCE BOOK**
Datasheet, Crystals, p. 91

SECTION
2.3 Substances' properties depend on their bonds.
pp. 56–61

Time: 4 periods (2 blocks)
R Lesson Plan, pp. 93–94

 • **RESOURCE CENTER,** Properties of Ionic and Covalent Compounds

 UNIT TRANSPARENCY BOOK
- Big Idea Flow Chart, p. T9
- Daily Vocabulary Scaffolding, p. T10
- 3-Minute Warm-Up, p. T13
- Chapter Outline, pp. T15–T16

PE • EXPLORE Bonds in Metals, p. 56
• CHAPTER INVESTIGATION, Chemical Bonds, pp. 60–61

R **UNIT RESOURCE BOOK**
- CHAPTER INVESTIGATION, Chemical Bonds, A, B, & C, pp. 110–118
- Additional INVESTIGATION, Weird Water, A, B, & C, pp. 119–127

KEY TO ICONS

 CD/CD-ROM

 Teacher Edition

 UNIT TRANSPARENCY BOOK

 SPANISH ASSESSMENT BOOK

 INTERNET

 Pupil Edition

 UNIT RESOURCE BOOK

UNIT ASSESSMENT BOOK

SCIENCE TOOLKIT

READING AND REINFORCEMENT

ASSESSMENT

STANDARDS

- Description Wheel, B20–21
- Main Idea and Detail Notes, C37
- Daily Vocabulary Scaffolding, H1–8

 UNIT RESOURCE BOOK
- Vocabulary Practice, pp. 105–106
- Decoding Support, p. 107
- Summarizing the Chapter, pp. 128–129

 Audio Readings CD
Listen to Pupil Edition.

Audio Readings in Spanish CD
Listen to Pupil Edition in Spanish.

- Chapter Review, pp. 63–64
- Standardized Test Practice, p. 65

 UNIT ASSESSMENT BOOK
- Diagnostic Test, pp. 20–21
- Chapter Test, A, B, & C, pp. 25–36
- Alternative Assessment, pp. 37–38

 Spanish Chapter Test, pp. 237–240

 Test Generator CD-ROM
Generate customized tests.

 Lab Generator CD-ROM
Rubrics for Labs

National Standards
A.2–8, A.9.a–c, A.9.e–f, B.1.b

See p. 38 for the standards.

 UNIT RESOURCE BOOK
- Reading Study Guide, A & B, pp. 72–75
- Spanish Reading Study Guide, pp. 76–77
- Challenge and Extension, p. 78
- Reinforcing Key Concepts, p. 81

 Ongoing Assessment, pp. 42, 44–45

 Section 2.1 Review, p. 45

UNIT ASSESSMENT BOOK
Section 2.1 Quiz, p. 22

National Standards
A.2–8, A.9.a–c, A.9.e–f, B.1.b

UNIT RESOURCE BOOK
- Reading Study Guide, A & B, pp. 84–87
- Spanish Reading Study Guide, pp. 88–89
- Challenge and Extension, p. 90
- Reinforcing Key Concepts, p. 92
- Challenge Reading, pp. 103–104

 Ongoing Assessment, pp. 47–48, 51–52, 54

 Section 2.2 Review, p. 54

 UNIT ASSESSMENT BOOK
Section 2.2 Quiz, p. 23

National Standards
A.2–7, A.9.a–b, A.9.e–f

UNIT RESOURCE BOOK
- Reading Study Guide, A & B, pp. 95–98
- Spanish Reading Study Guide, pp. 99–100
- Challenge and Extension, p. 101
- Reinforcing Key Concepts, p. 102

 Ongoing Assessment, pp. 56–58

 Section 2.3 Review, p. 59

 UNIT ASSESSMENT BOOK
Section 2.3 Quiz, p. 24

National Standards
A.2–7, A.9.a–b, A.9.e–f, B.1.b

Previewing Resources for Differentiated Instruction

CHAPTER INVESTIGATION

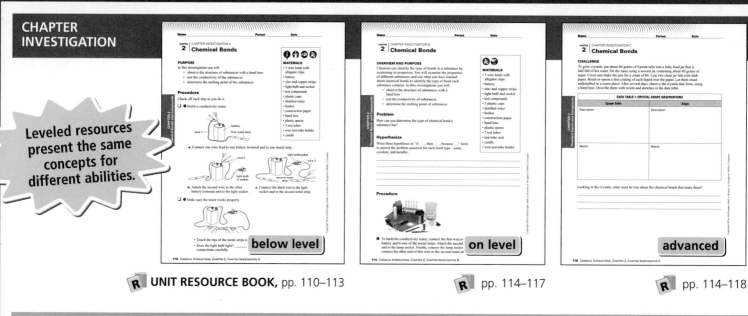

Leveled resources present the same concepts for different abilities.

R **UNIT RESOURCE BOOK,** pp. 110–113

R pp. 114–117

R pp. 114–118

READING STUDY GUIDE

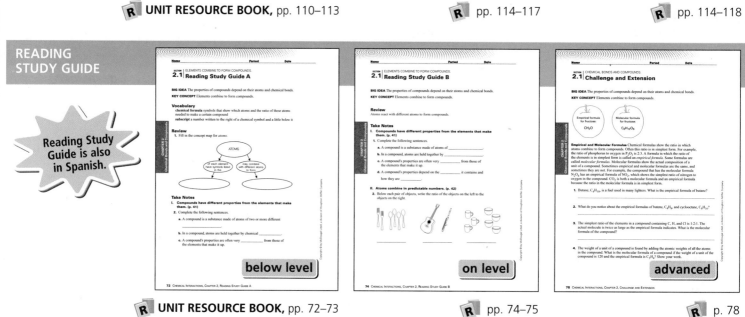

Reading Study Guide is also in Spanish.

R **UNIT RESOURCE BOOK,** pp. 72–73

R pp. 74–75

R p. 78

CHAPTER TEST

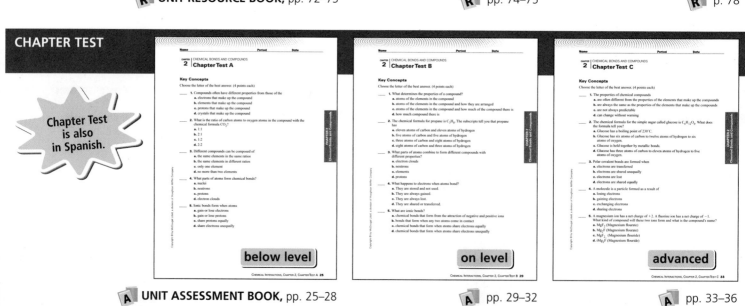

Chapter Test is also in Spanish.

A **UNIT ASSESSMENT BOOK,** pp. 25–28

A pp. 29–32

A pp. 33–36

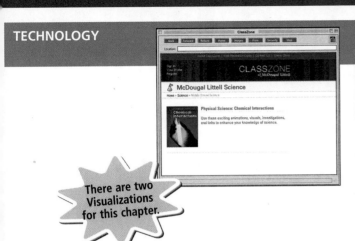

There are two Visualizations for this chapter.

AUDIO READINGS

McDOUGAL LITTELL

LAB GENERATOR

Customize and edit labs with this easy-to-use CD-ROM

- Searchable database of all labs from the program
- Additional lab options
- Template for creating your own labs
- Rubrics and other resources

Science

 **CLASSZONE.COM**

CD/CD-ROMS

 **CLASSZONE.COM**

VISUAL CONTENT

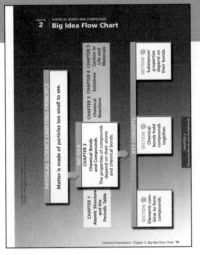

CHAPTER 2 CHEMICAL BONDS AND COMPOUNDS
2 Big Idea Flow Chart

 **UNIT TRANSPARENCY BOOK,** p. T9

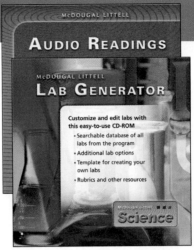

CHAPTER 2 CHEMICAL BONDS AND COMPOUNDS
2 Note-Taking Model

MAIN IDEAS	DETAIL NOTES
Chemical bonds between atoms involve electrons.	Chemical bonds hold the atoms of elements together in compounds.
	Chemical bonds form when the electrons in the electron clouds around two atoms interact.

Chemical Interactions, Chapter 2, Note-Taking Model T11

p. T11

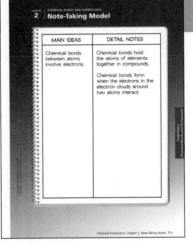

CHAPTER 2 CHEMICAL BONDS AND COMPOUNDS
2 Comparing Bonds

In Salar de Uyuni, Bolivia, salt is mined in great quantities from salt water. All types of chemical bonds are involved.

Ionic Bonds (salt) Covalent Bonds (air) Polar Covalent Bonds (water)

Sodium Chloride (NaCl)
A complete transfer of electrons produces ionic bonds.

Nitrogen (N_2) and Oxygen (O_2)
Some molecules contain multiple covalent bonds.

Water (H_2O)
The covalent bonds in water are very polar.

Chemical Interactions, Chapter 2, Teaching Visual T14

p. T14

MORE SUPPORT

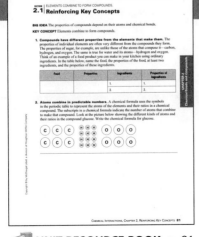

SECTION ELEMENTS COMBINE TO FORM COMPOUNDS
2.1 Reinforcing Key Concepts

BIG IDEA The properties of compounds depend on their atoms and chemical bonds.

KEY CONCEPT Elements combine to form compounds.

Reinforcing Key Concepts for each section

Chemical Interactions, Chapter 2, Reinforcing Key Concepts 81

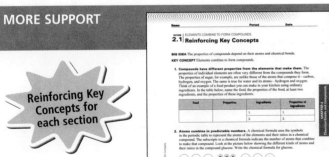 **UNIT RESOURCE BOOK,** p. 81

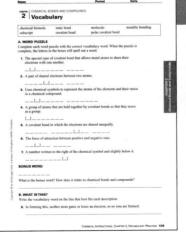

CHAPTER CHEMICAL BONDS AND COMPOUNDS
2 Vocabulary

Chemical Interactions, Chapter 2, Vocabulary Practice 105

 pp. 105–106

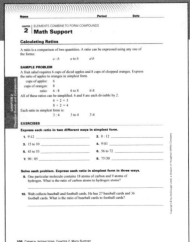

CHAPTER ELEMENTS COMBINE TO FORM COMPOUNDS
2 Math Support

Calculating Ratios

108 Chemical Interactions, Chapter 2, Math Support

 p. 108

INTRODUCE

Have students look at the photograph of skydivers diving as a group. Discuss how the question in the box links to the Big Idea:

- If you described the characteristics of the group of skydivers, how would the group characteristics compare to those of the individual skydivers?

- Based on the formation, to how many other skydivers can one skydiver connect?

National Science Education Standards

Content

B.1.b Substances react chemically in characteristic ways with other substances to form new substances (compounds) with different characteristic properties. In chemical reactions, the total mass is conserved. Substances often are placed in categories or groups if they react in similar ways; metals is an example of such a group.

Process

A.2–8 Design and conduct an investigation; use tools to gather and interpret data; use evidence to describe, predict, explain, model; think critically to make relationships between evidence and explanation; recognize different explanations and predictions; communicate scientific procedures and explanations; use mathematics.

A.9.a–c, A.9.e–f Understand scientific inquiry by using different investigations, methods, mathematics, and explanations based on logic, evidence, and skepticism.

CHAPTER

Chemical Bonds and Compounds

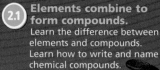

the BIG idea

The properties of compounds depend on their atoms and chemical bonds.

> How do these skydivers stay together? How is this similar to the way atoms stay together?

Key Concepts

SECTION
2.1 **Elements combine to form compounds.**
Learn the difference between elements and compounds. Learn how to write and name chemical compounds.

SECTION
2.2 **Chemical bonds hold compounds together.**
Learn about the different types of chemical bonds.

SECTION
2.3 **Substances' properties depend on their bonds.**
Learn how bonds give compounds certain properties.

Internet Preview

CLASSZONE.COM
Chapter 2 online resources: Content Review, two Visualizations, two Resource Centers, Math Tutorial, Test Practice

INTERNET PREVIEW

CLASSZONE.COM For student use with the following pages:

Review and Practice
- Content Review, pp. 40, 62
- Math Tutorial: Ratios, p. 46
- Test Practice, p. 65

Activities and Resources
- Internet Activity: Bonding, p. 39
- Resource Centers: Chemical Formulas, p. 44; Properties of Compounds, p. 58
- Visualization: Polar Electron Clouds, p. 51

NSTA *SCi*
scilinks.org *LINKS*
Compounds **Code: MDL023**

Mixing It Up

Get some red and yellow modeling compound. Make three red and two yellow balls, each about the diameter of a nickel. Blend one red and one yellow ball together. Blend one yellow and two red balls together.

Observe and Think How different do your combinations look from the original? from each other?

The Shape of Things

Pour some salt onto dark paper. Look at the grains through a hand lens. Try to observe a single grain.

Observe and Think What do you notice about the salt grains? What do you think might affect the way the grains look?

Internet Activity: Bonding

Go to **ClassZone.com** and watch the animation showing ionic and covalent bonding. Observe the differences in the two types of bonding.

Observe and Think What's the difference between an ionic and a covalent bond? Explain how covalent bonding can have different characteristics.

NSTA
scilinks.org
SCiLINKS
Compounds Code: MDL023

Chapter 2: **Chemical Bonds and Compounds** 39 **B**

TEACHING WITH TECHNOLOGY

Molecule Modeling Software As students discuss molecular shape, have them use molecule modeling software to examine the shapes of several different molecules. Two opportunities are when they read about carbon dioxide and the molecules shown on p. 54 and when you discuss the polyatomic ions described in Teach Difficult Concepts, p. 58.

EXPLORE the BIG idea

These inquiry-based activities are appropriate for use at home or as a supplement to classroom instruction.

Mixing It Up

PURPOSE To introduce how properties of a compound differ from the properties of its components. Students observe changes in properties when modeling the modeling compound alone and when combined.

TIP *10 min.* Different colors of food coloring in water can be used instead of modeling compound.

Answer: The resulting colors are different from the original colors. The second combination is darker.

REVISIT after p. 42.

The Shape of Things

PURPOSE To introduce the crystal structure of an ionic compound. Students observe the structure of salt.

TIP *10 min.* Have students draw what they see.

Answer: They are cubes with sharp edges. Size might affect how the crystals look, but the shape would be the same.

REVISIT after p. 49.

Internet Activity: Bonding

PURPOSE To introduce students to differences in ionic and covalent bonding.

TIP *20 min.* Have students list characteristics of each type of bond as they go through the examples.

Answer: In ionic compounds, electrons are transferred. In covalent compounds, electrons are shared.

REVISIT after p. 51.

◗ CONCEPT REVIEW
Activate Prior Knowledge

- Tell students they will draw a helium atom, which has two protons and two neutrons.
- Have students identify where to put the protons. Ask where they would put the neutrons.
- Ask students how many electrons are needed to balance the charges. Have students note the charge on the electron cloud.

◗ TAKING NOTES

Main Idea and Detail Notes

Have small groups of students compare their charts. Have them discuss similarities and differences.

Vocabulary Strategy

Encourage students to make at least three spokes describing each term. Examine their descriptions to see whether they are too general and should be more specific. If a description is too detailed, advise the student to split the information between two spokes.

Vocabulary and Note-Taking Resources

R
- Vocabulary Practice, pp. 105–106
- Decoding Support, p. 107

T
- Daily Vocabulary Scaffolding, p. T10
- Note-Taking Model, p. T11

🔧
- Description Wheel, B20–21
- Main Idea and Detail Notes, C37
- Daily Vocabulary Scaffolding, H1–8

CHAPTER 2
Getting Ready to Learn

◗ CONCEPT REVIEW

- Electrons occupy a cloud around an atom's nucleus.
- Atoms form ions by losing or gaining electrons.

◗ VOCABULARY REVIEW

electron p. 11
element *See Glossary.*

CONTENT REVIEW
CLASSZONE.COM
Review concepts and vocabulary.

▶ TAKING NOTES

MAIN IDEA AND DETAIL NOTES

Make a two-column chart. Write the main ideas, such as those in the blue headings, in the column on the left. Write details about each of those main ideas in the column on the right.

VOCABULARY STRATEGY

Place each vocabulary term at the center of a **description wheel** diagram. Write some words describing it on the spokes.

See the Note-Taking Handbook on pages R45–R51.

B 40 Unit: Chemical Interactions

SCIENCE NOTEBOOK

MAIN IDEAS	DETAIL NOTES
Atoms combine in predictable numbers.	• Each compound has a specific ratio of atoms. • A ratio is a comparison between two quantities.
Writing chemical formulas	• Find symbols on the periodic table. • Note ratio of atoms with subscripts.

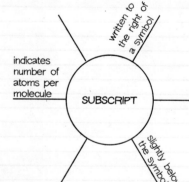

indicates number of atoms per molecule

written to the right of a symbol

SUBSCRIPT

slightly below the symbol

CHECK READINESS

Administer the Diagnostic Test to determine students' readiness for new science content and their mastery of requisite math skills.

 Diagnostic Test, pp. 20–21

Technology Resources

Students needing content and math skills should visit **ClassZone.com**.

- **CONTENT REVIEW**
- **MATH TUTORIAL**

- **CONTENT REVIEW CD-ROM**

2.1 Elements combine to form compounds.

◀ **BEFORE, you learned**

- Atoms make up everything on Earth
- Atoms react with different atoms to form compounds

▶ **NOW, you will learn**

- How compounds differ from the elements that make them
- How a chemical formula represents the ratio of atoms in a compound
- How the same atoms can form different compounds

VOCABULARY

chemical formula p. 43
subscript p. 43

EXPLORE Compounds

How are compounds different from elements?

PROCEDURE

1. Examine the lump of carbon, the beaker of water, and the sugar. Record your observations of each.

2. Pour some sugar into a test tube and heat it over a candle for several minutes. Record your observations.

MATERIALS

- carbon
- water
- sugar
- test tube
- test-tube holder
- candle

WHAT DO YOU THINK?

- The sugar is made up of atoms of the same elements that are in the carbon and water. How are sugar, carbon, and water different from one another?
- Does heating the sugar give you any clue that sugar contains more than one element?

Compounds have different properties from the elements that make them.

MAIN IDEA AND DETAILS
Make a two-column chart to start organizing information on compounds.

If you think about all of the different substances around you, it is clear that they cannot all be elements. In fact, while there are just over 100 elements, there are millions of different substances. Most substances are compounds. A compound is a substance made of atoms of two or more different elements. Just as the 26 letters in the alphabet can form thousands of words, the elements in the periodic table can form millions of compounds.

The atoms of different elements are held together in compounds by chemical bonds. Chemical bonds can hold atoms together in large networks or in small groups. Bonds help determine the properties of a compound.

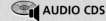

Chapter 2: **Chemical Bonds and Compounds** 41 **B**

2.1 FOCUS

◉ Set Learning Goals

Students will

- Describe how compounds are made from combinations of atoms.
- Explain how chemical formulas represent compounds.
- Model a compound in an experiment.

◑ 3-Minute Warm-Up

Display Transparency 12 or copy this exercise on the board:

Draw a diagram of a neutral carbon atom. A neutral carbon atom has six protons in its nucleus. On your diagram, label the nucleus and electron cloud, and indicate the total positive or negative charge on each. *Diagrams should show the nucleus as a dot within the electron cloud. The electron cloud should be labeled 6− and a line to the nucleus should be labeled 6+.*

Ⓣ 3-Minute Warm-Up, p. T12

2.1 MOTIVATE

EXPLORE Compounds

PURPOSE To compare the properties of a compound and those of its elements

TIP *10 min.* Extend the activity, using a piece of uncorroded iron and a piece of rusty iron.

WHAT DO YOU THINK? *They differ in color, texture, and state. If the sugar is heated enough, it turns black, like carbon, and droplets of water form at the top of the test tube.*

RESOURCES FOR DIFFERENTIATED INSTRUCTION

Below Level

UNIT RESOURCE BOOK
- Reading Study Guide A, pp. 72–73
- Decoding Support, p. 107

🔊 **AUDIO CDS**

Advanced

UNIT RESOURCE BOOK
Challenge and Extension, p. 78

English Learners

UNIT RESOURCE BOOK
Spanish Reading Study Guide, pp. 76–77

🔊 **AUDIO CDS**

- Audio Readings in Spanish
- Audio Readings (English)

History of Science

In the early 1800s, John Dalton and other scientists noticed that atoms combine in whole number volumes. Two liters of hydrogen, for example, always combined with one liter of oxygen to make two liters of water. This happens because these volumes contain the correct ratio of atoms to form the compound.

EXPLORE (the **BIG** idea)

Revisit "Mixing It Up" on p. 39. Have students make an analogy between their results and chemical compounds.

Integrate the Sciences

Although the air is approximately 78 percent nitrogen, this nitrogen is not in a form that plants can use. Nitrogen is quite unreactive in its elemental form. Things such as bacteria in the soil can change elemental nitrogen into compounds. These compounds have properties such that plants are able to absorb and use nitrogen.

Ongoing Assessment

Describe how compounds are made from combinations of atoms.

Ask: What can you predict about the properties of the compound potassium fluoride compared to the properties of the elements potassium and fluorine? *They are different.*

 Answer: They are often quite different.

The properties of a compound depend not only on which atoms the compound contains, but also on how the atoms are arranged. Atoms of carbon and hydrogen, for example, can combine to form many thousands of different compounds. These compounds include natural gas, components of automobile gasoline, the hard waxes in candles, and many plastics. Each of these compounds has a certain number of carbon and hydrogen atoms arranged in a specific way.

The properties of compounds are often very different from the properties of the elements that make them. For example, water is made from two atoms of hydrogen bonded to one atom of oxygen. At room temperature, hydrogen and oxygen are both colorless, odorless gases, and they remain gases down to extremely low temperatures. Water, however, is a liquid at temperatures up to 100°C (212°F) and a solid below 0°C (32°F). Sugar is a compound composed of atoms of carbon, hydrogen, and oxygen. Its properties, however, are unlike those of carbon, hydrogen, or oxygen.

calcium + **chlorine** = **calcium chloride**

The picture above shows what happens when the elements calcium and chlorine combine to form the compound calcium chloride. Calcium is a soft, silvery metallic solid. Chlorine is a greenish-yellow gas that is extremely reactive and poisonous to humans. Calcium chloride, however, is a nonpoisonous white solid. People who live in cold climates often use calcium chloride to melt the ice that forms on streets in the wintertime.

 CHECK YOUR READING How do the properties of a compound compare with the properties of the elements that make it?

Atoms combine in predictable numbers.

A given compound always contains atoms of elements in a specific ratio. For example, the compound ammonia always has three hydrogen atoms for every nitrogen atom—a 3 to 1 ratio of hydrogen to nitrogen. This same 3:1 ratio holds for every sample of ammonia, under all physical conditions. A substance with a different ratio of hydrogen to nitrogen atoms is not ammonia. For example, hydrazoic acid also contains atoms of hydrogen and nitrogen but in a ratio of one hydrogen atom to three nitrogen atoms, or 1:3.

READING TiP

A ratio is a numerical relationship between two values. If you had 3 apples for every 1 orange, you'd have a ratio of 3 to 1.

DIFFERENTIATE INSTRUCTION

? More Reading Support

A What two factors determine the properties of a compound? *the types of atoms it contains and how the atoms are arranged*

English Learners Put the definitions for the terms *compound, chemical formula,* and *subscript* on your classroom Science Word Wall or have English learners make flash cards for quick reference.

"Investigate Element Ratios" on p. 43 has directions that use the imperative mood, in which "you" is implied. Make sure English learners understand how to read and follow these directions.

INVESTIGATE Element Ratios

How can you model a compound?

PROCEDURE

1. Collect a number of nuts and bolts. The nuts represent hydrogen atoms. The bolts represent carbon atoms.

2. Connect the nuts to the bolts to model the compound methane. Methane contains four hydrogen atoms attached to one carbon atom. Make as many of these models as you can.

3. Count the nuts and bolts left over.

WHAT DO YOU THINK?

- What ratio of nuts to bolts did you use to make a model of a methane atom?
- How many methane models did you make? Why couldn't you make more?

CHALLENGE
The compound ammonia has one nitrogen atom and three hydrogen atoms. How would you use the nuts and bolts to model this compound?

SKILL FOCUS
Modeling

MATERIAL
nuts and bolts

TIME
20 minutes

Chemical Formulas

Remember that atoms of elements can be represented by their chemical symbols, as given in the periodic table. A **chemical formula** uses these chemical symbols to represent the atoms of the elements and their ratios in a chemical compound.

Carbon dioxide is a compound consisting of one atom of carbon attached by chemical bonds to two atoms of oxygen. Here is how you would write the chemical formula for carbon dioxide:

- Find the symbols for carbon (C) and oxygen (O) on the periodic table. Write these symbols side by side.
- To indicate that there are two oxygen atoms for every carbon atom, place the subscript 2 to the right of the oxygen atom's symbol. A **subscript** is a number written to the right of a chemical symbol and slightly below it.
- Because there is only one atom of carbon in carbon dioxide, you need no subscript for carbon. The subscript 1 is never used. The chemical formula for carbon dioxide is, therefore,

$$CO_2$$

The chemical formula shows one carbon atom bonded to two oxygen atoms.

VOCABULARY
Remember to create a description wheel for *chemical formula* and other vocabulary words.

READING TiP
The word *subscript* comes from the prefix *sub-*, which means "under," and the Latin word *scriptum*, which means "written." A subscript is something written under something else.

INVESTIGATE Element Ratios

PURPOSE To model a compound using representations of atoms in definite proportions

TIP *20 min.* Provide students with a random number of nuts and bolts allowing for some of one to be left over. An excess "reactant" encourages students to think in terms of proportions.

WHAT DO YOU THINK? *Four nuts were attached to one bolt. The number of models is limited by the number of either nuts or bolts. If you do not have exactly the right ratio, you cannot make the compound.*

CHALLENGE *You would use one bolt to represent nitrogen and three nuts to represent hydrogen.*

For additional practice, have students use the Modeling Compounds Datasheet.

R • Modeling Compounds Datasheet, p. 79
• Datasheet, Element Ratios, p. 80

Technology Resources

Customize this student lab as needed or look for an alternative. Print rubrics to assess student lab reports.

🧪 **Lab Generator CD-ROM**

Metacognitive Strategy

Ask students to discuss how performing this investigation helped them make generalizations about compounds and their element ratios. Students should remark that there need to be a certain number of atoms for compounds to form.

DIFFERENTIATE INSTRUCTION

More Reading Support

B How do we represent compounds? *by a chemical formula*

C $NaNO_3$ has how many oxygen atoms? *three*

Alternative Assessment Have students design a similar experiment, in which other materials are used to model carbon dioxide, CO_2, and phosphorus pentachloride, PCl_5.

Teach from Visuals

Have students review the table of chemical formulas. Ask:

- Phosphorus trichloride has the formula PCl_3. What would be the entries for the second and third columns if this compound were included in the table? *the diagram of one phosphorus atom and three chlorine atoms and the ratio 1:3*

- A unit of barium chloride contains one barium ion and two chloride ions. What are the atomic ratio and the chemical formula for this compound? *1:2, $BaCl_2$*

Mathematics Connection

In some compounds, the ratio of the atoms is not expressed in simplest terms. Ask:

- A butane molecule contains four carbon atoms, and the simplest ratio of carbon to hydrogen atoms is 2:5. What is the chemical formula for butane? *C_4H_{10}*

- A molecule of glucose has the formula $C_6H_{12}O_6$. What is the ratio of atoms in this molecule? *6:12:6* What is this ratio in simplest form? *1:2:1*

Ongoing Assessment

Explain how chemical formulas represent compounds.

Ask: How many carbon atoms are in a molecule of $C_{12}H_{22}O_{11}$? *twelve*

 Answer: four

 Answer: Different ratios of elements indicate different compounds.

Chemical Formulas

Chemical formulas show the ratios of atoms in a chemical compound.

Compound Name	Atoms	Atomic Ratio	Chemical Formula
Hydrogen chloride	H Cl	1:1	HCl
Water	H H O	2:1	H_2O
Ammonia	N H H H	1:3	NH_3
Methane	C H H H H	1:4	CH_4
Propane	C C C H H H H H H H H	3:8	C_3H_8

 READING VISUALS How many more hydrogen atoms does propane have than methane?

 RESOURCE CENTER
CLASSZONE.COM
Find out more about chemical formulas.

The chart above shows the names, atoms, ratios, and chemical formulas for several chemical compounds. The subscripts for each compound indicate the number of atoms that combine to make that compound. Notice how hydrogen combines with different atoms in different ratios. Notice in particular that methane and propane are made of atoms of the same elements, carbon and hydrogen, only in different ratios. This example shows why it's important to pay attention to ratios when writing chemical formulas.

CHECK YOUR READING Why is the ratio of atoms in a chemical formula so important?

Same Elements, Different Compounds

Even before chemists devised a way to write chemical formulas, they realized that different compounds could be composed of atoms of the same elements. Nitrogen and oxygen, for example, form several compounds. One compound consists of one atom of nitrogen attached to one atom of oxygen. This compound's formula is NO. A second compound has one atom of nitrogen attached to two atoms of oxygen, so its formula is NO_2. A third compound has two nitrogen atoms attached to one oxygen atom; its formula is N_2O. The properties of these compounds are different, even though they are made of atoms of the same elements.

DIFFERENTIATE INSTRUCTION

 More Reading Support

D What compounds contain the same two elements? *methane and propane*

E Write the formula for one atom of nitrogen and one atom of oxygen. *NO*

Advanced Have students research the work of John Dalton to find his symbols for the elements. Have students make a table comparing his symbols to the modern chemical symbols of the periodic table.

R Challenge and Extension, p. 78

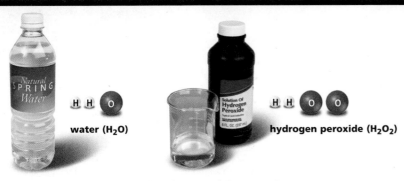

water (H_2O)   hydrogen peroxide (H_2O_2)

There are many other examples of atoms of the same elements forming different compounds. The photographs above show two bottles filled with clear, colorless liquids. You might use the liquid in the first bottle to cool off after a soccer game. The bottle contains water, which is a compound made from two atoms of hydrogen and one atom of oxygen (H_2O). You could not survive for long without water.

You definitely would not want to drink the liquid in the second bottle, although this liquid resembles water. This bottle also contains a compound of hydrogen and oxygen, hydrogen peroxide, but hydrogen peroxide has two hydrogen and two oxygen atoms (H_2O_2). Hydrogen peroxide is commonly used to kill bacteria on skin. One way to tell these two compounds apart is to test them using a potato. A drop of hydrogen peroxide on a raw potato will bubble; a drop of water on the potato will not.

The difference between the two compounds is greater than the labels or their appearance would indicate. The hydrogen peroxide that you buy at a drugstore is a mixture of hydrogen peroxide and water. In its concentrated form, hydrogen peroxide is a thick, syrupy liquid that boils at 150°C (302°F). Hydrogen peroxide can even be used as a fuel.

 CHECK YOUR READING What are the chemical formulas for water and hydrogen peroxide?

2.1 Review

KEY CONCEPTS

1. How do the properties of compounds often compare with the properties of the elements that make them?

2. How many atoms are in the compound represented by the formula $C_{12}H_{22}O_{11}$?

3. How can millions of compounds be made from the atoms of about 100 elements?

CRITICAL THINKING

4. **Apply** If a chemical formula has no subscripts, what can you conclude about the ratio of the atoms in it?

5. **Infer** How might you distinguish between hydrogen peroxide and water?

⬥ CHALLENGE

6. **Analyze** A chemist analyzes two compounds and finds that they both contain only carbon and oxygen. The two compounds, however, have different properties. How can two compounds made from the same elements be different?

Chapter 2: **Chemical Bonds and Compounds** 45 **B**

ANSWERS

1. In many cases they are quite different.

2. 12 carbon atoms + 22 hydrogen atoms + 11 oxygen atoms = 45 total atoms

3. Atoms can combine in many different ratios.

4. All atoms are in a 1:1 ratio.

5. Compare how they react with another substance.

6. The carbon and oxygen atoms are in different ratios in the compounds.

Set Learning Goal

To determine and use ratios to represent relative numbers of atoms in a chemical formula

Present the Science

A given compound contains the same elements in the same proportion by mass. For example, ammonia is always NH_3, with nitrogen and hydrogen in a 1:3 ratio. According to the law of multiple proportions, different compounds can form from the same elements, but the elements will be in different ratios in the different compounds. For example, copper and oxygen form CuO and Cu_2O. The ratios of elements in these compounds are 1:1 and 2:1, respectively.

Develop Number Sense

Students are used to writing ratios in simplest form where the numbers in the ratios have no common factor. Ask why representing ratios in simplest form may not always give accurate information. *Because atoms often combine in ratios that are not the lowest possible numerically.*

Close

Ask: How does the ratio of hydrogen to oxygen compare for water and glucose? *It is the same for both.* If this ratio is the same, why are they different compounds? *Glucose also contains carbon.*

- Math Support, p. 108
- Math Practice, p. 109

Technology Resources

Students can visit **ClassZone.com** for practice in calculating ratios.

 MATH TUTORIAL

MATH in SCIENCE

i **MATH TUTORIAL**
CLASSZONE.COM
Click on Math Tutorial for more help with ratios.

A good strikeout-to-walk ratio for a baseball pitcher is 2:1. This means that for every two strikeouts achieved, the pitcher only allows one walk.

B 46 Unit: Chemical Interactions

Regarding Ratios

No pitcher gets a batter out every time. Sometimes even the worst pitchers have spectacular games. If you're a fan of professional baseball, you've probably seen the quality of certain players rated by using a ratio. A ratio is a comparison of two quantities. For a major league baseball pitcher, for example, one ratio you might hear reported is the number of strikeouts to the number of walks during a season. Chemical formulas are also ratios—ratios that compare the numbers of atoms in a compound.

Example

Consider the chemical formula for the compound glucose:

$$C_6H_{12}O_6$$

From this formula you can write several ratios. To find the ratio of carbon atoms to hydrogen atoms, for instance, do the following:

(1) Find the number of each kind of atom by noting the subscripts.

6 carbon, 12 hydrogen

(2) Write the first number on the left and the second on the right, and place a colon between them.

6:12

(3) Reduce the ratio by dividing each side by the largest number that goes into each evenly, in this case 6.

1:2

ANSWER The ratio of carbon to hydrogen in glucose is 1:2.

Use the table below to answer the following questions.

Compounds and Formulas	
Compound Name	**Chemical Formula**
Carbon dioxide	CO_2
Methane	CH_4
Sulfuric acid	H_2SO_4
Glucose	$C_6H_{12}O_6$
Formic acid	CH_2O_2

1. In carbon dioxide, what is the ratio of carbon to oxygen?

2. What is the ratio of carbon to hydrogen in methane?

3. In sulfuric acid, what is the ratio of hydrogen to sulfur? the ratio of sulfur to oxygen?

CHALLENGE What two chemical compounds in the table have the same ratio of carbon atoms to oxygen atoms?

ANSWERS

1. 1:2
2. 1:4
3. 2:1, 1:4
CHALLENGE *carbon dioxide and formic acid*

KEY CONCEPT

2.2 Chemical bonds hold compounds together.

BEFORE, you learned

- Elements combine to form compounds
- Electrons are located in a cloud around the nucleus
- Atoms can lose or gain electrons to form ions

NOW, you will learn

- How electrons are involved in chemical bonding
- About the different types of chemical bonds
- How chemical bonds affect structure

VOCABULARY

ionic bond p. 48
covalent bond p. 50
molecule p. 51
polar covalent bond p. 51

THINK ABOUT

How do you keep things together?

Think about the different ways the workers at this construction site connect materials. They may use nails, screws, or even glue, depending on the materials they wish to keep together. Why would they choose the method they do? What factors do you consider when you join two objects?

Chemical bonds between atoms involve electrons.

MAIN IDEA AND DETAILS Make a two-column chart to organize information on chemical bonds.

Water is a compound of hydrogen and oxygen. The air you breathe, however, contains oxygen gas, a small amount of hydrogen gas, as well as some water vapor. How can hydrogen and oxygen be water sometimes and at other times not? The answer is by forming chemical bonds.

Chemical bonds are the "glue" that holds the atoms of elements together in compounds. Chemical bonds are what make compounds more than just mixtures of atoms.

Remember that an atom has a positively charged nucleus surrounded by a cloud of electrons. Chemical bonds form when the electrons in the electron clouds around two atoms interact. How the electron clouds interact determines the kind of chemical bond that is formed. Chemical bonds have a great effect on the chemical and physical properties of compounds. Chemical bonds also influence how different substances interact. You'll learn more about how substances interact in a later chapter.

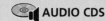

RESOURCES FOR DIFFERENTIATED INSTRUCTION

Below Level

UNIT RESOURCE BOOK
- Reading Study Guide A, pp. 84–85
- Decoding Support, p. 107

AUDIO CDS

Advanced

UNIT RESOURCE BOOK
- Challenge and Extension, p. 90
- Challenge Reading, pp. 103–104

English Learners

UNIT RESOURCE BOOK
Spanish Reading Study Guide, pp. 88–89

AUDIO CDS
- Audio Readings in Spanish
- Audio Readings (English)

2.2 FOCUS

Set Learning Goals

Students will
- Explain how electrons are involved in chemical bonding.
- Describe what the different types of chemical bonds are.
- Determine how chemical bonds affect structure.
- Observe how a crystal grows in an experiment.

3-Minute Warm-Up

Display Transparency 12 or copy this exercise on the board:

Match each definition to a term.

Definitions

1. a negatively charged particle that moves around an atom's nucleus *a*
2. an atom that has lost an electron *c*
3. an atom that has gained an electron *b*

Terms

a. electron
b. negative ion
c. positive ion

T 3-Minute Warm-Up, p. T12

2.2 MOTIVATE

THINK ABOUT

PURPOSE To understand how things can be connected in different ways

DEMONSTRATE Show students several connections on clothing, such as buttons, zippers, and Velcro. Ask them to explain why they would use one method and not another, based on the situation. (Snaps may be too bulky for some materials; buttons may not stay attached well on certain other materials.)

Ongoing Assessment

Explain how electrons are involved in chemical bonding.

Ask: What determines the kind of chemical bond formed? *how the electron clouds interact*

Teach from Visuals

Have students examine the table of groups 1, 2, and 17. Ask what shows the charges on individual elements. +/− signs to the right of each element's symbol. Remind them that compounds have no overall charge. Point out that an element from group 1 can combine with an element from group 17 in single molecules: Li + Br = LiBr. An element from group 2 combines with an element from group 17 by doubling the molecules of the element from group 17: Ca + F = CaF$_2$.

Address Misconceptions

IDENTIFY Ask students what would happen if they were to mix a bottle of hydrogen gas with a bottle of oxygen gas. If they answer, "Water would form," they hold the misconception that chemical combination is no different from physical combination.

CORRECT Take either the graphite from a pencil or a piece of charcoal (carbon) and place it in a glass of water. Tell students that here they have all the chemical ingredients for sugar but they have no sugar, only a mess. To make a chemical compound, like sugar, chemical bonds must be formed.

REASSESS Ask students why simply mixing the eggs, flour, water, and other ingredients for a cake does not make a cake. *Because bonds need to form. Baking is what forms the bonds.*

Technology Resources

Visit **ClassZone.com** for background on common student misconceptions.

MISCONCEPTION DATABASE

Ongoing Assessment

Answer: positive

Atoms can transfer electrons.

REMINDER Remember that elements in columns show similar chemical properties.

Ions are formed when atoms gain or lose electrons. Gaining electrons changes an atom into a negative ion. Losing electrons changes an atom into a positive ion. Individual atoms do not form ions by themselves. Instead, ions typically form in pairs when one atom transfers one or more electrons to another atom.

An element's location on the periodic table can give a clue as to the type of ions the atoms of that element will form. The illustration to the left shows the characteristic ions formed by several groups. Notice that all metals lose electrons to form positive ions. Group 1 metals commonly lose only one electron to form ions with a single positive charge. Group 2 metals commonly lose two electrons to form ions with two positive charges. Other metals, like the transition metals, also always form positive ions, but the number of electrons they may lose varies.

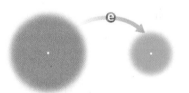

Nonmetals form ions by gaining electrons. Group 17 nonmetals, for example, gain one electron to form ions with a 1− charge. The nonmetals in Group 16 gain two electrons to form ions with a 2− charge. The noble gases do not normally gain or lose electrons and so do not normally form ions.

CHECK YOUR READING What type of ions do metals form?

Ionic Bonds

What happens when an atom of an element from Group 1, like sodium, meets an atom of an element from Group 17, like chlorine? Sodium is likely to lose an electron to form a positive ion. Chlorine is likely to gain an electron to form a negative ion. An electron, therefore, moves from the sodium atom to the chlorine atom.

sodium atom (Na)   chlorine atom (Cl)   sodium ion (Na$^+$)   chloride ion (Cl$^-$)

Remember that particles with opposite electrical charges attract one another. When the ions are created, therefore, they are drawn toward one another by electrical attraction. This force of attraction between positive and negative ions is called an **ionic bond.**

DIFFERENTIATE INSTRUCTION

More Reading Support

A What do you get when an atom loses or gains an electron? *an ion*

B What do you call the force of attraction between positive and negative ions? *an ionic bond*

English Learners English learners may need help understanding different word forms. Look at the words *ionic* and *ionize.* Adding *-ic* to the end of *ion,* produces an adjective. Adding *-ize* to the end of the same word produces a verb. Help English learners recognize indicators such as these when learning new terms.

Electrical forces act in all directions. Each ion, therefore, attracts all other nearby ions with the opposite charge. The next illustration shows how this all-around attraction produces a network of sodium and chloride ions known as a sodium chloride crystal.

Notice how each positive ion is surrounded by six negative ions, and each negative ion is surrounded by six positive ions. This regular arrangement gives the sodium chloride crystal its characteristic cubic shape. You can see this distinctive crystal shape when you look at table salt crystals through a magnifying glass.

C

Ionic bonds form between all nearby ions of opposite charge. These interactions make ionic compounds very stable and their crystals very strong. Although sodium chloride crystals have a cubic shape, other ionic compounds form crystals with different regular patterns. The shape of the crystals of an ionic compound depends, in part, on the ratio of positive and negative ions and the sizes of the ions.

The cubic shape of sodium chloride crystals is a result of how the ions form crystals.

Names of Ionic Compounds

The name of an ionic compound is based on the names of the ions it is made of. The name for a positive ion is the same as the name of the atom from which it is formed. The name of a negative ion is formed by dropping the last part of the name of the atom and adding the suffix -ide. To name an ionic compound, the name of the positive ion is placed first, followed by the name of the negative ion. For example, the chemical name for table salt is sodium chloride. *Sodium* is the positive sodium ion and *chloride* is the negative ion formed from chlorine.

Therefore, to name the compound with the chemical formula BaI_2

- First, take the name of the positive metal element: barium.
- Second, take the name of the negative, nonmetal element, iodine, and give it the ending -ide: iodide.
- Third, combine the two names: barium iodide.

Similarly, the name for KBr is potassium bromide, and the name for MgF_2 is magnesium fluoride.

Teacher Demo

To illustrate how opposite charges attract, use two pieces of cellophane tape. Tape the smooth side of one piece to the sticky side of the other. When pulled apart rapidly the tapes should attract. Have two students do the activity at the same time and bring each piece of tape together. They should see pieces repel and attract. Tell them this is the same type of attraction and repulsion that exists among ions.

Teach Difficult Concepts

Have students look closely at the illustration of the sodium chloride crystal. Ask them if each positive ion belongs with a particular negative ion. Emphasize how in the crystal, ions are attracted to all nearby ions of the opposite charge, not a particular, single ion. There are no individual NaCl units.

Real World Example

Explain to students that a water softener is used to deal with ionic compounds in water. The water softener in a home replaces calcium and magnesium ions in water with sodium ions. Sodium ions are less likely to clog pipes.

EXPLORE (the BIG idea)

Revisit "The Shape of Things" on p. 39. Have students compare their drawings to the photograph of salt crystals on this page.

DIFFERENTIATE INSTRUCTION

More Reading Support

C What type of bond holds a crystal together? *ionic*

D What is the name of MgO? *magnesium oxide*

Below Level Provide students with symbols of common ions that are not in group 1, 2, or 17. Have students prepare two stacks of index cards, one with symbols of metallic ions and the other with symbols of nonmetallic ions. Have students randomly choose a metallic ion and a nonmetallic ion and name the compound that they would combine to form.

Before the work of chemist Gilbert Newton Lewis in the early 1900s, bonding theory was limited to electron transfer. Lewis was the first to suggest that atoms share electrons when forming a bond. His theory was that atoms could only have a certain number of electrons around them. Atoms would share electrons so they would have this maximum number.

Teach Difficult Concepts

Have students investigate the role of electronegativity in determining the nature of bonds. Electronegativity is a measure of the attraction of an atom for electrons in a chemical bond. Some periodic tables list electronegativity. Tell students that the highest electronegativities are associated with the atoms that have the greatest attraction for electrons.

Atoms can share electrons.

In general, an ionic bond forms between atoms that lose electrons easily to form positive ions, such as metals, and atoms that gain electrons easily to form negative ions, such as nonmetals. Another way in which atoms can bond together is by sharing electrons. Nonmetal atoms usually form bonds with each other by sharing electrons.

Covalent Bonds

VOCABULARY
Make a description wheel for *covalent bond* and other vocabulary words.

? E

A pair of shared electrons between two atoms is called a **covalent bond.** In forming a covalent bond, neither atom gains or loses an electron, so no ions are formed. The shared electrons are attracted to both positively charged nuclei. The illustrations below show a covalent bond between two iodine atoms. In the first illustration, notice how the electron clouds overlap. A covalent bond is also often represented as a line between the two atoms, as in the second illustration.

Iodine (I_2)

electron cloud model   ball-and-stick model

READING TiP
To help yourself remember that a covalent bond involves a sharing of electrons, remember that the prefix *co-* means "partner."

The number of covalent bonds that an atom can form depends on the number of electrons that it has available for sharing. For example, atoms of the halogen group and hydrogen can contribute only one electron to a covalent bond. These atoms, therefore, can form only one covalent bond. Atoms of group 16 elements can form two covalent bonds. Atoms of the elements of Group 15 can form three bonds. Carbon and silicon in Group 14 can form four bonds. For example, in methane (CH_4), carbon forms four covalent bonds with four hydrogen atoms, as shown below.

Methane (CH_4)

ball-and-stick model   space-filling model

We don't always show the lines representing the covalent bonds between the atoms. The space-filling model still shows the general shape of the bonded atoms, but occupies far less space on the page.

DIFFERENTIATE INSTRUCTION

? **More Reading Support**

E What type of bond is formed when electrons are shared? *covalent*

Below Level Have students model a single bond by using a toothpick and two gumdrops. Have students test how easily they can turn the gumdrops. Have them model a double bond by using two toothpicks. Again test how easily the "atoms" can spin. Then have them model a triple bond by using three toothpicks. Test again how easily the gumdrops can spin. Ask students how the three bonds compare and which bond would be the strongest? *Each additional bond is more rigid. A triple bond would be the strongest.*

Each carbon-hydrogen bond in methane is a single bond because one pair of electrons is shared between the atoms. Sometimes atoms may share more than one pair of electrons with another atom. For example, the carbon atom in carbon dioxide (CO_2) forms double bonds with each of the oxygen atoms. A double bond consists of four (two pairs of) shared electrons. Two nitrogen atoms form a triple bond, meaning that they share six (three pairs of) electrons.

Carbon Dioxide (CO_2)

Nitrogen (N_2)

READING TiP
Remember that each line in the model stands for a covalent bond—one shared pair of electrons.

A group of atoms held together by covalent bonds is called a **molecule.** A molecule can contain from two to many thousand atoms. Most molecules contain the atoms of two or more elements. For example, water (H_2O), ammonia (NH_3), and methane (CH_4) are all compounds made up of molecules. However, some molecules contain atoms of only one element. The following elements exist as two-atom molecules: H_2, N_2, O_2, F_2, Cl_2, Br_2, and I_2.

 CHECK YOUR READING What is a molecule?

Polar Covalent Bonds

In an iodine molecule, both atoms are exactly the same. The shared electrons therefore are attracted equally to both nuclei. If the two atoms involved in a covalent bond are very different, however, the electrons have a stronger attraction to one nucleus than to the other and spend more time near that nucleus. A covalent bond in which the electrons are shared unequally is called a **polar covalent bond.** The word *polar* refers to anything that has two extremes, like a magnet with its two opposite poles.

READING TiP
To remind yourself that polar covalent bonds have opposite partial charges, remember that Earth has both a North Pole and a South Pole.

Water (H_2O)

ball-and-stick model

space-filling model

In a water molecule (H_2O), the oxygen atom attracts electrons far more strongly than the hydrogen atoms do. The oxygen nucleus has eight protons, and the hydrogen nucleus has only one proton. The oxygen atom pulls the shared electrons more strongly toward it. In a water molecule, therefore, the oxygen side has a slightly negative charge, and the hydrogen side has a slightly positive charge.

 VISUALIZATION CLASSZONE.COM
Examine how electrons move in a polar covalent molecule.

Teach Difficult Concepts

What makes a molecule is difficult to explain completely. A molecule is often described as the "smallest part of a substance that retains all the properties of that substance." In this sense, one NaCl pair would be considered a molecule. We do not use the term *molecule* to refer to ionic compounds simply because it is rare to find a single pair of sodium and chloride ions.

Real World Example

In the water molecule, oxygen attracts electrons more strongly than hydrogen. The molecule is so polar that the negative end of one water molecule is attracted to the positive end of another molecule to form a bond called a hydrogen bond. Hydrogen bonding is what is responsible for water's high surface tension and higher than expected boiling point.

EXPLORE (the BIG idea)

Revisit "Internet Activity: Bonding" on p. 39. Have students revise their answers if necessary.

Ongoing Assessment

CHECK YOUR READING *Answer: a group of atoms held together by covalent bonds*

More Reading Support

F What type of bond forms from unequal sharing of electrons? *polar covalent*

Inclusion Prepare physical models of the molecules on pp. 50 and 51 by using small Styrofoam balls and toothpicks or molecular model kits. Have these molecules available for students who need a tactile model.

Teach Difficult Concepts

Students may think that scientists can always classify chemical bonds with certainty. Explain that the bond between two atoms is not always completely ionic or completely covalent. To illustrate this, hang a piece of white paper and a piece of black paper on the board. Ask students to pick out black and white objects in the room. Then hold up several objects or pieces of paper that are various shades of gray. Point out that although some things are entirely black or white, many objects are in the range that goes from black to white. Point out that only identical atoms bond by 100 percent covalent bonds.

Teach from Visuals

Ask: Why is there no line between the sodium and chloride ions? *There is no line because electrons are not shared.*

 This visual is also available as T14 in the Unit Transparency Book.

Ongoing Assessment

Describe the different types of chemical bonds.

Ask: What type of bond has molecules with negative and positive ends? *polar covalent*

 Answer: nitrogen, oxygen, and water

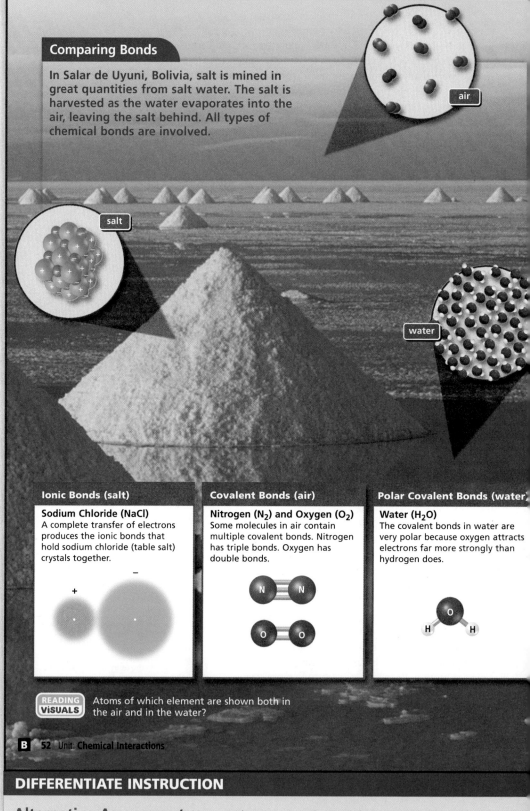

Comparing Bonds

In Salar de Uyuni, Bolivia, salt is mined in great quantities from salt water. The salt is harvested as the water evaporates into the air, leaving the salt behind. All types of chemical bonds are involved.

air

salt

water

Ionic Bonds (salt)

Sodium Chloride (NaCl)
A complete transfer of electrons produces the ionic bonds that hold sodium chloride (table salt) crystals together.

Covalent Bonds (air)

Nitrogen (N_2) and Oxygen (O_2)
Some molecules in air contain multiple covalent bonds. Nitrogen has triple bonds. Oxygen has double bonds.

Polar Covalent Bonds (water)

Water (H_2O)
The covalent bonds in water are very polar because oxygen attracts electrons far more strongly than hydrogen does.

READING VISUALS Atoms of which element are shown both in the air and in the water?

DIFFERENTIATE INSTRUCTION

Alternative Assessment Have students prepare bar graphs showing the mixture of compounds in air in terms of the types of bonds that make up the compounds. Each graph should contain four bars, one each for the three types of bonds and one for no bonding. Have them include the covalent bonds of nitrogen at 78% and oxygen at 21%, and polar water vapor at 4%. (The amount of water vapor in air can vary from 0% to 4%.) The bar for ionic compounds will be at 0. Argon has no bonds, just single atoms.

Chemical bonds give all materials their structures.

The substances around you have many different properties. The structure of the crystals and molecules that make up these substances are responsible for many of these properties. For example, crystals bend rays of light, metals shine, and medications attack certain diseases in the body because their atoms are arranged in specific ways.

Ionic Compounds

Most ionic compounds have a regular crystal structure. Remember how the size, shape, and ratio of the sodium ions and chloride ions give the sodium chloride crystal its shape. Other ionic compounds, such as calcium chloride, have different but equally regular structures that depend upon the ratio and sizes of the ions. One consequence of such rigid structures is that, when enough force is applied to the crystal, it shatters rather than bends.

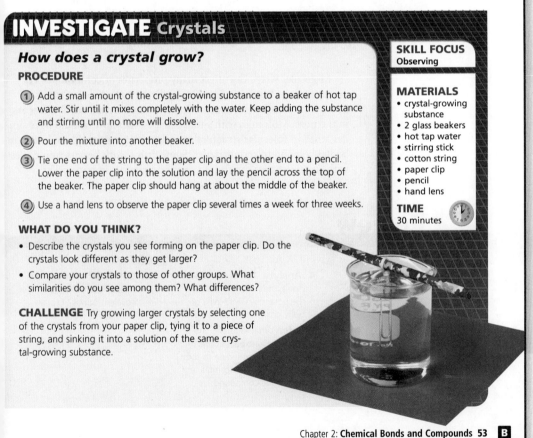

INVESTIGATE Crystals

How does a crystal grow?

PROCEDURE

① Add a small amount of the crystal-growing substance to a beaker of hot tap water. Stir until it mixes completely with the water. Keep adding the substance and stirring until no more will dissolve.

② Pour the mixture into another beaker.

③ Tie one end of the string to the paper clip and the other end to a pencil. Lower the paper clip into the solution and lay the pencil across the top of the beaker. The paper clip should hang at about the middle of the beaker.

④ Use a hand lens to observe the paper clip several times a week for three weeks.

WHAT DO YOU THINK?

• Describe the crystals you see forming on the paper clip. Do the crystals look different as they get larger?

• Compare your crystals to those of other groups. What similarities do you see among them? What differences?

CHALLENGE Try growing larger crystals by selecting one of the crystals from your paper clip, tying it to a piece of string, and sinking it into a solution of the same crystal-growing substance.

SKILL FOCUS
Observing

MATERIALS
• crystal-growing substance
• 2 glass beakers
• hot tap water
• stirring stick
• cotton string
• paper clip
• pencil
• hand lens

TIME
30 minutes

INVESTIGATE Crystals

PURPOSE To observe how a crystal grows

TIPS *30 min.*

• Check toxicity and other safety factors before using a laboratory chemical.

• If possible, have different groups use different crystal-growing substances. Use either sugar, salt, alum, or copper sulfate.

• Alum, which can be purchased at a pharmacy or some grocery stores, grows good crystals and is inexpensive.

• Hot distilled water will produce more regular crystals because of the lack of impurities.

• While crystals are growing, it is important to keep the beaker free of dust and undisturbed. You might try putting a coffee filter over the cup.

WHAT DO YOU THINK? *The shape of the crystal will depend on the solute chosen. The shape remains the same as the crystal grows.*

CHALLENGE *As the water evaporates, the crystal should remain the same shape but grow larger.*

 Datasheet, Crystals, p. 91

Technology Resources

Customize this student lab as needed or look for an alternative. Print rubrics to assess student lab reports.

 Lab Generator CD-ROM

Metacognitive Strategy

Ask students to think about how often they should take a picture of the crystal growth if they wanted to make a presentation of the investigation.

DIFFERENTIATE INSTRUCTION

More Reading Support

G What is responsible for many of the properties of substances? *their structure*

Advanced Have students research the uses for crystals in the electronics industry.

Have students who are interested in nuclear magnetic resonance (NMR), which can detect how atoms are bonded together, read the following article:

R Challenge Reading, pp. 103–104

Covalent Compounds

Unlike ionic compounds, covalent compounds exist as individual molecules. Chemical bonds give each molecule a specific, three-dimensional shape called its molecular structure. Molecular structure can influence everything from how a specific substance feels to the touch to how well it interacts with other substances.

A few basic molecular structures are shown below. Molecules can have a simple linear shape, like iodine (I_2), or they can be bent, like a water molecule (H_2O). The atoms in an ammonia molecule (NH_3) form a pyramid, and methane (CH_4) molecules even have a slightly more complex shape. The shape of a molecule depends on the atoms it contains and the bonds holding it together.

READING TiP
To help yourself appreciate the differences among these structures, try making three-dimensional models of them.

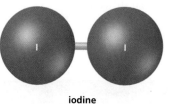

iodine (I_2)   water (H_2O)   ammonia (NH_3)   methane (CH_4)

Molecular shape can affect many properties of compounds. For example, there is some evidence to indicate that we detect scents because molecules with certain shapes fit into certain smell receptors in the nose. Molecules with similar shapes, therefore, should have similar smells. Molecular structure also plays an essential role in how our bodies respond to certain drugs. Some drugs work because molecules with certain shapes can fit into specific receptors in body cells.

2.2 Review

KEY CONCEPTS

1. What part of an atom is involved in chemical bonding?

2. How are ionic bonds and covalent bonds different?

3. Describe two ways that crystal and molecular structures affect the properties of ionic and covalent compounds.

CRITICAL THINKING

4. **Analyze** Would you expect the bonds in ammonia to be polar covalent? Why or why not?

5. **Infer** What kind of bond would you expect atoms of strontium and iodine to form? Why? Write the formula and name the compound.

CHALLENGE

6. **Conclude** Is the element silicon likely to form ionic or covalent bonds? Explain.

ANSWERS

1. the electron cloud

2. An ionic bond results from the attraction of oppositely charged ions. Covalent bonds result from atoms sharing electrons.

3. Their crystal structures make ionic compounds shatter when enough force is

applied. Molecular structures can influence how a compound smells and how it reacts with other compounds.

4. Yes; nitrogen has more protons in its nucleus and thus has more attraction for the shared electrons.

5. Strontium is in Group 2 and forms Sr^{2+} ions. Iodine is in Group 17 and forms I- ions. Strontium iodide is ionic, SrI_2

6. Silicon is in the same group as carbon. Carbon forms covalent bonds, so silicon probably does, too.

SKILL: ISOLATING VARIABLES

Stick to It

Glues join objects by forming something like chemical bonds between their surfaces. While glue manufacturers try to make glues as strong as possible, simply being strong does not mean that a glue will join all surfaces equally well. For example, a glue that will hold two pieces of wood together very well may not be able to form a lasting bond between two pieces of plastic piping or two metal sheets.

Variables

When testing a new glue, a scientist wants to know exactly how that glue will perform under all conditions. In any test, however, there are a number of variables that could affect the quality of the bonds formed by the glue. The scientist needs to discover exactly which of these variables most affects the glue's ability to form lasting bonds. Identifying these variables and the effects each has on the glue's strength and lifetime enables glue makers to recommend the best uses for the glue. Following are a few of the variables a glue maker may consider when testing a glue.

- What surfaces the glue is being used to join
- How much glue is used in a test
- How evenly the glue is applied to the surface
- How much force the glue can withstand
- Over how long a time the force is applied
- The environment the glue is used in (wet, dry, or dusty)

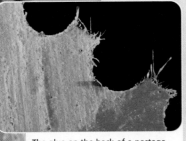

The glue on the back of a postage stamp must be activated somehow. This scanning electron microscope photo shows postage stamp glue before (green) and after (blue) it has been activated by moisture.

This highly magnified photograph shows the attachment formed by a colorless, waterproof wood glue.

Variables to Test

On Your Own You are a scientist at a glue company. You have developed a new type of glue and need to know how specific conditions will affect its ability to hold surfaces together. First, select one variable you wish to test. Next, outline how you would ensure that only that variable will differ in each test. You might start out by listing all the variables you can think of and then put checks by each one and describe how you are controlling it.

As a Group Discuss the outlines of your tests with others. Are there any variables you haven't accounted for?

CHALLENGE Adhesive tapes come in many different types. Outline how you would test how well a certain tape holds in a wet environment and in a dry environment.

THINK SCIENCE
Scientific Methods of Thinking

Set Learning Goal

To isolate and identify the variables involved in testing a glue

Present the Science

- One very strong glue is made from an acrylic resin. The glue bonds with whatever it touches when it comes in contact with water. Almost any surface contains at least trace amounts of water, so this type of glue bonds readily. Acrylic glues can do more than repair broken vases. They are sometimes used to close wounds and detect fingerprints.
- White glues bond when the solvent they contain evaporates. The solvent in most white glues is water, so when the glue dries, it bonds to the object it contacts.
- The glue that is used on repositionable papers forms tiny spheres. Therefore, the overall hold of the glue is weak.

Guide the Activity

- As students are making their lists of variables, ask them to think of what they do when they apply a glue.
- Next to each variable, have students state exactly how it is accounted for in their experiments.

Close

Ask students why doing tests at the same time is better than doing them days or weeks apart. *You can be more familiar with the variables and so control them better.*

ANSWERS

Check student lists for numbers of variables. More important than the number of variables identified, however, is the number students have successfully accounted for in their tests.
CHALLENGE Check to ensure students have addressed the special effects of moisture on tape.

Students will

- Describe how metal atoms form chemical bonds with one another.
- Analyze how ionic and covalent bonds influence substances' properties.
- Identify different forms of the same element.

3-Minute Warm-Up

Display Transparency 13 or copy this exercise on the board:

Decide if these statements are true. If they are not true, correct them.

1. An atom can form more than one covalent bond with another atom. *true*

2. Covalent bonds always have equal sharing of electrons. *usually unequal sharing of electrons*

3. A molecule's structure can influence how it reacts with other molecules. *true*

T 3-Minute Warm-Up, p. T13

2.3 MOTIVATE

EXPLORE Bonds in Metals

PURPOSE To determine what objects will conduct an electric current

TIP *10 min.* Ask students to bring a D-cell battery from home.

WHAT DO YOU THINK? *Most metal objects will conduct a current.*

Ongoing Assessment

Describe how metal atoms form chemical bonds with one another.

Ask: Why are metals able to carry an electrical current? *The electrons in a metal are free to move.*

KEY CONCEPT

2.3 Substances' properties depend on their bonds.

◄ **BEFORE, you learned**

- Chemical bonds hold the atoms of compounds together
- Chemical bonds involve the transfer or sharing of electrons
- Molecules have a structure

► **NOW, you will learn**

- How metal atoms form chemical bonds with one another
- How ionic and covalent bonds influence substances' properties

VOCABULARY

metallic bond p. 56

▼ **REMINDER**

Chemical bonds involve the sharing of or transfer of electrons.

EXPLORE Bonds in Metals

What objects conduct electricity?

PROCEDURE

1. Tape one end of a copper wire to one terminal of the battery. Attach the other end of the copper wire to the light bulb holder. Attach a second wire to the holder. Tape a third wire to the other terminal of the battery.

2. Touch the ends of both wires to objects around the classroom. Notice if the bulb lights or not.

WHAT DO YOU THINK?
- Which objects make the bulb light?
- How are these objects similar?

MATERIALS
- masking tape
- 3 pieces of copper wire (15 cm)
- D cell (battery)
- light bulb and holder
- objects to test

Metals have unique bonds.

Metal atoms bond together by sharing their electrons with one another. The atoms share the electrons equally in all directions. The equal sharing allows the electrons to move easily among the atoms of the metal. This special type of bond is called a **metallic bond.**

The properties of metals are determined by metallic bonds. One common property of metals is that they are good conductors of electric current. The electrons in a metal flow through the material, carrying the electric current. The free movement of electrons among metal atoms also means that metals are good conductors of heat. Metals also typically have high melting points. Except for mercury, all metals are solids at room temperature.

RESOURCES FOR DIFFERENTIATED INSTRUCTION

Below Level
UNIT RESOURCE BOOK
- Reading Study Guide A, pp. 95–96
- Decoding Support, p. 107

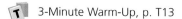 **AUDIO CDS**

R **Additional INVESTIGATION,**
Weird Water, A, B, & C, pp. 119–127;
Teacher Instructions, pp. 329–330

Advanced
UNIT RESOURCE BOOK
Challenge and Extension, p. 101

English Learners
UNIT RESOURCE BOOK
Spanish Reading Study Guide, pp. 99–100

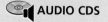 **AUDIO CDS**

- Audio Readings in Spanish
- Audio Readings (English)

Metallic Properties

Copper and other metals get their properties from metallic bonds.

The ability of electrons to move freely makes metals
- good conductors of electricity
- good conductors of heat
- easy to shape

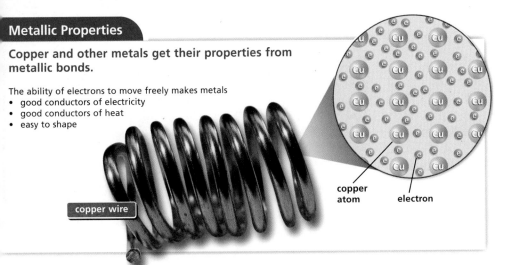

copper wire

copper atom

electron

Two other properties of metals are that they are easily shaped by pounding and can be drawn into a wire. These properties are also explained by the nature of the metallic bond. In metallic compounds, atoms can slide past one another. It is as if the atoms are swimming in a pool of surrounding electrons. Pounding the metal simply moves these atoms into other positions. This property makes metals ideal for making coins.

 CHECK YOUR READING What three properties do metals have because of metallic bonds?

Ionic and covalent bonds give compounds certain properties.

The properties of a compound depend on the chemical bonds that hold its atoms together. For example, you can be pretty certain an ionic compound will be a solid at room temperature. Ionic compounds, in fact, usually have extremely high melting and boiling points because it takes a lot of energy to break all the bonds between all the ions in the crystal. The rigid crystal network also makes ionic compounds hard, brittle, and poor conductors of electricity. No moving electrical charges means no current will flow.

Ionic compounds, however, often dissolve easily in water, separating into positive ions and negative ions. The separated ions can move freely, so solutions of ionic compounds are good conductors of electricity. Your body, in fact, uses ionic solutions to help transmit impulses between nerve and muscle cells. Exercise can rapidly deplete the body of these ionic solutions, and so a good sports drink contains ionic

MAIN IDEA AND DETAILS Make a two-column chart to organize information about ionic and covalent bonds.

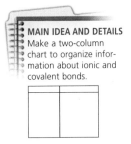

Real World Example

Most metals you use are not pure elements but mixtures of metals called alloys. Brass, bronze, and steel are all alloys. Alloys are most commonly formed from two metals that have similar atom sizes.

Teach Difficult Concepts

Students may have trouble with the distinction between chemical and physical properties. Consider copper and iron, two common metals used extensively in many everyday objects. Both elements have melting points over 1000°C, are easily shaped, and are good conductors of heat and electricity. All these properties are physical properties. The chemical properties of these elements are quite different.

- Iron is relatively reactive. It rusts and reacts with acids.
- Copper is relatively unreactive. It slowly reacts with materials in the air, as demonstrated by discolored pennies and the green patina that appears on copper items.

Ongoing Assessment

CHECK YOUR READING *Answer: Metals are good conductors of heat and electricity and are easy to shape.*

DIFFERENTIATE INSTRUCTION

More Reading Support

A What property of the electrons in a metal gives the metal its properties?
They move easily.

English Learners Students may not understand how bulleted text can be read as three endings to the same sentence, as in the graphic at the top of this page.

The ability of electrons to move freely makes metals

- good conductors of electricity

- good conductors of heat

- easy to shape

Point out that each bulleted line completes the sentence.

Real World Example

Petroleum is a mixture of hydrocarbons, all of which are covalent compounds. The component parts of this mixture can be separated because some of the molecules are larger than others. The smaller molecules, such as those in gasoline, have lower boiling points. The larger molecules, such as those in lubricating oil, have higher boiling points. In fractional distillation, the different parts of petroleum are heated until they reach their boiling points. The vapors rise through a fractionating tower, where they cool. As each component liquifies, it is removed from the mixture.

Teach Difficult Concepts

Sometimes both covalent and ionic bonds occur in the same compound. Some ionic compounds contain ions called polyatomic ions. Polyatomic ions are covalently bonded groups of atoms that act as a single ion. Nitrate (NO_3^-), sulfate (SO_4^{2-}), and ammonium (NH_4^+) are examples of polyatomic ions. Explain that chemists often treat these groups as single ions because they commonly appear together.

Teaching with Technology

Have students use software to model polyatomic ions. Using one of the many periodic tables available on the Web, have students note common compounds formed by various elements. Challenge students to identify other elements that have different forms.

Ongoing Assessment

Analyze how ionic and covalent bonds influence substances' properties.

Ask: You can see that a specific compound comes in crystal form, but it melts at a fairly low temperature. What can you infer? *The compound probably has covalent bonds.*

A hot pool in Yellowstone Park's Upper Geyser Basin. These pools are often characterized by their striking colors.

RESOURCE CENTER
CLASSZONE.COM

Find out more about the properties of ionic and covalent compounds.

B

compounds like potassium chloride that replace the ions lost during physical activity.

Mineral hot springs, like those found in Yellowstone National Park, are another example of ionic solutions. Many of the ionic compounds dissolved in these hot springs contain the element sulfur, which can have an unpleasant odor. Evidence of these ionic compounds can be seen in the white deposits around the pool's rim.

Covalent compounds have almost the exact opposite properties of ionic compounds. Since the atoms are organized as individual molecules, melting or boiling a covalent compound does not require breaking chemical bonds. Therefore, covalent compounds often melt and boil at lower temperatures than ionic compounds. Unlike ionic compounds, molecules stay together when dissolved in water, which means covalent compounds are poor conductors of electricity. Table sugar, for example, does not conduct an electric current when in solution.

Bonds can make the same element look different.

Covalent bonds do not always form small individual molecules. This explains how the element carbon can exist in three very different forms—diamond, graphite, and fullerene. The properties of each form depend on how the carbon atoms are bonded to each other.

Diamond is the hardest natural substance. This property makes diamond useful for cutting other substances. Diamonds are made entirely of carbon. Each carbon atom forms covalent bonds with four other carbon atoms. The pattern of linked atoms extends throughout the entire volume of a diamond crystal. This three-dimensional structure of carbon atoms gives diamonds their strength—diamond bonds do not break easily.

DIFFERENTIATE INSTRUCTION

More Reading Support

B Are covalent compounds good conductors of electricity when dissolved? Why or why not? *No, molecules stay together and do not conduct an electric current.*

Advanced Have students investigate buckminsterfullerene. Why do scientists want to make it? What uses could it have?

Challenge and Extension, p. 101

Additional Investigation To reinforce Section 2.3 learning goals, use the following full-period investigation:

Additional INVESTIGATION, Weird Water, A, B, & C, pp. 119–127, 329–330
(Advanced students should complete Levels B and C.)

Another form of carbon is graphite. Graphite is the dark, slippery component of pencil "lead." Graphite has a different structure from diamond, although both are networks of interconnected atoms. Each carbon atom in graphite forms covalent bonds with three other atoms to form two-dimensional layers. These layers stack on top of one another like sheets of paper. The layers can slide past one another easily. Graphite feels slippery and is used as a lubricant to reduce friction between metal parts of machines.

graphite

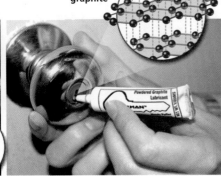

diamond

A third form of carbon, fullerene, contains large molecules. One type of fullerene, called buckminsterfullerene, has molecules shaped like a soccer ball. In 1985 chemists made a fullerene molecule consisting of 60 carbon atoms. Since then, many similar molecules have been made, ranging from 20 to more than 100 atoms per molecule.

buckminsterfullerene

2.3 Review

KEY CONCEPTS

1. How do metal atoms bond together?

2. Why do ionic compounds have high melting points?

3. What are three forms of the element carbon?

CRITICAL THINKING

4. **Apply** A compound known as cubic boron nitride has a structure similar to that of a diamond. What properties would you expect it to have?

5. **Infer** Sterling silver is a combination of silver and copper. How are the silver and copper atoms held together?

◔ CHALLENGE

6. **Infer** Why might the water in mineral springs be a better conductor of electricity than drinking water?

Integrate the Sciences

The ability of carbon to form different kinds of bonds is crucial for life on Earth. Compounds called organic compounds contain carbon and other elements such as nitrogen and oxygen and play essential roles in living things. Organic chemistry and biochemistry are subdisciplines of chemistry that study these compounds.

Reinforce (the **BIG** idea)

Have students relate the section to the Big Idea.

R Reinforcing Key Concepts, p. 102

2.3 ASSESS & RETEACH

Assess

A Section 2.3 Quiz, p. 24

Reteach

Show students samples of graphite and, if possible, diamond. Have them examine the samples, list the properties for each, and compare them. *Diamond is a clear, hard crystal. Graphite is a grayish black slippery powder.*

Technology Resources

Have students visit **ClassZone.com** for reteaching of Key Concepts.

 CONTENT REVIEW

 CONTENT REVIEW CD-ROM

ANSWERS

1. Metal atoms share electrons with many other metal atoms at the same time.

2. Each ion in a crystal is rigidly bonded to the several ions that surround it. Because each ion is strongly bonded to other ions, much energy has to be added to break the bonds.

3. graphite, diamond, and fullerene

4. hard, high melting and boiling points

5. The metal atoms are attracted to the electrons of all the other metal atoms around them.

6. Mineral springs water would be more likely to have ionic materials in it, so it would be more likely to conduct a current.

Focus

PURPOSE To investigate properties of substances that have different types of bonds

OVERVIEW Students will examine ionic and covalent compounds and metals to determine their properties. They will use the properties to characterize three compounds. Students will find the following:

- Sugar is a covalent compound.
- Epsom salts is an ionic compound.
- Iron is a metal.

Lab Preparation

- Epsom salts and sugar can be purchased from local sources.
- If time is limited, assemble the conductivity testers ahead of time and place the samples in labeled cups.
- Prior to the investigation, have students read through the investigation and prepare their data tables. Or you may wish to copy and distribute datasheets and rubrics.

 UNIT RESOURCE BOOK, pp. 110–118

 SCIENCE TOOLKIT, F14

Lab Management

- Have students use laboratory scoops or small pieces of paper to add the materials to the test tubes.
- Any solutions can be poured down the drain with plenty of water. Save uncontaminated iron for reuse.

SAFETY Students should wear safety goggles and aprons throughout the investigation. Be sure proper fire safety is observed when candles are being used.

INCLUSION For students who are visually impaired, wire the conductivity tester to a low-voltage buzzer instead of a light bulb.

Chemical Bonds

OVERVIEW AND PURPOSE Chemists can identify the type of bonds in a substance by examining its properties. In this investigation you will examine the properties of different substances and use what you have learned about chemical bonds to identify the type of bond each substance contains. You will

- observe the structure of substances with a hand lens
- test the conductivity of substances
- determine the melting point of substances

Problem

How can you determine the type of chemical bond a substance has?

Hypothesize

Write three hypotheses in "if . . . , then . . . , because . . ." form to answer the problem question for each bond type—ionic, covalent, and metallic.

Procedure

MATERIALS
- 3 wire leads with alligator clips
- battery
- zinc and copper strips
- light bulb and socket
- test compounds
- 3 plastic cups
- distilled water
- beaker
- construction paper
- hand lens
- plastic spoon
- 3 test tubes
- test-tube rack
- candle
- wire test-tube holder

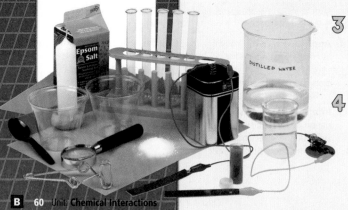

1. Create a data table similar to the one shown on the sample notebook page.

2. To build the conductivity tester, connect the first wire to one terminal of the battery and to one of the metal strips. Attach the second wire to the other terminal and to the lamp socket. Finally, connect the lamp socket to the third wire, and connect the other end of this wire to the second metal strip.

3. To make sure your tester works properly, touch the tips of the metal strips together. If the bulb lights, the tester is working properly. If not, check the connections carefully.

4. Get the following test compounds from your teacher: Epsom salts ($MgSO_4$), sugar ($C_{12}H_{22}O_{11}$), and iron filings (Fe). For each substance, put about 20 grams in a cup and label it.

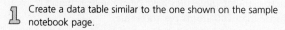
INVESTIGATION RESOURCES

 CHAPTER INVESTIGATION, Chemical Bonds
- Level A, pp. 110–113
- Level B, pp. 114–117
- Level C, p. 118

Advanced students should complete Levels B & C.

 Writing a Lab Report, D12–13

Technology Resources

Customize this student lab as needed or look for an alternative. Print rubrics to assess student lab reports.

 Lab Generator CD-ROM

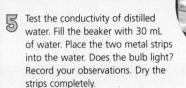

5. Test the conductivity of distilled water. Fill the beaker with 30 mL of water. Place the two metal strips into the water. Does the bulb light? Record your observations. Dry the strips completely.

6. Place dry Epsom salts on dark paper. Observe them with a hand lens. Do you see any kind of patterns in the different grains? Put the salts between the metal strips. Can you get the bulb to light by bringing the strips closer together? Record your observations.

7. Add all but a small amount of the Epsom salts to the beaker of water. Stir well. Repeat the conductivity test. What happens when you put the metal strips into the solution? Record your results.

8. Rinse and dry the beaker. Repeat steps 6–7 with other test substances. Record your results.

9. Put the remainder of each test substance into its own clean, dry test tube. Label the tubes. Light the candle. Use a test tube holder to hold each compound over the candle flame for 2 minutes. Do you notice any signs of melting? Record your observations.

▶ Observe and Analyze

1. **RECORD OBSERVATIONS** Be sure you have entered all your observations in your data table.

2. **CLASSIFY** Using the periodic table, find the elements these compounds contain. How might consulting the periodic table help you determine what type of bond exists in the compound?

▶ Conclude

1. **INTERPRET** Review your recorded observations. Classify the compounds as having ionic, covalent, or metallic bonds. Fill in the last column of the the data table with your conclusions.

2. **INFER** Compare your results with your hypotheses. Did your results support your hypotheses?

3. **EVALUATE** Describe possible limitations, errors, or places where errors might have occurred.

4. **APPLY** Electrocardiograms are graphs that show the electrical activity of the heart. When an electrocardiogram is made, a paste of sodium chloride is used to hold small metal discs on the patient's skin. What property of ionic compounds does this medical test make use of?

▶ INVESTIGATE Further

CHALLENGE To grow crystals, put about 60 grams of Epsom salts into a baby-food jar that is half full of hot water. Do the same using a second jar containing about 60 grams of sugar. Cover and shake the jars for a count of 60. Line two clean jar lids with dark paper. Brush or spoon a thin coating of each liquid over the paper. Let them stand in a warm place. After several days, observe the crystals that form, using a hand lens.

Chemical Bonds
Problem How can you determine the type of chemical bond a substance has?

Hypothesize

Observe and Analyze
Table 1: Properties of Bonds

Property	Epsom Salts (MgSO₄)	Sugar (C₁₂H₂₂O₁₁)	Iron Filings (Fe)	Bond Type
Crystal structure				
Conductivity of solid				
Conductivity in water				
Melting				

Conclude

▶ Observe and Analyze

1. *The solid iron and the Epsom-salts solution conduct a current. Only the sugar melts. Sugar and Epsom salts have a crystal structure. The sugar crystal and Epsom salts have a regular shape.*

2. *Epsom salts contains magnesium, which is a group 2 metal, so it is probably ionic. Sugar ($C_{12}H_{22}O_{11}$) contains no metals, so it is probably covalent. Iron is a transition metal, so it has metallic bonds.*

▶ Conclude

1. *Epsom salts is ionic, sugar is covalent, and iron contains metallic bonds.*

2. *Answers will vary.*

3. *Sample answer: The water used might have contained ions, the solutions might not have contained enough solute, or the hand lens may not have been powerful enough to reveal crystal structure.*

4. *They conduct electric current.*

▶ INVESTIGATE Further

CHALLENGE *The crystals will look irregular.*

Post-Lab Discussion

Ask: Why did you need to do more than one test to determine sugar was a covalent compound? *Sugar was soluble in water and had a crystal shape just like ionic compounds.*

BACK TO

the BIG idea

Have students look back at the photograph on pp. 38–39. Ask them to use the photograph to summarize what they have learned about chemical bonding and to use the skydiving analogy to describe bonding. *Just as the skydivers joined hands, atoms connect by chemical bonds. Skydivers hold hands just as atoms share electrons.*

◯ KEY CONCEPTS SUMMARY

SECTION 2.1
Ask: At room temperature, the element nitrogen is a gas. Can you assume that all compounds containing nitrogen are gases? *No. Compounds often have very different physical properties.*

SECTION 2.2
Ask: How is a polar covalent bond similar to both an ionic bond and a covalent bond? *Although electrons are still being shared by the atoms, they spend more time nearer to one atom.*

SECTION 2.3
Ask: What kind of compound would you expect to find in a battery, ionic or covalent? Why? *ionic; ionic compounds conduct electricity in solution.*

Review Concepts

- Big Idea Flow Chart, p. T9
- Chapter Outline, pp. T15–T16

 Chapter Review

the BIG idea

The properties of compounds depend on their atoms and chemical bonds.

CONTENT REVIEW
CLASSZONE.COM

◀ KEY CONCEPTS SUMMARY

 Elements combine to form compounds.

- Compounds have different properties from the elements that made them.
- Atoms combine in predictable numbers.

calcium (Ca) +	chlorine (Cl$_2$) =	calcium chloride (CaCl$_2$)

VOCABULARY
chemical formula p. 43
subscript p. 43

 Chemical bonds hold compounds together.

- Chemical bonds between atoms involve electrons.
- Atoms can transfer electrons.
- Atoms can share electrons.
- Chemical bonds give all materials their structure.

ionic bond   covalent bond

VOCABULARY
ionic bond p. 48
covalent bond p. 50
molecule p. 51
polar covalent bond
  p. 51

 Substances' properties depend on their bonds.

- Metals have unique bonds.
- Ionic and covalent bonds give compounds certain properties.
- Bonds can make the same element look different.

  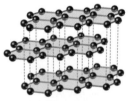

copper   diamond fragment   graphite fragment

VOCABULARY
metallic bond p. 56

B 62 Unit: Chemical Interactions

Technology Resources

Have students visit **ClassZone.com** or use the CD-ROM for a cumulative review of concepts.

 CONTENT REVIEW

 CONTENT REVIEW CD-ROM

Engage students in a whole-class interactive review of Key Concepts. Edit content as you wish.

 POWER PRESENTATIONS

Reviewing Vocabulary

Copy and complete the table below. Under each bond type, describe
- how electrons are distributed
- how the compound is structured
- one of the properties of the compound containing this type of bond

Some of the table has been filled out for you.

Ionic Bonds	Covalent Bonds	Metallic Bonds
1.	shared electron pair	**2.**
3.	**4.**	close-packed atoms in sea of electrons
have high melting points	**5.**	**6.**

Reviewing Key Concepts

Multiple Choice *Choose the letter of the best answer.*

7. Most substances are
- **a.** elements
- **b.** compounds
- **c.** metals
- **d.** nonmetals

8. All compounds are made of
- **a.** atoms of two or more elements
- **b.** two or more atoms of the same element
- **c.** atoms arranged in a crystal
- **d.** atoms joined by covalent bonds

9. The chemical formula for a compound having one barium (Ba) ion and two chloride (Cl) ions is
- **a.** BCl
- **b.** BaCl
- **c.** $BaCl_2$
- **d.** Ba_2Cl_2

10. The 4 in the chemical formula CH_4 means there are
- **a.** four carbon atoms to one hydrogen atom
- **b.** four carbon and four hydrogen atoms
- **c.** four hydrogen atoms to one carbon atom
- **d.** four total carbon CH combinations

11. The compound KBr has the name
- **a.** potassium bromide
- **b.** potassium bromine
- **c.** bromide potassium
- **d.** bromine potassium

12. An atom becomes a positive ion when it
- **a.** is attracted to all nearby atoms
- **b.** gains an electron from another atom
- **c.** loses an electron to another atom
- **d.** shares an electron with another atom

13. A polar covalent bond forms when two atoms
- **a.** share one electron equally
- **b.** share two electrons equally
- **c.** share one electron unequally
- **d.** share two electrons unequally

14. Metallic bonds make many metals
- **a.** poor conductors of heat
- **b.** liquid at room temperature
- **c.** difficult to shape
- **d.** good conductors of electricity

15. Three forms of carbon are
- **a.** diamond, graphite, and salt
- **b.** diamond, graphite, and fullerene
- **c.** graphite, salt, and carbonate
- **d.** diamond, salt, and fullerene

Short Answer *Write a short answer to each question.*

16. Why does a mixture of sodium chloride and water conduct electricity but a sodium chloride crystal does not?

17. Describe what makes diamond and graphite, two forms of the element carbon, so different.

Reviewing Vocabulary

1. transferred electrons
2. sea of electrons
3. crystal
4. individual molecules
5. have low melting points
6. conduct an electric current

Reviewing Key Concepts

7. b
8. a
9. c
10. c
11. a
12. c
13. d
14. d
15. b
16. Charged particles are free to move in a solution of sodium and chloride ions in water.
17. The carbon atoms are bonded differently, giving them different structures.

ASSESSMENT RESOURCES

UNIT ASSESSMENT BOOK
- Chapter Test A, pp. 25–28
- Chapter Test B, pp. 29–32
- Chapter Test C, pp. 33–36
- Alternative Assessment, pp. 37–38

SPANISH ASSESSMENT BOOK
Spanish Chapter Test, pp. 237–240

Technology Resources

Edit test items and answer choices.

 Test Generator CD-ROM

Visit **ClassZone.com** to extend test practice.

 Test Practice

Thinking Critically

18. NH_3

19. There should be a negative charge by nitrogen and positive charge by hydrogen.

20. $MgCl_2$

21. Temperature and heat cause bonds to be rearranged.

22. Seawater has a greater concentration of ions.

23. The attraction among the metal atoms is the same in all directions so they can slide past each other when pressure is applied.

24. All elements detected on Mars are present on Earth. Thus, it is unlikely that different bonds exist there.

25. Ions attract other ions from all sides, and so do not form individual molecules.

26. potassium chloride; ionic

27. I. potassium, sulfur, and oxygen; 7

    II. carbon and fluorine; 5

    III. carbon and hydrogen; 14

    IV. potassium and chlorine; 2

28. II. 1:4; III. 4:10; IV. 1:1; I. potassium to sulfur 2:1, sulfur to oxygen 1:4, potassium to oxygen 2:4

29. The metals gain and lose electrons relatively easily. Both metals and saltwater conduct an electric current well.

30. Hydrogen can form only one bond. It has only one electron.

the BIG idea

31. Skydivers hold hands much as atoms share electrons to form bonds. Skydivers might also hold onto a rope; they can share a rope.

32. Phosphorus can have different patterns of bonding.

UNIT PROJECTS

Collect schedules, materials lists, and questions. Be sure dates and materials are obtainable and questions are focused.

R Unit Projects, pp. 5–10

Thinking Critically

Use the illustration above to answer the next two questions.

18. **IDENTIFY** Write the chemical formula for the molecule pictured above.

19. **ANALYZE** The nitrogen atom has a far greater attraction for electrons than hydrogen atoms. Copy the molecule pictured above and indicate which parts of the molecule have a slightly positive charge and which parts have a slightly negative charge.

20. **PREDICT** The chemical formula for calcium chloride is $CaCl_2$. What would you predict the formula for magnesium chloride to be? [**Hint:** Find magnesium on the periodic table.]

21. **INFER** When scientists make artificial diamonds, they sometimes subject graphite to very high temperatures and pressures. What do you think happens to change the graphite into diamond?

22. **SYNTHESIZE** Why would seawater be a better conductor of electricity than river water?

23. **ANALYZE** How does the nature of the metallic bond explain the observation that most metals can be drawn into a wire?

24. **EVALUATE** Do you think the types of bonds you've studied occur on the planet Mars? Explain.

25. **INFER** Why don't we use the term *ionic molecule?*

Use the chemical formulas below and a periodic table to answer the next three questions.

Compound
I. K_2SO_4
II. CF_4
III. C_4H_{10}
IV. KCl

26. **APPLY** Name compound IV. Does this compound have ionic or covalent bonds?

27. **ANALYZE** Name the elements in each compound. Tell how many atoms are in each compound.

28. **CALCULATE** Express the ratio of atoms in compounds II, III, and IV. For compound I, express all three ratios.

29. **APPLY** By 1800 Alessandro Volta had made the first electric battery. He placed pieces of cardboard soaked in saltwater in between alternating zinc and silver discs. What properties of the metals and the saltwater made them good materials for a battery?

30. **PREDICT** What is the maximum number of covalent bonds that a hydrogen atom can form? Explain your answer.

the BIG idea

31. **DRAW CONCLUSIONS** Look at the photograph on pages 38–39 again. Can you now recognize any similarities between how the skydivers stay together and how atoms stay together?

32. **APPLY** Phosphorus can be a strange element. Pure phosphorus is sometimes white, black, or red. What can account for the differences in appearance?

UNIT PROJECTS

If you need to create graphs or other visuals for your project, be sure you have graph paper, poster board, markers, or other supplies.

MONITOR AND RETEACH

If students have trouble applying the concepts in items 26–28, have them make a labeled diagram of groups 1, 2, and 17 from the periodic table. Their diagrams need to include only the symbols of the elements. Students can use these diagrams to familiarize themselves with what elements are in the same group and thus form similar bonds.

Students may benefit from summarizing one or more sections of the chapter.

 Summarizing the Chapter, pp. 128–129

Interpreting Tables

The table below lists some of the characteristics of substances that contain different types of bonds. Use the table to answer the questions.

Bond Type	Usually Forms Between	Electrons	Properties	Examples
Ionic	an atom of a metal and an atom of a nonmetal	transferred between atoms	• high melting points • conducts electricity when in water	BaS, $BaBr_2$, Ca_3N_2, LiCl, ZnO
Covalent	atoms of nonmetallic elements	shared between atoms but often not equally	• low melting points • does not conduct electricity	C_2H_6, C, Cl_2, H_2, $AsCl_3$
Metallic	atoms of metallic elements	freely moving about the atoms	• high melting points • conducts electricity at all times • easily shaped	Ca, Fe, Na, Cu, Zn

1. Which of these compounds would you expect to have the highest melting point?
- **a.** C_2H_6
- **b.** Cl_2
- **c.** $AsCl_3$
- **d.** $BaBr_2$

2. Which substance is likely to be easily shaped?
- **a.** $BaBr_2$
- **b.** LiCl
- **c.** Na
- **d.** C

3. In the compound LiCl, electrons are
- **a.** shared equally
- **b.** shared but not equally
- **c.** transferred between atoms to form ions
- **d.** freely moving among the atoms

4. Which of the following is an ionic compound?
- **a.** C_2H_6
- **b.** Cl_2
- **c.** $AsCl_3$
- **d.** ZnO

5. Which of the following compounds has a low melting point?
- **a.** Cl_2
- **b.** ZnO
- **c.** Cu
- **d.** $BaBr_2$

6. A solid mass of which substance would conduct electricity?
- **a.** Ca_3N_2
- **b.** LiCl
- **c.** Cu
- **d.** $AsCl_3$

Extended Response

Answer the next two questions in detail.
Include some of the terms from the list in the box.
Underline each term you use in your answer.

share electron	transfer electron
freely moving electrons	charge
compound	chemical formula

7. Compare how electrons are involved in making the three main types of bonds: ionic, covalent, and metallic.

8. Just about 100 elements occur naturally. There are, however, millions of different materials. How can so few basic substances make so many different materials?

Chapter 1: **Chemical Bonds and Compounds 65** **B**

Interpreting Tables

1. d   3. c   5. a
2. c   4. d   6. c

Extended Response

7. RUBRIC

4 points for a response that correctly describes all three types of bonds and uses the following terms accurately:
- share electron
- freely moving electrons
- compound
- transfer electron
- charge

An atom might <u>transfer</u> one or more <u>electrons</u> to another atom to form <u>charged</u> particles. An ionic bond forms when these oppositely charged particles attract each other. Atoms <u>share electrons</u> when covalent bonds form. Both ionic and covalent bonds form compounds. Metallic bonds form when <u>freely moving electrons</u> surround metal ions and are easily attracted to all of them.

3 points correctly describes two types and uses four terms accurately
2 points correctly describes two types and uses two terms accurately
1 point correctly describes one type and uses one term accurately

8. RUBRIC

4 points for a response that answers the question completely and uses the following term accurately:
- compound

Sample: Atoms of one element can combine with the atoms of different elements. The same two atoms can sometimes combine in different ratios. Just as the same letters may form several different words, the same atoms may form several different <u>compounds</u>.

3 points incomplete answer to the question and uses the term accurately
2 points incomplete answer to the question and uses the term inaccurately
1 point incomplete answer to the question and does not use the term

---

METACOGNITIVE ACTIVITY

Have students answer the following questions in their **Science Notebook:**

1. What did you find the most challenging to understand about chemical bonds?

2. What questions do you still have about chemical bonds?

3. How have you solved a problem while working on your Unit Project?

CHAPTER
3 Chemical Reactions

Physical Science
UNIFYING PRINCIPLES

PRINCIPLE 1
Matter is made of particles too small to see.

PRINCIPLE 2
Matter changes form and moves from place to place.

PRINCIPLE 3
Energy changes from one form to another, but it cannot be created or destroyed.

PRINCIPLE 4
Physical forces affect the movement of all matter on Earth and throughout the universe.

Unit: Chemical Interactions
BIG IDEAS

CHAPTER 1
Atomic Structure and the Periodic Table
A substance's atomic structure determines its physical and chemical properties.

CHAPTER 2
Chemical Bonds and Compounds
The properties of compounds depend on their atoms and chemical bonds.

CHAPTER 3
Chemical Reactions
Chemical reactions form new substances by breaking and making chemical bonds.

CHAPTER 4
Solutions
When substances dissolve to form a solution, the properties of the mixture change.

CHAPTER 5
Carbon in Life and Materials
Carbon is essential to living things and to modern materials.

CHAPTER 3
KEY CONCEPTS

SECTION (3.1)	SECTION (3.2)	SECTION (3.3)	SECTION (3.4)
Chemical reactions alter arrangements of atoms.	**The masses of reactants and products are equal.**	**Chemical reactions involve energy changes.**	**Life and industry depend on chemical reactions.**
1. Atoms interact in chemical reactions.	1. Careful observations led to the discovery of the conservation of mass.	1. Chemical reactions release or absorb energy.	1. Living things require chemical reactions.
2. Chemical reactions can be classified.	2. Chemical reactions can be described by chemical equations.	2–3. Exothermic reactions release energy; endothermic reactions absorb energy.	2. Chemical reactions are used in technology.
3. The rates of chemical reactions can vary.	3. Chemical equations must be balanced.	4. Exothermic and endothermic reactions work together to supply energy.	3. Industry uses chemical reactions to make useful products.

 The Big Idea Flow Chart is available on p. T17 in the **UNIT TRANSPARENCY BOOK**.

Chemical reactions alter arrangements of atoms. pp. 69–77

1. Atoms interact in chemical reactions.
Substances change in two ways.
- In physical changes, the substance itself does not change, although its appearance or some of its properties may change.
- In chemical changes, a substance changes into different substances. A **chemical reaction** rearranges atoms. Bonds are broken in **reactants,** and new bonds are formed in the **products.**

Evidence of a chemical reaction includes a change in color or temperature or the formation of a **precipitate** or a gas.

2. Chemical reactions can be classified.
A synthesis reaction combines two or more simpler reactants to form a new, more complex product. A decomposition reaction breaks a reactant into two or more simpler products. A combustion reaction always involves oxygen. The other reactant often contains carbon and hydrogen.

3. The rates of chemical reactions can vary.
Four factors can change the rate of a chemical reaction:
- the concentration of reactants
- the surface area of reactants
- the temperature of the reaction mixture
- the presence of a catalyst

A **catalyst** takes part in a reaction but is not consumed during the reaction. It decreases the energy needed to start a reaction, and it increases the reaction rate. The diagram below represents the effect of a catalyst called an enzyme on a reaction.

Reactants   **Reactants combined**   **New product**

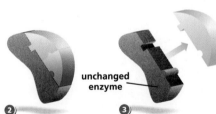

enzyme (catalyst)

unchanged enzyme

① An enzyme is a catalyst for chemical reactions in living things.

② Enzymes allow reactions that would not normally take place to occur.

③ A new product is made, but the enzyme is not changed by the reaction.

The masses of reactants and products are equal. pp. 78–85

1. Careful observations led to the discovery of the conservation of mass.
Antoine Lavoisier's careful quantitative experiments showed that in a chemical reaction, the total mass of reactants is always equal to the total mass of products. In other words, mass is neither created nor destroyed during chemical reactions.

2. Chemical reactions can be described by chemical equations.
A chemical equation represents the way in which a reaction rearranges the atoms in chemicals. To write an equation, you must know the reactants and products, their chemical formulas, and the direction of the reaction. The arrow in an equation indicates the direction of the reaction.

$$C + O_2 \rightarrow CO_2$$

3. Chemical equations must be balanced.
A chemical equation must reflect the law of conservation of mass, so each side of an equation must have the same number of atoms of each element.

- An equation is balanced by changing the number of molecules of reactants or products represented. This is done by adding coefficients in front of some of the chemical formulas, as shown in the equations below.

$N_2 + H_2 \rightarrow NH_3$
Unbalanced Equation

$N_2 + 3H_2 \rightarrow 2NH_3$
Balanced Equation

- When balancing an equation, subscripts in the chemical formulas of the reactants and products cannot be changed. Changing a subscript in a chemical formula changes the substance represented by the formula. Therefore, equations must be balanced by changing coefficients, which changes only the amounts of the reactants and products represented by the equation.

Common Misconceptions

BURNING IS A CHEMICAL CHANGE Students often think that burning is a physical change. Burning, a combustion reaction, is a chemical change that requires oxygen and usually produces water, carbon, and carbon dioxide.

 This misconception is addressed on p. 71.

 MISCONCEPTION DATABASE
CLASSZONE.COM Background on student misconceptions

THE INTRINSIC MOTION OF PARTICLES IN MATTER Students may not understand that all particles in matter, including those in solid objects, have kinetic energy and are in constant, random motion.

TE This misconception is addressed on p. 75.

Previewing Content

3.3 Chemical reactions involve energy changes. pp. 86–93

1. Chemical reactions release or absorb energy.
Chemical reactions break the chemical bonds in reactants and make new bonds in the products. Breaking bonds requires energy; forming bonds releases energy. The energy in chemical bonds is called **bond energy.**
- If more energy is released when the bonds in products form than is used to break the bonds in reactants, energy is released by the **exothermic reaction.**
- If more energy is needed to break the bonds in reactants than is released when the bonds in products form, energy is absorbed by the **endothermic reaction.**

2. Exothermic reactions release energy.
In exothermic reactions, the reactants have lower bond energies than the products, so energy is released, often as heat and light. All common combustion reactions are exothermic. Below is an energy diagram showing an exothermic reaction.

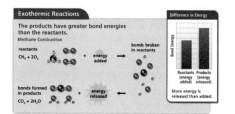

3. Endothermic reactions absorb energy.
In endothermic reactions, the reactants have higher bond energies than the products, so energy is absorbed. Below is an energy diagram showing an endothermic reaction.

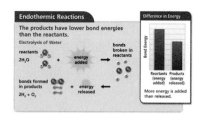

4. Exothermic and endothermic reactions work together to supply energy.
Endothermic and exothermic reactions can form a cycle. For example, the energy stored by the series of endothermic reactions in photosynthesis can be released by exothermic reactions such as combustion.

3.4 Life and industry depend on chemical reactions. pp. 94–99

1. Living things require chemical reactions.
Photosynthesis is an endothermic process. During photosynthesis, plants absorb energy from sunlight and store this energy in the chemical bonds of sugars. These sugars are broken down during the exothermic reactions of **respiration,** the process that produces energy for living organisms. The processes of photosynthesis and respiration are essentially the reverse of one another.

Photosynthesis:
$$6CO_2 + 6H_2O + energy \rightarrow C_6H_{12}O_6 + 6O_2$$
Respiration:
$$C_6H_{12}O_6 + 6O_2 \rightarrow 6CO_2 + 6H_2O + energy$$

Most reactions that take place in living organisms use enzymes, catalysts that cause reactions to take place at the relatively low temperatures of living tissue.

2. Chemical reactions are used in technology.
Combustion engines use gasoline in a chemical reaction that releases energy. Catalytic converters are technological devices that remove unwanted pollutants from the burning of gasoline in automobile engines. These devices use metals as catalysts. The metal catalysts in a catalytic converter, which include platinum, palladium, and rhodium, allow reactions between exhaust gases to occur. These reactions change the exhaust gases into gases that are typical parts of Earth's atmosphere, such as oxygen, nitrogen, water vapor, and carbon dioxide.

3. Industry uses chemical reactions to make useful products.
The electronics industry produces silicon for microchips by refining SiO_2 (quartz) into pure silicon. Silicon treated with photoresist and light produces miniature electronic circuits.

Common Misconceptions

BALANCING EQUATIONS Some students may try to balance chemical equations by changing the chemical formulas rather than adding coefficients. Changing the formulas changes the substances involved, not the amounts.

 This misconception is addressed on p. 82.

MISCONCEPTION DATABASE
CLASSZONE.COM Background on student misconceptions

BREAKING BONDS A very common misconception is that energy is released when bonds are broken. The overall reaction may release energy, but energy is always required to break bonds.

 This misconception is addressed on p. 87.

Previewing Labs

EXPLORE (the BIG idea)

Changing Steel Wool, p. 67
Students observe a chemical reaction that involves the rusting of steel wool.

TIME 10 minutes
MATERIALS small lump of steel wool, cup, vinegar, tongs, small plastic bottle, balloon

---

A Different Rate, p. 67
Students vary the temperature of a reaction mixture to explore how reaction rate is affected by temperature.

TIME 10 minutes
MATERIALS 2 plastic cups, hot and cold tap water, 2 seltzer tablets, stopwatch

---

Internet Activity: Reactions, p. 67
Students explore chemical reactions and how to balance chemical equations.

TIME 20 minutes
MATERIALS computer with Internet access

SECTION 3.1

EXPLORE Chemical Changes, p. 69
Students examine the evidence for a chemical change caused by a chemical reaction.

TIME 10 minutes
MATERIALS 25 mL vinegar, clear bowl, plastic spoon, spoonful of table salt, 20 pennies, large ungalvanized iron nail

---

INVESTIGATE Chemical Reactions, p. 74
Students vary the surface area of a reactant in a chemical reaction to infer how surface area affects reaction rate.

TIME 15 minutes
MATERIALS 2 seltzer tablets, 2 plastic cups, warm tap water, stopwatch

---

SECTION 3.2

INVESTIGATE Conservation of Mass, p. 79
Students measure the mass of the reactants, products, and experimental setup in a chemical reaction in order to observe the law of conservation of mass.

TIME 35 minutes
MATERIALS teaspoon, 2 tsp baking soda, funnel, balloon, 2 tsp vinegar, plastic bottle, balance

---

SECTION 3.3

EXPLORE Energy Changes, p. 86
Students measure changes in temperature during a chemical process in order to identify a transfer of energy.

TIME 10 minutes
MATERIALS graduated cylinder, hot tap water, plastic cup, thermometer, stopwatch, plastic spoon, 5 tsp Epsom salts

---

CHAPTER INVESTIGATION
Exothermic or Endothermic?, pp. 92–93
Students classify each of two processes as exothermic or endothermic by measuring the change in temperature of reaction mixtures.

TIME 40 minutes
MATERIALS graduated cylinder, 30 mL hydrogen peroxide, 2 100 mL beakers, 2 thermometers, stopwatch, measuring spoons, 1 g yeast, balance, plastic spoon, large plastic cup, hot tap water, 30 mL vinegar, 1 g baking soda

---

SECTION 3.4

INVESTIGATE Sugar Combustion, p. 95
Students infer that different experimental conditions can determine whether a reaction happens.

TIME 20 minutes
MATERIALS candle, matches, tongs, 2 sugar cubes, stopwatch, ashes

---

R **Additional INVESTIGATION,** Modeling Chemical Reactions, A, B, & C, pp. 189–197; Teacher Instructions, pp. 329–330

Previewing Chapter Resources

	INTEGRATED TECHNOLOGY	LABS AND ACTIVITIES

CHAPTER 3
Chemical Reactions

 CLASSZONE.COM
- eEdition Plus
- EasyPlanner Plus
- Misconception Database
- Content Review
- Test Practice
- Visualizations
- Resource Centers
- Internet Activity: Reactions
- Math Tutorial

 SCILINKS.ORG

 CD-ROMS
- eEdition
- EasyPlanner
- Power Presentations
- Content Review
- Lab Generator
- Test Generator

 AUDIO CDS
- Audio Readings
- Audio Readings in Spanish

 EXPLORE the Big Idea, p. 67
- Changing Steel Wool
- A Different Rate
- Internet Activity: Reactions

 UNIT RESOURCE BOOK
Unit Projects, pp. 5–10

 Lab Generator CD-ROM
Generate customized labs.

SECTION
3.1 Chemical reactions alter arrangements of atoms. pp. 69–77

Time: 2 periods (1 block)
 Lesson Plan, pp. 130–131

- **VISUALIZATION,** Concentration and Reaction Rate
- **RESOURCE CENTER,** Catalysts
- **MATH TUTORIAL**

 UNIT TRANSPARENCY BOOK
- Big Idea Flow Chart, p. T17
- Daily Vocabulary Scaffolding, p. T18
- Note-Taking Model, p. T19
- 3-Minute Warm-Up, p. T20

- EXPLORE Chemical Changes, p. 69
- INVESTIGATE Chemical Reactions, p. 74
- Math in Science, p. 77

 UNIT RESOURCE BOOK
- Datasheet, Chemical Reactions, p. 139
- Additional INVESTIGATION, Modeling Chemical Reactions, A, B, & C, pp. 189–197
- Math Support & Practice, pp. 178–179

SECTION
3.2 The masses of reactants and products are equal. pp. 78–85

Time: 2 periods (1 block)
 Lesson Plan, pp. 141–142

 UNIT TRANSPARENCY BOOK
- Daily Vocabulary Scaffolding, p. T18
- 3-Minute Warm-Up, p. T20

- INVESTIGATE Conservation of Mass, p. 79
- Science on the Job, p. 85

 UNIT RESOURCE BOOK
Datasheet, Conservation of Mass, p. 150

SECTION
3.3 Chemical reactions involve energy changes. pp. 86–93

Time: 3 periods (1.5 blocks)
 Lesson Plan, pp. 152–153

 VISUALIZATION, Endothermic and Exothermic Reactions

 UNIT TRANSPARENCY BOOK
- Daily Vocabulary Scaffolding, p. T18
- 3-Minute Warm-Up, p. T21

- EXPLORE Energy Changes, p. 86
- CHAPTER INVESTIGATION, Exothermic or Endothermic?, pp. 92–93

UNIT RESOURCE BOOK
CHAPTER INVESTIGATION, Exothermic or Endothermic?, A, B, & C, pp. 180–188

SECTION
3.4 Life and industry depend on chemical reactions. pp. 94–99

Time: 3 periods (1.5 blocks)
 Lesson Plan, pp. 162–163

 UNIT TRANSPARENCY BOOK
- Big Idea Flow Chart, p. T17
- Daily Vocabulary Scaffolding, p. T18
- 3-Minute Warm-Up, p. T21
- "Chemical Reactions in Catalytic Converters" Visual, p. T22
- Chapter Outline, pp. T23–T24

 INVESTIGATE Sugar Combustion, p. 95

 UNIT RESOURCE BOOK
Datasheet, Sugar Combustion, p. 171

READING AND REINFORCEMENT

ASSESSMENT

STANDARDS

 (Science Toolkit)
- Four Square, B22–23
- Combination Notes, C36
- Daily Vocabulary Scaffolding, H1–8

 UNIT RESOURCE BOOK
- Vocabulary Practice, pp. 175–176
- Decoding Support, p. 177
- Summarizing the Chapter, pp. 198–199

Audio Readings CD
Listen to Pupil Edition.

Audio Readings in Spanish CD
Listen to Pupil Edition in Spanish.

 (PE)
- Chapter Review, pp. 101–102
- Standardized Test Practice, p. 103

 (A) **UNIT ASSESSMENT BOOK**
- Diagnostic Test, pp. 39–40
- Chapter Test, A, B, & C, pp. 45–56
- Alternative Assessment, pp. 57–58

 (SP A)
Spanish Chapter Test, pp. 241–244

Test Generator CD-ROM
Generate customized tests.

Lab Generator CD-ROM
Rubrics for Labs

National Standards
A.2–8, A.9.a–f, B.1.b, B.3.e, E.6.c, F.5.c, G.1.b

See p. 66 for the standards.

 UNIT RESOURCE BOOK
- Reading Study Guide, A & B, pp. 132–135
- Spanish Reading Study Guide, pp. 136–137
- Challenge and Extension, p. 138
- Reinforcing Key Concepts, p. 140

 (TE) Ongoing Assessment, pp. 70–71, 73, 75–76

 (PE) Section 3.1 Review, p. 76

 (A) **UNIT ASSESSMENT BOOK**
Section 3.1 Quiz, p. 41

National Standards
A.2–8, A.9.a, A.9.c–e, G.1.b

 UNIT RESOURCE BOOK
- Reading Study Guide, A & B, pp. 143–146
- Spanish Reading Study Guide, pp. 147–148
- Challenge and Extension, p. 149
- Reinforcing Key Concepts, p. 151

 (TE) Ongoing Assessment, pp. 79–80, 82–84

 (PE) Section 3.2 Review, p. 84

(A) **UNIT ASSESSMENT BOOK**
Section 3.2 Quiz, p. 42

National Standards
A.2–8, A.9.a, A.9.c–e, B.1.b, G.1.b

UNIT RESOURCE BOOK
- Reading Study Guide, A & B, pp. 154–157
- Spanish Reading Study Guide, pp. 158–159
- Challenge and Extension, p. 160
- Reinforcing Key Concepts, p. 161
- Challenge Reading, pp. 173–174

(TE) Ongoing Assessment, pp. 86–91

(PE) Section 3.3 Review, p. 91

(A) **UNIT ASSESSMENT BOOK**
Section 3.3 Quiz, p. 43

National Standards
A.2–8, A.9.a, A.9.c–e, B.3.e, G.1.b

UNIT RESOURCE BOOK
- Reading Study Guide, A & B, pp. 164–167
- Spanish Reading Study Guide, pp. 168–169
- Challenge and Extension, p. 170
- Reinforcing Key Concepts, p. 172

(TE) Ongoing Assessment, pp. 95–99

(PE) Section 3.4 Review, p. 99

(A) **UNIT ASSESSMENT BOOK**
Section 3.4 Quiz, p. 44

National Standards
A.2–8, A.9.a, A.9.c–e, E.6.c, F.5.c, G.1.b

Previewing Resources for Differentiated Instruction

CHAPTER INVESTIGATION

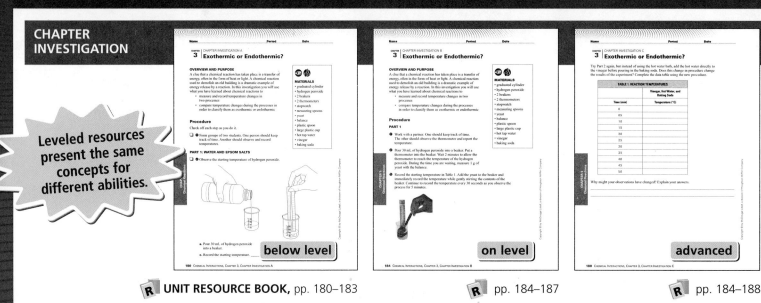

CHAPTER 3 | CHAPTER INVESTIGATION A
Exothermic or Endothermic?

below level

UNIT RESOURCE BOOK, pp. 180–183

CHAPTER 3 | CHAPTER INVESTIGATION B
Exothermic or Endothermic?

on level

R pp. 184–187

CHAPTER 3 | CHAPTER INVESTIGATION C
Exothermic or Endothermic?

advanced

R pp. 184–188

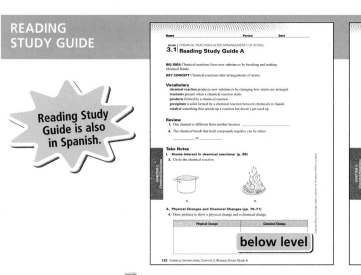

Leveled resources present the same concepts for different abilities.

READING STUDY GUIDE

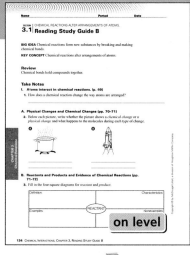

SECTION 3.1 | CHEMICAL REACTIONS ALTER ARRANGEMENTS OF ATOMS
Reading Study Guide A

below level

UNIT RESOURCE BOOK, pp. 132–133

SECTION 3.1 | CHEMICAL REACTIONS ALTER ARRANGEMENTS OF ATOMS
Reading Study Guide B

on level

R pp. 134–135

SECTION 3.1 | CHEMICAL REACTIONS ALTER ARRANGEMENTS OF ATOMS
Challenge and Extension

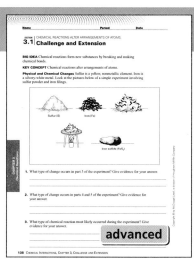

advanced

R p. 138

Reading Study Guide is also in Spanish.

CHAPTER TEST

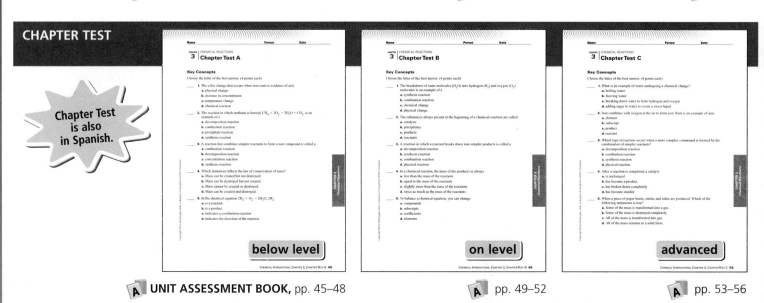

CHAPTER 3 | CHEMICAL REACTIONS
Chapter Test A

below level

UNIT ASSESSMENT BOOK, pp. 45–48

CHAPTER 3 | CHEMICAL REACTIONS
Chapter Test B

on level

A pp. 49–52

CHAPTER 3 | CHEMICAL REACTIONS
Chapter Test C

advanced

A pp. 53–56

Chapter Test is also in Spanish.

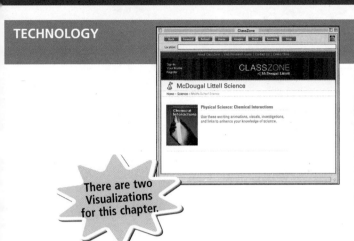

There are two Visualizations for this chapter.

CLASSZONE.COM

CD/CD-ROMS

CLASSZONE.COM

VISUAL CONTENT

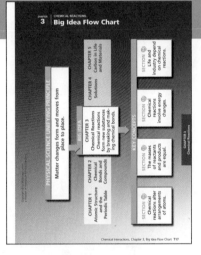

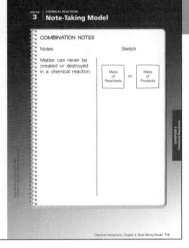

T UNIT TRANSPARENCY BOOK, p. T17

T p. T19

T p. T22

MORE SUPPORT

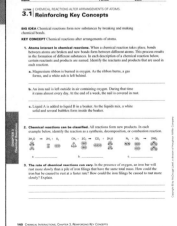

Reinforcing Key Concepts for each section

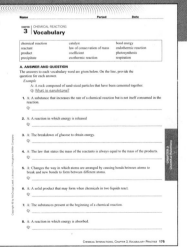

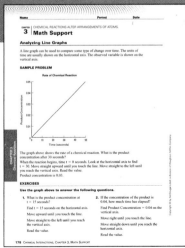

R UNIT RESOURCE BOOK, p. 140

R pp. 175–176

R p. 178

INTRODUCE

the BIG idea

Have students look at the photograph of the chemical reaction and discuss how the question in the box links to the Big Idea. For further discussion:

- How might new substances be formed by a chemical reaction?
- What types of changes might indicate that a chemical reaction has occurred?

National Science Education Standards

Content

B.1.b Substances react chemically in characteristic ways with other substances to form new substances (compounds) with different characteristic properties. In chemical reactions, the total mass is conserved.

B.3.e In most chemical and nuclear reactions, energy is transferred into or out of a system.

Process

A.2–8 Design and conduct an investigation; use tools to gather and interpret data; use evidence to describe, predict, explain, model; think critically to make relationships between evidence and explanation; recognize different explanations and predictions; communicate scientific procedures and explanations; use mathematics.

A.9.a–f Understand scientific inquiry by using different investigations, methods, mathematics, technology, explanations based on logic, evidence, and skepticism.

E.6.c Science drives technology; technology drives science.

F.5.c Technology influences society through its products and processes.

G.1.b Science requires different abilities.

CHAPTER

3 Chemical Reactions

the BIG idea

Chemical reactions form new substances by breaking and making chemical bonds.

What changes are happening in this chemical reaction?

Key Concepts

SECTION 3.1
Chemical reactions alter arrangements of atoms. Learn how chemical reactions are identified and controlled.

SECTION 3.2
The masses of reactants and products are equal. Learn how chemical equations show the conservation of mass.

SECTION 3.3
Chemical reactions involve energy changes. Learn how energy is absorbed or released by chemical reactions.

SECTION 3.4
Life and industry depend on chemical reactions. Learn about some chemical reactions in everyday life.

Internet Preview

CLASSZONE.COM
Chapter 3 online resources: Content Review, two Visualizations, two Resource Centers, Math Tutorial, Test Practice

INTERNET PREVIEW

CLASSZONE.COM   For student use with the following pages:

Review and Practice
- Content Review, pp. 68, 100
- Math Tutorial: Interpreting Line Graphs, p. 77
- Test Practice, p. 103

Activities and Resources
- Internet Activity: p. 67
- Visualizations: Concentration and Reaction Rate, p. 74; Endothermic and Exothermic Reactions, p. 90
- Resource Center: Catalysts in Living Things, p. 76

NSTA scilinks.org **SCiLINKS**

Chemical Reactions
Code: MDL024

Changing Steel Wool

Place a small lump of steel wool in a cup. Pour in enough vinegar to cover the steel wool. After five minutes, take the steel wool out of the vinegar. Shake the steel wool to remove any excess vinegar. Place the steel wool in a small plastic bottle, and cover the mouth of the bottle with a balloon. Observe the steel wool and balloon after one hour.

Observe and Think What happened to the steel wool and balloon? What might have caused this to occur?

A Different Rate

Half fill one cup with hot tap water and a second cup with cold tap water. Drop a seltzer tablet into each cup at the same time. Time how long it takes for each tablet to stop fizzing.

Observe and Think Which tablet fizzed for a longer period of time? How might you explain any differences?

Internet Activity: Reactions

Go to **ClassZone.com** to explore chemical reactions and chemical equations. Learn how a chemical equation can be balanced.

Observe and Think How do chemical equations show what happens during a chemical reaction?

NSTA
scilinks.org
SCILINKS
Chemical Reactions Code: MDL024

Chapter 3: Chemical Reactions 67 **B**

TEACHING WITH TECHNOLOGY

Video Camera Use a video camera to record sugar combustion (p. 95). The reaction takes place quickly; videotaping it will allow students to observe it again.

CBL and Probeware If you have probeware, you may use a temperature probe to record temperature changes in the Chapter Investigation on pp. 92–93. Students can then use a graphing calculator (or graphing software) to make graphs of the temperatures they record.

EXPLORE (the BIG idea)

These inquiry-based activities are appropriate for use at home or as a supplement to classroom instruction.

Changing Steel Wool

PURPOSE To observe a chemical change. Students observe a chemical reaction that involves the rusting of steel wool.

TIP *10 min.* Do not use scouring pads that contain soap.

Answer: The steel wool rusted, and the balloon was pulled into the bottle because oxygen from the air in the bottle reacted with the metal.

REVISIT after p. 72.

A Different Rate

PURPOSE To explore how reaction rate is affected by temperature. Students vary the temperature of a reaction mixture and measure changes in reaction rate.

TIP *10 min.* Use effervescent tablets.

Answer: The tablet in cold water fizzed longer because the reaction proceeded more slowly.

REVISIT after p. 75.

Internet Activity: Reactions

PURPOSE To examine chemical equations and learn how to balance them.

TIP *20 min.* Students should check their work by tallying the number of atoms of each element on each side of the equation.

Answer: The same number of atoms of each element in the reaction is on each side of the equation, which indicates that atoms are neither created nor destroyed during a chemical reaction.

REVISIT after p. 83.

Chapter 3 **67** **B**

PREPARE

◑ CONCEPT REVIEW

Activate Prior Knowledge

Make a model of a water molecule by attaching two identical gumdrops to a marshmallow with toothpicks. The angle between the toothpicks should be slightly greater than a right angle (about 105°).

- Ask: What do the toothpicks represent? *covalent bonds* What does the marshmallow represent? *an oxygen atom* The gumdrops? *hydrogen atoms*
- Ask: Are electrons shared or are they gained and lost by the atoms within the molecule? *shared*

◑ TAKING NOTES

Combination Notes

Combining a sketch with notes will help students to visualize a new concept and connect it with an example. Students should use a two-column format, writing their notes in one column and drawing their sketch in the other.

Vocabulary Strategy

The four square diagram organizes all aspects of a term into a coherent pattern. By filling in their own words, students personalize their understanding. Point out that it's okay to leave a blank square if there is no clear nonexample for the term.

Vocabulary and Note-Taking Resources

- Vocabulary Practice, pp. 175–176
- Decoding Support, p. 177

- Daily Vocabulary Scaffolding, p. T18
- Note-Taking Model, p. T19

- Four Square, B22–23
- Combination Notes, C36
- Daily Vocabulary Scaffolding, H1–8

CHAPTER 3
Getting Ready to Learn

◀ CONCEPT REVIEW

- Atoms combine to form compounds.
- Atoms gain or lose electrons when they form ionic bonds.
- Atoms share electrons in covalent bonds.

◀ VOCABULARY REVIEW

electron p. 11
ionic bond p. 48
covalent bond p. 50
See Glossary for definitions.
atom, chemical change

CONTENT REVIEW
CLASSZONE.COM
Review concepts and vocabulary.

▶ TAKING NOTES

COMBINATION NOTES

To take notes about a new concept, first make an informal outline of the information. Then make a sketch of the concept and label it so you can study it later.

VOCABULARY STRATEGY

Write each new vocabulary term in the center of a **four square** diagram. Write notes in the squares around each term. Include a definition, some characteristics, and some examples of the term. If possible, write some things that are not examples of the term.

See the Note-Taking Handbook on pages R45–R51.

B 68 Unit: Chemical Interactions

SCIENCE NOTEBOOK

NOTES

Chemical reactions
- cause chemical changes
- make new substances
- change reactants into products

Evidence of Chemical Reactions

before   after

increase in temperature

Definition substance present before a chemical reaction occurs	Characteristics its bonds are broken during a reaction
	REACTANT	
Examples oxygen in a combustion reaction	Nonexample carbon dioxide in a combustion reaction

CHECK READINESS

Administer the Diagnostic Test to determine students' readiness for new science content and their mastery of requisite math skills.

 Diagnostic Test, pp. 39–40

Technology Resources

Students needing content and math skills should visit **ClassZone.com**.

- **CONTENT REVIEW**
- **MATH TUTORIAL**

 CONTENT REVIEW CD-ROM

B 68   Unit: **Chemical Interactions**

KEY CONCEPT

Chemical reactions alter arrangements of atoms.

◀ **BEFORE,** you learned

- Atoms of one element differ from atoms of all other elements
- Chemical bonds hold compounds together
- Chemical bonds may be ionic or covalent

▶ **NOW,** you will learn

- About chemical changes and how they occur
- About three types of chemical reactions
- How the rate of a chemical reaction can be changed

VOCABULARY

chemical reaction p. 69
reactant p. 71
product p. 71
precipitate p. 72
catalyst p. 76

EXPLORE Chemical Changes

How can you identify a chemical change?

PROCEDURE

1. Pour about 3 cm (1 in.) of vinegar into the bowl. Add a spoonful of salt. Stir until the salt dissolves.

2. Put the pennies into the bowl. Wait two minutes, and then put the nail into the bowl.

3. Observe the nail after five minutes and record your observations.

WHAT DO YOU THINK?

- What did you see on the nail? Where do you think it came from?
- Did a new substance form? What evidence supports your conclusion?

MATERIALS

- vinegar
- clear bowl
- plastic spoon
- table salt
- 20 pennies
- large iron nail

Atoms interact in chemical reactions.

COMBINATION NOTES
Use combination notes to organize information about how atoms interact during chemical reactions.

You see substances change every day. Some changes are physical, such as when liquid water changes to water vapor during boiling. Other changes are chemical, such as when wood burns to form smoke and ash, or when rust forms on iron. During a chemical change, substances change into one or more different substances.

A **chemical reaction** produces new substances by changing the way in which atoms are arranged. In a chemical reaction, bonds between atoms are broken and new bonds form between different atoms. This breaking and forming of bonds takes place when particles of the original materials collide with one another. After a chemical reaction, the new arrangements of atoms form different substances.

Chapter 3: **Chemical Reactions** 69 **B**

RESOURCES FOR DIFFERENTIATED INSTRUCTION

Below Level

UNIT RESOURCE BOOK
Reading Study Guide A, pp. 132–133
Decoding Support, p. 177

 AUDIO CDS

R **Additional INVESTIGATION,**
Modeling Chemical Reactions, A, B, & C, pp. 189–197;
Teacher Instructions, pp. 329–330

Advanced

UNIT RESOURCE BOOK
Challenge and Extension, p. 138

English Learners

UNIT RESOURCE BOOK
Spanish Reading Study Guide, pp. 136–137

AUDIO CDS

- Audio Readings in Spanish
- Audio Readings (English)

Teach from Visuals

To help students interpret the blow-up diagrams, ask what happens to the water molecules when water changes from the solid state to the liquid state. *The water molecules from the ice are no longer locked in place, allowing the molecules to move freely past each other.*

Real World Example

Making popcorn involves a physical change of water. Unpopped popcorn kernels contain about 14 percent water. When you heat the kernels, the water expands and turns to steam. Pressure builds up inside the kernel. Eventually, the kernel explodes. As it explodes, the soft starch inside the kernel inflates and bursts, turning the kernel inside out. The kernel releases the steam, and the popcorn is popped.

Ongoing Assessment

CHECK YOUR READING *Answer: Its physical form changes, but the substance itself remains unchanged.*

Physical Changes

 A A change in the state of a substance is an example of a physical change. The substance may have some different properties after a physical change, but it is still the same substance. For example, you know that water can exist in three different physical states: the solid state (ice), the liquid state (water), and the gas state (water vapor). However, regardless of what state water is in, it still remains water, that is, H_2O molecules. As ice melts, the molecules of water move around more quickly, but the molecules do not change. As water vapor condenses, the molecules of water move more slowly, but they are still the same molecules.

Substances can undergo different kinds of physical changes. For example, sugar dissolves in water but still tastes sweet because the molecules that make up sugar do not change when it dissolves. The pressure of helium changes when it is pumped from a high-pressure tank into a balloon, but the gas still remains helium.

CHECK YOUR READING What happens to a substance when it undergoes a physical change?

When water changes from a liquid to a solid, it undergoes a physical change.

Ice is composed of water molecules that are locked together.

Liquid water is composed of molecules that move freely past each other.

DIFFERENTIATE INSTRUCTION

? **More Reading Support**

A What kind of change occurs when a substance changes state? *a physical change*

English Learners Have students write the definitions for *chemical reaction, precipitate,* and *catalyst* in their Science Word Dictionaries. Help English learners understand reactants and products by using an analogy. For example, reactants are like the ingredients in a cake recipe, and the product is like the cake. Encourage students to think of other analogies for reactants and products.

Chemical Changes

Water can also undergo a chemical change. Water molecules can be broken down into hydrogen and oxygen molecules by a chemical reaction called electrolysis. When an electric current is passed through liquid water (H_2O), it changes the water into two gases—hydrogen and oxygen. The molecules of water break apart into individual atoms, which then recombine into hydrogen molecules (H_2) and oxygen molecules (O_2). The original material (water) changes into different substances through a chemical reaction.

Hydrogen and oxygen are used as rocket fuel for the space shuttle. During liftoff, liquid hydrogen and liquid oxygen are combined in a reaction that is the opposite of electrolysis. This reaction produces water and a large amount of energy that helps push the shuttle into orbit.

CHECK YOUR READING How does a chemical change differ from a physical change?

Electrolysis of Water

- hydrogen gas (H_2)
- oxygen gas (O_2)
- water (H_2O)

Water molecules can be split apart to form separate hydrogen and oxygen molecules.

Reactants and Products

Reactants are the substances present at the beginning of a chemical reaction. In the burning of natural gas, for example, methane (CH_4) and oxygen (O_2) are the reactants in the chemical reaction. **Products** are the substances formed by a chemical reaction. In the burning of natural gas, carbon dioxide (CO_2) and water (H_2O) are the products formed by the reaction. Reactants and products can be elements or compounds, depending on the reaction taking place.

During a chemical reaction, bonds between atoms in the reactants are broken and new bonds are formed in the products. When natural gas is burned, bonds between the carbon and hydrogen atoms in methane are broken, as are the bonds between the oxygen atoms in oxygen molecules. New bonds are formed between carbon and oxygen in carbon dioxide gas and between hydrogen and oxygen in water vapor.

Reactants—bonds broken		Products—new bonds formed	
methane +	oxygen	carbon dioxide +	water
(CH_4)	(O_2)	(CO_2)	(H_2O)

CHECK YOUR READING What must happen for reactants to be changed into products?

To help students interpret the electrolysis diagram, ask how the contents of each test tube differ from the contents of the beaker. *One test tube contains hydrogen gas, and the other contains oxygen gas. Both gases came from the water in the beaker.*

Address Misconceptions

IDENTIFY Ask: Is burning a physical change or a chemical change? If students say that it is a physical change, they may hold the misconception that combustion is a physical change because no chemicals are involved.

CORRECT Place a small piece of paper in a beaker. Set the paper on fire and immediately cover the beaker with a sheet of glass. When the flames go out, have students observe the underside of the glass. Caution: Use heat-resistant glass, gloves, and tongs, and make sure students are at a safe distance from the fire.

REASSESS Ask students what products burning paper produces. *carbon (soot), carbon dioxide, and water* Point out that burning is a chemical change because it produces new substances.

Technology Resources

Visit **ClassZone.com** for background on common student misconceptions.

MISCONCEPTION DATABASE

Ongoing Assessment

Recognize chemical changes and describe how they occur.

Ask: How can you tell that splitting water to form hydrogen and oxygen is a chemical change? How does this change occur? *New substances form. Bonds break, and new bonds form.*

CHECK YOUR READING *Answer: During a chemical change, the original substances change into different substances.*

CHECK YOUR READING *Answer: The bonds in the reactants must be broken, and new bonds must form in the products.*

DIFFERENTIATE INSTRUCTION

More Reading Support

B What products are formed when water is broken down during a chemical change? *hydrogen and oxygen molecules*

Below Level Have students compare and contrast physical and chemical changes by making a table listing the features of each. Students' tables should also give at least one example of each kind of change.

Set up chemical reactions to show evidence of each type of chemical change.

- To demonstrate a color change, place some dry cornstarch in a small dish. Put a few drops of iodine solution on the starch. A dark blue iodine-starch complex will form. Caution: Iodine solution can stain hands and clothing.

- To produce a precipitate, pour about 5 mL of 0.1M silver nitrate solution (2 g $AgNO_3$ dissolved in 100 mL water) into a flask. Slowly add about 5 mL of 0.1M potassium iodide solution (2 g KI dissolved in 100 mL water). A bright yellow precipitate of silver iodide (AgI) will form. Caution: Silver nitrate solution can stain hands and clothing.

- To produce a gas, place a spoonful of baking soda ($NaHCO_3$) in a small beaker. Add about 10 mL of vinegar (acetic acid). Bubbles of carbon dioxide will appear.

- To demonstrate a temperature change, burn a small amount of rubbing alcohol in an evaporating dish. Tell students that heat is produced during the combustion reaction. Caution: Make sure students maintain a safe distance.

EXPLORE (the BIG idea)

Revisit "Changing Steel Wool" on p. 67. Have students explain their results.

Real World Example

The rusting of iron and corrosion of other metals are very costly chemical reactions. Millions of dollars are spent worldwide each year to paint building parts, bridges, ships, machinery, storage tanks, and other metal structures. In addition, rust takes an enormous toll on automobiles. The salt used to melt snow and ice on roads enhances the reaction that produces rust.

VOCABULARY
Remember to use a four square diagram for *precipitate* and other vocabulary terms.

?
C

?
D

Evidence of Chemical Reactions

Some chemical changes are easy to observe—the products formed by the rearrangement of atoms look different than the reactants. Other changes are not easy to see but can be detected in other ways.

Color Change Substances often change color during a chemical reaction. For example, when gray iron rusts the product that forms is brown, as shown in the photograph below.

Formation of a Precipitate Many chemical reactions form products that exist in a different physical state from the reactants. A solid product called a **precipitate** may form when chemicals in two liquids react, as shown in the photograph below. Seashells are often formed this way when a sea creature releases a liquid that reacts with seawater.

Color Change

Formation of a Precipitate

Formation of a Gas Chemical reactions may produce a gas, like that often formed when antacid pills are mixed with excess stomach acid. The photograph below shows an example in which carbon dioxide gas is produced by a chemical reaction.

Temperature Change Most chemical reactions involve a temperature change. Sometimes this change can be inferred from the observation of a flame, as in the burning of the metal magnesium in the photograph below. Other temperature changes are not immediately obvious. If you have touched concrete before it hardens, you may have noticed that it felt warm. This warmth is due to a chemical reaction.

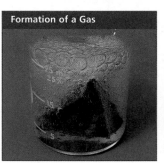
Formation of a Gas

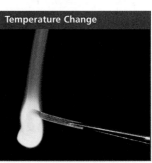
Temperature Change

DIFFERENTIATE INSTRUCTION

 More Reading Support

C What is a precipitate? *a solid product*

D How can a temperature change in a reaction be observed? *seeing a flame or feeling warmth*

Advanced Have students investigate the reaction of various antacids with excess stomach acid. What gas is sometimes produced? *carbon dioxide* What happens to the excess stomach acid? *It changes to a harmless salt and water.*

R Challenge and Extension, p. 138

Chemical reactions can be classified.

Scientists classify chemical reactions in several ways to help make the different types of reactions easier to understand. All reactions form new products, but the ways in which products are made can differ.

Synthesis In a synthesis reaction, a new compound is formed by the combination of simpler reactants. For example, nitrogen dioxide (NO_2), a component of smog, forms when nitrogen and oxygen combine in the air.

READING TiP

Synthesis means "making a substance from simpler substances."

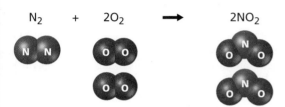

$$N_2 \quad + \quad 2O_2 \quad \longrightarrow \quad 2NO_2$$

Decomposition In a decomposition reaction, a reactant breaks down into simpler products, which could be elements or other compounds. Decomposition reactions can be thought of as being the reverse of synthesis reactions. For example, water can be decomposed into its elements—hydrogen and oxygen.

READING TiP

Decomposition means "separation into parts."

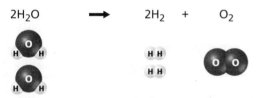

$$2H_2O \quad \longrightarrow \quad 2H_2 \quad + \quad O_2$$

Combustion In a combustion reaction, one reactant is always oxygen and another reactant often contains carbon and hydrogen. The carbon and hydrogen atoms combine with oxygen, producing carbon dioxide and water. The burning of methane is a combustion reaction.

READING TiP

Combustion is the process of burning with oxygen.

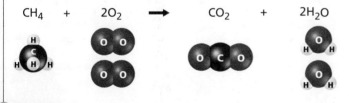

$$CH_4 \quad + \quad 2O_2 \quad \longrightarrow \quad CO_2 \quad + \quad 2H_2O$$

CHECK YOUR READING How are synthesis reactions different from decomposition reactions?

Teach from Visuals

To help students interpret the three diagrams of chemical reactions, have them examine the reactants and products of each reaction. Ask:

- Do any of the reactants remain unchanged? How can you tell? *No; the atoms in the reactants are reorganized to form the products.*
- What products form from the combustion of substances containing carbon and hydrogen atoms? *water and carbon dioxide*
- Atoms of which elements make up a water molecule? *hydrogen and oxygen*

Ongoing Assessment

Identify three types of chemical reactions.

Ask: What types of chemical reactions are discussed on this page? *synthesis, decomposition, and combustion*

CHECK YOUR READING *Answer: Synthesis reactions make more complex products from simpler reactants; decomposition reactions break down complex reactants into simpler products.*

DIFFERENTIATE INSTRUCTION

More Reading Support

E What reactant is always in a combustion reaction? *oxygen*

Additional Investigation To reinforce Section 3.1 learning goals, use the following full-period investigation:

R **Additional INVESTIGATION,** Modeling Chemical Reactions, A, B, & C, pp. 189–197, 329–330
(Advanced students should complete Levels B and C.)

Inclusion Cardboard cutouts or ball-and-stick models of the atoms and molecules shown on this page can be used by students with visual impairments to envision the reactants and products in the three types of reactions.

INVESTIGATE Chemical Reactions

PURPOSE To infer how the surface area of reactants affects reaction rate

TIP *15 min.* Only antacids containing carbonate or a bicarbonate will fizz. Use effervescent seltzer tablets.

WHAT DO YOU THINK *The whole tablet should fizz longer than the crushed tablet. The crushed tablet has a greater total surface area, so the reaction proceeds more rapidly.*

CHALLENGE *A greater surface area means that more particles are available to react, so they collide more frequently.*

 Datasheet, Chemical Reactions, p. 139

Technology Resources

Customize this student lab as needed or look for an alternative. Print rubrics to assess student lab reports.

 Lab Generator CD-ROM

Teach Difficult Concepts

Students often have trouble visualizing that a large particle has less surface area than many small particles with an equal mass or volume. Have students calculate the volume and surface area of a cube 4 cm on each side. $V = 64$ cm^3, area = 96 cm^2 Then have them calculate the volume and surface area of 64 cubes, each is 1 cm on each side. Point out that these 64 cubes would fit inside the larger cube. $V = 64$ cm^3, area = 384 cm^2

The rates of chemical reactions can vary.

Most chemical reactions take place when particles of reactants collide with enough force to react. Chemical reactions can occur at different rates. Striking a match causes a very quick chemical reaction, while the rusting of an iron nail may take months. However, the rate of a reaction can be changed. For instance, a nail can be made to rust more quickly. Three physical factors—concentration, surface area, and temperature—and a chemical factor—a catalyst—can greatly affect the rate of a chemical reaction.

Concentration

 Observe how changing the concentration of a reactant can change the rate of a reaction.

Concentration measures the number of particles present in a certain volume. A high concentration of reactants means that there is a large number of particles that can collide and react. Turning the valve on a gas stove to increase the flow of gas increases the concentration of methane molecules that can combine with oxygen in the air. The result is a bigger flame and a faster combustion reaction.

Surface Area

Suppose one of the reactants in a chemical reaction is present as a single large piece of material. Particles of the second reactant cannot get inside the large piece, so they can react only with particles on the surface. To make the reaction go faster, the large piece of material could be broken into smaller pieces before the reaction starts.

INVESTIGATE Chemical Reactions

How can the rate of a reaction be changed?

PROCEDURE

(1) Place a whole seltzer tablet in one cup. Crush the second tablet and place it in the second cup.

(2) Fill each cup halfway with water.

(3) Time how long the tablet in each cup fizzes.

WHAT DO YOU THINK?

• How long did the whole tablet fizz? What about the crushed tablet?

• How are these results related to the rate of a chemical reaction? Explain.

CHALLENGE How might your results be related to collisions between particles during a chemical reaction?

SKILL FOCUS
Inferring

MATERIALS
• 2 seltzer tablets
• 2 plastic cups
• tap water
• stopwatch

TIME
15 minutes

DIFFERENTIATE INSTRUCTION

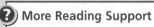

(?) More Reading Support

F What three physical factors affect reaction rate? *concentration, surface area, and temperature*

Alternative Assessment Students can answer and extend the Challenge question visually by making a drawing or model. Students should understand that more collisions between reactant particles take place when more surface area is exposed.

Breaking a large piece of material into smaller parts increases the surface area of the material. All of the inner material has no surface when it is inside a larger piece. Each time the large piece is broken, however, more surfaces are exposed. The amount of material does not change, but breaking it into smaller parts increases its surface area. Increasing the surface area increases the rate of the reaction.

CHECK YOUR READING Why does a reaction proceed faster when the reactants have greater surface areas?

Temperature

The rate of a reaction can be increased by making the particles move faster. The result is that more collisions take place per second and occur with greater force. The most common way to make the particles move faster is to add energy to the reactants, which will raise their temperature.

REMINDER
Temperature is the average amount of kinetic energy of the particles in a substance.

Many chemical reactions during cooking go very slowly, or do not take place at all, unless energy is added to the reactants. Too much heat can make a reaction go too fast, and food ends up burned. Chemical reactions can also be slowed or stopped by decreasing the temperature of the reactants. Again, think about cooking. The reactions that take place during cooking can be stopped by removing the food from the heat source.

Particles and Reaction Rates

Changes in Reactants	Normal Reaction Rate	Increased Reaction Rate
Concentration An increase in concentration of the reactants increases the number of particles that can interact.		
Surface area An increase in the surface area of the reactants increases the number of particles that can interact.		
Temperature Adding energy makes particles move faster and increases temperature. The increase in motion allows reactants to collide and react more frequently.		

To help students interpret the table of factors that affect reaction rate, ask: Do the three factors shown in the table affect the reaction rate through physical or chemical changes? Explain. *physical changes; they increase reaction rate through physical changes to the reactants*

Address Misconceptions

IDENTIFY Ask students to describe the motion of particles in matter. If students describe anything other than random, straight-line motion, or do not know that all particles in matter are in constant motion, they may hold a misconception about the intrinsic motion of particles in matter.

CORRECT Model the kinetic theory of matter in relation to solids, liquids, and gases. Fill a clear plastic container with marbles or other small objects. Cover it with a lid, and shake it to demonstrate a solid. Remove a portion of the marbles to demonstrate a liquid, and leave only a few to demonstrate a gas.

REASSESS Ask students to draw the path of a particle in a solid. *Students' diagrams should show the particle vibrating around the same position. Every time the particle collides with another particle, its direction changes.*

Technology Resources
Visit **ClassZone.com** for background on common student misconceptions.
 MISCONCEPTION DATABASE

EXPLORE (the BIG idea)
Revisit "A Different Rate" on p. 67. Have students explain their results.

Ongoing Assessment
CHECK YOUR READING *Answer: More particles are available to collide and react when the surface area is greater.*

More Reading Support
G Why are collisions between particles important in chemical reactions?
Particles of reactants must collide to react.

English Learners Point out that the word *rate* has several meanings. Have students look up the meanings in a dictionary. Then have them use each meaning in a sentence and indicate which meaning is used in this section. *Rate here is the speed of reaction in relation to time.*

Ongoing Assessment

Describe how the rate of a chemical reaction can be changed.

Ask: Why does increased temperature increase the rate of a reaction? *The particles of reactants move faster, and they collide harder and more often.*

 CHECK YOUR READING *Answer: They increase the rate of chemical reactions. Some reactions would proceed slowly or not at all without a catalyst.*

Teach from Visuals

To help students interpret the sequence of enzyme action, have them examine the diagram. Ask: How do you think an enzyme might increase the rate of a reaction? *An enzyme precisely lines up two molecules so they can react quickly.*

Reinforce (the **BIG** idea)

Have students relate the section to the Big Idea.

 Reinforcing Key Concepts, p. 140

3.1 ASSESS & RETEACH

Assess

[A] Section 3.1 Quiz, p. 41

Reteach

Have students make a set of flash cards. On one side of each card, they should write one type of chemical reaction. On the other side, they should write the definition, important criteria, and an example of that type of reaction. Have pairs of students trade cards to learn from each other's ideas.

Technology Resources

Have students visit **ClassZone.com** for reteaching of Key Concepts.

 CONTENT REVIEW

CONTENT REVIEW CD-ROM

Catalysts

 **RESOURCE CENTER**
CLASSZONE.COM

Learn more about catalysts and how they work in living things.

The rate of a reaction can be changed chemically by adding a catalyst. A **catalyst** is a substance that increases the rate of a chemical reaction but is not itself consumed in the reaction. This means that after the reaction is complete, the catalyst remains unchanged. Catalysts are very important for many industrial and biological reactions. In fact, many chemical reactions would proceed slowly or not take place at all without catalysts.

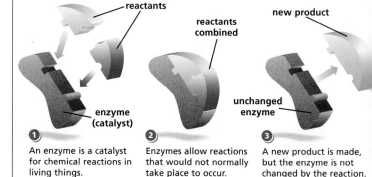

reactants

reactants combined

new product

enzyme (catalyst)

unchanged enzyme

① An enzyme is a catalyst for chemical reactions in living things.

② Enzymes allow reactions that would not normally take place to occur.

③ A new product is made, but the enzyme is not changed by the reaction.

In living things, catalysts called enzymes are absolutely necessary for life. Without them, many important reactions could not take place under the conditions within your body. In fact, in 2003, scientists reported that they had discovered the slowest known chemical reaction in living things. This reaction would normally take one trillion years. Enzymes, though, allow the reaction to occur in 0.01 seconds.

 CHECK YOUR READING Why are catalysts important in chemical reactions?

3.1 Review

KEY CONCEPTS

1. How do physical changes differ from chemical changes? Explain.

2. Describe four types of evidence of a chemical reaction.

3. Describe the ways in which the rate of a chemical reaction can be changed.

CRITICAL THINKING

4. **Synthesize** What evidence shows that the burning of methane is a chemical reaction?

5. **Compare** What about combustion reactions makes them different from either synthesis or decomposition reactions?

⚠ CHALLENGE

6. **Apply** How might the chewing of food be related to the rate of a chemical reaction—digestion—that occurs in your body? Explain.

ANSWERS

1. A physical change alters the characteristics of a substance, but not the substance itself. A chemical change produces different substances.

2. a color change, formation of a precipitate or a gas, a temperature change

3. Change the concentration, surface area, or temperature of reactants, or add a catalyst.

4. a change in temperature and the production of gases

5. Oxygen is always a reactant, and another reactant often contains carbon and hydrogen.

6. Chewing increases the surface area of food, which speeds up chemical reactions involved in digestion.

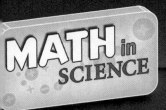

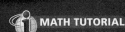

MATH TUTORIAL
CLASSZONE.COM

Click on Math Tutorial for more help with interpreting line graphs.

Before | After

The reactants in the iodine clock reaction produce a sudden color change several seconds after the reactants are mixed.

The Iodine Clock

Can a chemical reaction be timed? In the iodine clock reaction, a sudden color change indicates that the reaction has occurred. The length of time that passes before the color changes depends on the concentration ratios of the reactants. As shown in the graph below, the greater the concentration of the reactants, the faster the reaction.

Example

Suppose you are given an unknown iodine concentration to test in the iodine clock reaction. What is the concentration ratio of the iodine if it takes 40 seconds for the color change to occur?

(1) Find 40 seconds on the *x*-axis of the graph below and follow the vertical line up to the plotted data.

(2) Draw a horizontal line from that point on the curve to the *y*-axis to find the iodine concentration ratio in your sample.

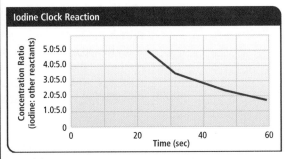

Iodine Clock Reaction

y-axis: Concentration Ratio (iodine: other reactants) — 5.0:5.0, 4.0:5.0, 3.0:5.0, 2.0:5.0, 1.0:5.0, 0
x-axis: Time (sec) — 0, 20, 40, 60

ANSWER The unknown concentration ratio is approximately 3.0:5.0.

Answer the following questions using the information in the graph above.

1. Approximately how long will it take for the reaction to occur if the concentration ratio is 4.0:5.0? 2.0:5.0?

2. Suppose you could extend the curve on the graph. If the reaction took 70 seconds to occur, what would be the approximate iodine concentration ratio?

CHALLENGE Using the following concentration ratios and times for another reactant, draw a reaction rate graph similar to the one shown above.

Concentration Ratios = 5.0:5.0, 4.0:5.0, 3.0:5.0, 2.0:5.0
Times = 24 sec, 25 sec, 43 sec, 68 sec

ANSWERS

1. approximately 28 sec; approximately 55 sec

2. 1.5 : 5.0

CHALLENGE Check students' graphs.

MATH IN SCIENCE
Math Skills Practice for Science

Set Learning Goal

To use a line graph to find reactant concentration

Present the Science

Two reactions are happening simultaneously in the iodine clock. A slow reaction produces triiodide, and a fast one makes the triiodide disappear. When the reactant for the fast reaction is used up, the triiodide no longer disappears but suddenly colors the solution dark blue as it reacts with starch.

Develop Graph Skills

• Point out the use and labeling of the two axes. The *y*-axis usually represents the dependent variable (the variable being studied), and the *x*-axis the independent variable (variable that is known or manipulated).

• Ask: Why might it be difficult to estimate untested values outside the data endpoints? *The curve outside the endpoints might suddenly drop or rise.*

DIFFERENTIATION TIP For students with visual impairments who have trouble reading the *x* and *y* values, provide a transparency of a grid that will help them determine alignment of a point with the axes.

Close

Ask students how they could make sure that the line of the graph continues in the same arc below 20 seconds and above 60. *Measure the reaction rate when the concentration ratio is less than 1.0:5.0 and greater than 5.0:5.0.*

• Math Support, p. 178
• Math Practice, p. 179

Technology Resources

Students can visit **ClassZone.com** for practice in interpreting line graphs.

 MATH TUTORIAL

● Set Learning Goals

Students will

- Explain why total mass does not change in a chemical reaction.
- Recognize how a chemical equation represents a chemical reaction.
- Outline how to balance a simple chemical equation.
- Measure in an experiment the mass of reactants and products in a chemical reaction.

◐ 3-Minute Warm-Up

Display Transparency 20 or copy this exercise on the board:

Decide if these statements are true. If not, correct them.

1. Reactants are changed into products by a physical change. *chemical change*

2. Changing the concentration of reactants can change the reaction rate. *true*

3. A color change in a reaction mixture is evidence that a chemical reaction has taken place. *true*

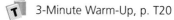 3-Minute Warm-Up, p. T20

3.2 MOTIVATE

THINK ABOUT

PURPOSE To examine how substances change during a chemical reaction

DISCUSS Brainstorm with students what happens to the matter that undergoes a chemical reaction. Ask: What happens to matter that seems to disappear? *A gas may form and escape into the air.*

Integrate the Sciences

Matter can, in fact, be destroyed, but not through chemical reactions. Matter is destroyed and converted into energy through nuclear reactions. This is the principle behind nuclear power.

KEY CONCEPT

3.2 The masses of reactants and products are equal.

◀ **BEFORE, you learned**

- Chemical reactions turn reactants into products by rearranging atoms
- Chemical reactions can be observed and identified
- The rate of chemical reactions can be changed

▶ **NOW, you will learn**

- About the law of conservation of mass
- How a chemical equation represents a chemical reaction
- How to balance a simple chemical equation

VOCABULARY

law of conservation of mass p. 79
coefficient p. 82

THINK ABOUT

What happens to burning matter?

You have probably watched a fire burn in a fireplace, a campfire, or a candle flame. It looks as if the wood or candle disappears over time, leaving a small pile of ashes or wax when the fire has finished burning. But does matter really disappear? Combustion is a chemical reaction, and chemical reactions involve rearrangements of atoms. The atoms do not disappear, so where do they go?

Careful observations led to the discovery of the conservation of mass.

COMBINATION NOTES
Take notes on the conservation of mass using combination notes.

The ashes left over from a wood fire contain less mass than the wood. In many other chemical reactions, mass also appears to decrease. That is, the mass of the products appears to be less than the mass of the reactants. In other reactions, the products appear to gain mass. For example, plants grow through a complex series of reactions, but where does their extra mass come from? At one time, scientists thought that chemical reactions could create or destroy matter.

During the 1780s the French chemist Antoine Lavoisier (luh-VWAH-zee-ay) showed that matter can never be created or destroyed in a chemical reaction. Lavoisier emphasized the importance of making very careful measurements in his experiments. Because of his methods, he was able to show that reactions that seem to gain mass or lose mass actually involve reactions with gases in the air. These gases could not be seen, but their masses could be measured.

RESOURCES FOR DIFFERENTIATED INSTRUCTION

Below Level
UNIT RESOURCE BOOK
- Reading Study Guide A, pp. 143–144
- Decoding Support, p. 177

 **AUDIO CDS**

Advanced
UNIT RESOURCE BOOK
Challenge and Extension, p. 149

English Learners
UNIT RESOURCE BOOK
Spanish Reading Study Guide, pp. 147–148

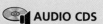 **AUDIO CDS**

- Audio Readings in Spanish
- Audio Readings (English)

An example of Lavoisier's work is his study of the reaction of the metal mercury when heated in air. In this reaction, the reddish-orange product formed has more mass than the original metal. Lavoisier placed some mercury in a jar, sealed the jar, and recorded the total mass of the setup. After the mercury had been heated in the jar, the total mass of the jar and its contents had not changed.

Lavoisier showed that the air left in the jar would no longer support burning—a candle flame was snuffed out by this air. He concluded that a gas in the air, which he called oxygen, had combined with the mercury to form the new product.

Lavoisier conducted many experiments of this type and found in all cases that the mass of the reactants is equal to the mass of the products. This conclusion, called the **law of conservation of mass,** states that in a chemical reaction atoms are neither created nor destroyed. All atoms present in the reactants are also present in the products.

Lavoisier carefully measured both the reactants and the products of chemical reactions.

 CHECK YOUR READING How did Lavoisier investigate the conservation of mass?

INVESTIGATE Conservation of Mass

Why is it important to measure the masses of reactants and products?

PROCEDURE

① Measure 2 tsp of baking soda. Use a funnel to put the baking soda in a balloon.

② Pour 2 tsp of vinegar into the plastic bottle.

③ Secure the balloon over the mouth of the bottle with the balloon hanging to the side of the bottle. Find and record the mass of the experimental setup.

④ Lift the balloon so that the baking soda drops into the bottle. Observe for five minutes, and then find and record the mass of the setup again.

WHAT DO YOU THINK?

- Did the mass of the experimental setup change?
- How do your observations demonstrate the conservation of mass?

CHALLENGE What do you think you would have observed if you had not used the balloon? Explain.

SKILL FOCUS
Measuring

MATERIALS
- teaspoon
- baking soda
- funnel
- balloon
- vinegar
- plastic bottle
- balance

TIME
35 minutes

IFFERENTIATE INSTRUCTION

) More Reading Support

A What scientific law states that the mass of the reactants is always equal to the mass of the products? *the law of conservation of mass*

English Learners Help English learners clearly understand the law of conservation of mass. If two or more students speak the same first language, encourage them to discuss the law of conservation of mass together in their native language. Then, encourage them to try the conversation in English, or pair them with more advanced students for help.

3.2 INSTRUCT

INVESTIGATE
Conservation of Mass

PURPOSE To show that mass is conserved in a chemical reaction

TIPS *35 min.*

- Different amounts of vinegar and baking soda can be used, but they should be kept in the same proportions.
- Students with small motor-skill disabilities may need help from a partner in manipulating the balloon.

WHAT DO YOU THINK? *No; all of the products were collected. The equal masses of the setup before and after the reaction show the conservation of mass.*

CHALLENGE *The mass would have been different after the reaction because any gas produced would have escaped.*

R Datasheet, Conservation of Mass, p. 150

Technology Resources

Customize this student lab as needed or look for an alternative. Print rubrics to assess student lab reports.

🔬 **Lab Generator CD-ROM**

Ongoing Assessment

Explain why total mass does not change in a chemical reaction.

Ask: Why is mass conserved in a chemical reaction? *Atoms are not created or destroyed in a chemical reaction.*

CHECK YOUR READING *Answer: By carefully observing and measuring all of the reactants and products in chemical reactions, he found that the total masses were always equal.*

Mathematics Connection

Ask: How many atoms are present in each of the following compounds? CH_4, C_2H_5OH, $MgSO_4$, $NaCl$ *5, 9, 6, 2*

Ask: How many atoms of hydrogen are in a molecule of C_2H_5OH? *6*

Develop Critical Thinking

PREDICT Although equations are shown in this chapter with the arrow pointing to the right, many chemical reactions are reversible. Have students predict what will happen if a reaction is reversible. *Many reactions will continue in both directions until they reach a point of equilibrium, where the formation of products from the reactants equals the formation of reactants from products.*

Teach Difficult Concepts

Chemical reactions rarely occur in the one step shown by a chemical equation. Rather, the chemical equation shows the overall process that occurs during the reaction.

Ongoing Assessment

Recognize that a chemical equation represents a chemical reaction.

Ask: What information is presented in a chemical equation? *reactants, products, direction of the reaction, the atomic symbols and chemical formulas of the reactants and products*

CHECK YOUR READING *Answer: by showing on each side of the equation the same number of atoms of each element involved in the reaction*

Chemical reactions can be described by chemical equations.

The law of conservation of mass states that in a chemical reaction, the total mass of reactants is equal to the total mass of products. For example, the mass of sodium plus the mass of chlorine that reacts with the sodium equals the mass of the product sodium chloride. Because atoms are rearranged in a chemical reaction, there must be the same number of sodium atoms and chlorine atoms in both the reactants and products.

Chemical equations represent how atoms are rearranged in a chemical reaction. The atoms in the reactants are shown on the left side of the equation. The atoms in the products are shown on the right side of the equation. Because atoms are rearranged and not created or destroyed, the number of atoms of each different element must be the same on each side of the equation.

CHECK YOUR READING How does a chemical equation show the conservation of mass?

In order to write a chemical equation, the information that you need to know is
- the reactants and products in the reaction
- the atomic symbols and chemical formulas of the reactants and products in the reaction
- the direction of the reaction

Carbon dioxide is a gas that animals exhale.

The following equation describes the formation of carbon dioxide from carbon and oxygen. In words, this equation says "Carbon reacts with oxygen to yield carbon dioxide." Notice that instead of an equal sign, an arrow appears between the reactants and the products. The arrow shows which way the reaction proceeds—from reactants on the left to the product or the products on the right.

reactants	direction of reaction	product
$C + O_2$	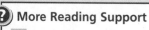	CO_2

Remember, the numbers below the chemical formulas for oxygen and carbon dioxide are called subscripts. A subscript indicates the number of atoms of an element in a molecule. You can see in the equation above that the oxygen molecule has two oxygen atoms, and the carbon dioxide molecule also has two oxygen atoms. If the chemical formula of a reactant or product does not have a subscript, it means that only one atom of each element is present in the molecule.

DIFFERENTIATE INSTRUCTION

(?) More Reading Support

B What does an arrow show in a chemical equation? *which way the reaction proceeds*

Advanced Chemical equations show the overall process that occurs during a reaction. Reactions, however, rarely occur in the single step shown by an equation. Have interested students do research to find a step reaction. Have them write the step reaction on the board and then discuss each step in order to learn more about how each one affects the overall reaction.

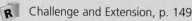 Challenge and Extension, p. 149

Chemical equations must be balanced.

Remember, chemical reactions follow the law of conservation of mass. Chemical equations show this conservation, or equality, in terms of atoms. The same number of atoms of each element must appear on both sides of a chemical equation. However, simply writing down the chemical formulas of reactants and products does not always result in equal numbers of atoms. You have to balance the equation to make the number of atoms equal on each side of an equation.

Balancing Chemical Equations

To learn how to balance an equation, look at the example of the combustion of natural gas, which is mostly methane (CH_4). The reactants are methane and oxygen. The products are carbon dioxide and water. You can write this reaction as the following equation.

REMINDER
Oxygen is always a reactant in a combustion reaction.

Unbalanced Equation

$$CH_4 + O_2 \longrightarrow CO_2 + H_2O$$

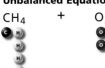

This equation is not balanced. There is one C on each side of the equation, so C is balanced. However, on the left side, H has a subscript of 4, which means there are four hydrogen atoms. On the right side, H has a subscript of 2, which means there are two hydrogen atoms. Also, there are two oxygen atoms on the left and three oxygen atoms on the right. Because of the conservation of mass, you know that hydrogen atoms do not disappear and oxygen atoms do not suddenly appear.

READING TIP
As you read how to balance the equation, look at the illustrations and count the atoms. The number of each type of atom is shown below the formula.

You can balance a chemical equation by changing the amounts of reactants or products represented.

- To balance H first, add another H_2O molecule on the right. Now, both C and H are balanced.
- There are now two oxygen atoms on the left side and four oxygen atoms on the right side. To balance O, add another O_2 molecule on the left.

Balanced Equation

$$CH_4 + O_2 + O_2 \longrightarrow CO_2 + H_2O + H_2O$$

Teach from Visuals

To help students determine if a chemical equation is balanced, ask: How many atoms of each element are on each side of the first equation on the page? (Note that the atoms underneath the equations are shown strictly for counting purposes, not as a model of the molecules.) *1 carbon on both sides; 4 hydrogen on left side, 2 hydrogen on right side; 2 oxygen on left side, 3 oxygen on right side.*

To help students balance the equation, ask:

- Atoms of which elements need to be added to the equation? *hydrogen, oxygen*
- On which side of the equation do atoms need to be added?*both*
- How has the second equation been changed so that it is balanced? *Another molecule of oxygen has been added to the left side, and another molecule of water has been added to the right side.*

Real World Example

Methane is the main component of natural gas that is used as fuel. It is a colorless, odorless, clean-burning fossil fuel that provides about 24 percent of the energy used in the United States. Natural gas formed millions of years ago when the remains of plants and algae were buried and compacted in sediment under intense heat and pressure. After natural gas is extracted from the ground by drilling, a substance called mercaptan is added to give the gas an odor. This odor makes it easier to detect gas leaks.

DIFFERENTIATE INSTRUCTION

More Reading Support

C Why does a chemical equation have to be balanced? *to make the number of atoms equal on both sides of the equation*

English Learners Point out to English learners that the word *subscript* comes from two Latin root words meaning "to write under." Subscripts are always written next to and slightly below the symbol for an atom. Tell English learners that word parts can often serve as clues to a word's meaning.

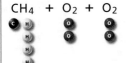

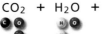

Using Coefficients to Balance Equations

The balanced equation for the combustion of methane shows that one molecule of methane reacts with two molecules of oxygen to produce one molecule of carbon dioxide and two molecules of water. The equation can be simplified by writing $2O_2$ instead of $O_2 + O_2$, and $2H_2O$ instead of $H_2O + H_2O$.

D

The numbers in front of the chemical formulas are called coefficients. **Coefficients** indicate how many molecules take part in the reaction. If there is no coefficient, then only one molecule of that type takes part in the reaction. The balanced equation, with coefficients, for the combustion of methane is shown below.

Balanced Equation with Coefficients

$$CH_4 \quad + \quad 2O_2 \quad \longrightarrow \quad CO_2 \quad + \quad 2H_2O$$
coefficient · · · · · subscript

Chemical formulas can have both coefficients and subscripts. In these cases, multiply the two numbers together to find the number of atoms involved in the reaction. For example, two water molecules ($2H_2O$) contain $2 \cdot 2 = 4$ hydrogen atoms and $2 \cdot 1 = 2$ oxygen atoms. Remember, coefficients in a chemical equation indicate how many molecules of each type take part in the reaction.

E

Only coefficients can be changed in order to balance a chemical equation. Subscripts are part of the chemical formula for reactants or products and cannot be changed to balance an equation. Changing a subscript changes the substance represented by the formula.

For example, the equation for the combustion of methane cannot be balanced by changing the formula CO_2 to CO. The formula CO_2 represents carbon dioxide gas, which animals exhale when they breathe. The formula CO represents carbon monoxide gas, which is a very different compound from CO_2. Carbon monoxide gas is poisonous, and breathing too much of it can be fatal.

 CHECK YOUR READING Why are coefficients used to balance equations?

> **REMINDER**
> A subscript shows the number of atoms in a molecule. If a subscript is changed, the molecule represented by the formula is changed.

The combustion of methane (CH_4) is used to melt glass.

B 82 Unit: **Chemical Interactions**

Balancing Equations with Coefficients

The steps below show how to balance the equation for the synthesis reaction between nitrogen (N_2) and hydrogen (H_2), which produces ammonia (NH_3).

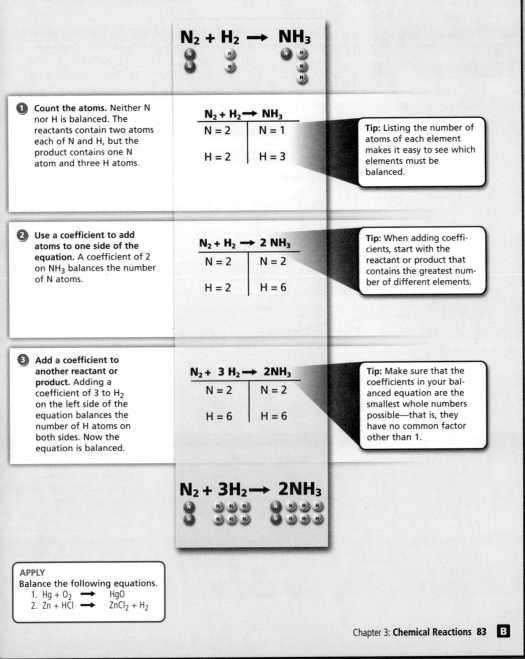

1 **Count the atoms.** Neither N nor H is balanced. The reactants contain two atoms each of N and H, but the product contains one N atom and three H atoms.

$N_2 + H_2 \rightarrow NH_3$	
N = 2	N = 1
H = 2	H = 3

Tip: Listing the number of atoms of each element makes it easy to see which elements must be balanced.

2 **Use a coefficient to add atoms to one side of the equation.** A coefficient of 2 on NH_3 balances the number of N atoms.

$$N_2 + H_2 \rightarrow 2\,NH_3$$

N = 2	N = 2
H = 2	H = 6

Tip: When adding coefficients, start with the reactant or product that contains the greatest number of different elements.

3 **Add a coefficient to another reactant or product.** Adding a coefficient of 3 to H_2 on the left side of the equation balances the number of H atoms on both sides. Now the equation is balanced.

$$N_2 + 3\,H_2 \rightarrow 2NH_3$$

N = 2	N = 2
H = 6	H = 6

Tip: Make sure that the coefficients in your balanced equation are the smallest whole numbers possible—that is, they have no common factor other than 1.

$$N_2 + 3H_2 \rightarrow 2NH_3$$

APPLY
Balance the following equations.
1. $Hg + O_2 \rightarrow HgO$
2. $Zn + HCl \rightarrow ZnCl_2 + H_2$

DIFFERENTIATE INSTRUCTION

Inclusion Kinesthetic learners and students with visual impairments can use three-dimensional objects, such as small foam balls, cardboard cutouts, or marshmallows, to create chemical equations. Distinguish for students what kind of atom each object represents.

Teach Difficult Concepts

To help students balance chemical equations, write the following equation on the board: $NO + H_2 \rightarrow NH_3 + H_2O$. Use red, blue, and white construction paper to represent oxygen, nitrogen, and hydrogen atoms respectively. Have students cut out circles. (four blue, four red, and twenty white).

Have students tape the circles together to model molecules of the reactants and products. Have them continue until they have used all the circles and have the same number of each color on both sides of the equation. The equation is balanced once they have two blue (nitrogen) circles, two red (oxygen) circles, and ten white (hydrogen) circles on each side. $2NO + 5H_2 \rightarrow 2NH_3 + 2H_2O$

Metacognitive Strategy

Have students write a short paragraph explaining whether a visual model helps them learn how to balance equations.

Develop Algebra Skills

When coefficients are used to balance an equation, the least common multiple of the coefficients should be used. Ask: Is $2N_2 + 6H_2 \rightarrow 4NH_3$ balanced? *yes*

Ask: How can the equation be expressed in a simpler form? $N_2 + 3H_2 \rightarrow 2NH_3$

Point out that the smallest whole numbers possible should be used.

EXPLORE (the **BIG** idea)

Revisit "Internet Activity: Reactions" on p. 67. Have students explain the steps they use to balance equations.

Ongoing Assessment

Outline how to balance a simple chemical equation.

Ask: What steps do you follow to balance a chemical equation? *Count the atoms of each element. Use coefficients to add atoms to one side at a time. Repeat until balanced.*

CAPTION Answers:
1. $2Hg + O_2 \rightarrow 2HgO$
2. $Zn + 2HCl \rightarrow ZnCl_2 + H_2$

Ongoing Assessment

Reinforce (the BIG idea)

Have students relate the section to the Big Idea.

 Reinforcing Key Concepts, p. 151

3.2 ASSESS & RETEACH

Assess

A Section 3.2 Quiz, p. 42

Reteach

Remind students about the law of conservation of mass and the concept of using chemical equations to represent chemical reactions. List the information needed in order to write a chemical equation, and then write the equation $2Li + H_2O \rightarrow LiOH + H_2$ on the board. Ask students to identify the key points of the equation (reactants on the left and products on right) and how many atoms are on each side of the equation. Then have students balance the equation by using coefficients to show that the number of atoms of each element is the same on both sides of the equation. $2Li + 2H_2O \rightarrow 2LiOH + H_2$

Technology Resources

Have students visit **ClassZone.com** for reteaching of Key Concepts.

 CONTENT REVIEW

CONTENT REVIEW CD-ROM

The decomposition of sodium azide is used to inflate air bags in automobiles.

Using the Conservation of Mass

A balanced chemical equation shows that no matter how atoms are rearranged during a chemical reaction, the same number of atoms must be present before and after the reaction. The following example demonstrates the usefulness of chemical equations and the conservation of mass.

The decomposition of sodium azide (NaN_3) is used to inflate automobile air bags. Sodium azide is a solid, and the amount of sodium azide needed in an air bag fills only a small amount of space. In fact, the amount of sodium azide used in air bags is only about 130 grams—an amount that would fit in a large spoon. An inflated air bag, though, takes up much more space even though it contains the same number of atoms that entered the reaction. The reason is illustrated by the chemical equation for this reaction.

Balanced Equation

$$2NaN_3 \longrightarrow 2Na + 3N_2$$

According to the balanced equation shown above, three molecules of nitrogen gas are formed for every two molecules of sodium azide that decompose. Because the nitrogen is a gas, it fills a much greater volume than the original sodium azide. In fact, 67 liters of nitrogen gas are produced by the 130 grams of sodium azide in the reaction. This amount of nitrogen is enough to quickly inflate the air bag during a collision—the decomposition of sodium azide to sodium and nitrogen takes 0.03 seconds.

CHECK YOUR READING Why must chemical equations be balanced?

3.2 Review

KEY CONCEPTS

1. State the law of conservation of mass.
2. Write the chemical equation that shows sodium (Na) and chlorine (Cl_2) combining to form table salt (NaCl).
3. Is the following equation balanced? Why or why not?
   $$CO \longrightarrow C + O_2$$

CRITICAL THINKING

4. **Communicate** Describe Lavoisier's experiment with mercury. How does this experiment show the law of conservation of mass?
5. **Synthesize** Suppose a log's mass is 5 kg. After burning, the mass of the ash is 1 kg. Explain what may have happened to the other 4 kg of mass.

CHALLENGE

6. **Synthesize** Suppose a container holds 1000 hydrogen molecules (H_2) and 1000 oxygen molecules (O_2) that react to form water. How many water molecules will be in the container? Will anything else be in the container? If so, what?

B 84 Unit: **Chemical Interactions**

ANSWERS

1. During a chemical reaction, matter is neither created nor destroyed; the mass of reactants always equals the mass of the products.

2. $2Na + Cl_2 \rightarrow 2NaCl$

3. No; one oxygen atom is on the left side of the arrow, and two are on the right side.

4. Mass in the system as a whole did not increase. This indicates that the mass of the product had to come from within the system,

presumably from the air within the jar.

5. It was released into the air as a gas because it was not collected during the reaction.

6. 1000 water molecules; yes, 500 O_2 molecules

Chemistry in Firefighting

A firefighter's job may seem simple: to put out fires. However, a firefighter needs to know about chemicals and chemical reactions. A fire is a combustion reaction that requires oxygen as a reactant. Without oxygen, a fire will normally burn itself out, so firefighters try to prevent oxygen from reaching the burning substances. Firefighters often use water or carbon dioxide for this purpose, but these materials make some types of fires more dangerous.

Grease Fires

Some fires can be extinguished by a chemical reaction. In kitchen grease fires, the chemicals that are used to fight the fire react with the grease. The reaction produces a foam that puts out the fire.

Metal Fires

Some fires involve metals such as magnesium. This metal burns at a very high temperature and reacts violently with water. Firefighters try to smother metal fires with a material such as sand.

The fire shown above is a magnesium fire in Chicago in 1998. Firefighters used water to protect surrounding buildings, but dumped road salt on the burning magnesium.

Hazardous Reactions

Chemicals may react with water to form poisonous gases or acids. Firefighters might use a foam that extinguishes the fire, cools the area around the fire, and traps gases released by the fire. The symbols shown on the left are among several that show firefighters what chemical dangers may be present.

EXPLORE

Build a carbon dioxide fire extinguisher.

1. Put 3 tsp of baking soda on a tissue and roll it into a tube. Tie the ends and middle of the tube with thread. Leave extra thread at one end of the tube.
2. Mold clay tightly around the straw.
3. Pour some vinegar into a bottle.
4. Hold the thread to suspend the tissue tube above the vinegar. Place the straw inside the bottle. Use the clay molded around the straw to hold the thread in place. Be sure that the straw is not touching the vinegar.
5. Shake and observe the fire extinguisher.

Chapter 3: **Chemical Reactions** 85 **B**

EXPLORE

To avoid submerging the bottom of the straw in the vinegar, students should measure the distance between the top of the bottle and the vinegar before figuring out where to put the clay on the straw. After they create the fire extinguisher, make sure they understand that when they shake it up, the baking soda and vinegar combine to make carbon dioxide, which will smother a fire (i.e., keep it from getting oxygen).

Set Learning Goal

To understand why firefighters need a knowledge of chemistry

Present the Science

Remind students that there are three elements to almost every fire: heat, a fuel source, and oxygen. Extinguishing a fire requires the removal of at least one of these elements. Water is useful for extinguishing fires in materials such as paper and wood, but can make other fires more dangerous.

GREASE FIRES Dry chemical extinguishers contain bases such as sodium bicarbonate (baking soda) and ammonium phosphate that react with fatty acids in grease. The resulting foam cuts off the fire's oxygen supply. Water may cause a grease fire to spread rather than extinguish it.

METAL FIRES The fire's oxygen supply must be cut off. Usually it will be smothered with a nonflammable material. When water is added directly on burning magnesium, massive fireballs can be produced.

HAZARDOUS MATERIALS Discuss the word *corrosive*. Water reacts with many chemicals, including cyanide salts. These chemicals contain a metal and a cyanide (CN^-) group. When most cyanide salts combine with water, deadly hydrogen cyanide (HCN) is produced.

Discussion Question

Ask: Why do most home fire extinguishers use dry chemicals instead of water? *Water can cause grease or chemical fires to spread, and water should never be used on an electrical fire. Dry chemicals are a safer choice.*

Close

Ask: Why do firefighters need a knowledge of chemistry? *to understand which materials will put out different types of fires; to avoid making a fire worse*

▶ Set Learning Goals

Students will

• Describe how energy changes in a chemical reaction.

• Explain how some chemical reactions release energy.

• Explain how some chemical reactions absorb energy.

○ 3-Minute Warm-Up

Display Transparency 21 or copy this exercise on the board:

Balance the equation.

$Al + Br_2 \rightarrow AlBr_3$ $2Al + 3Br_2 \rightarrow 2AlBr_3$

How can you tell if an equation is balanced? *An equation is balanced if the number of atoms of each element is the same on both sides of the equation.*

How does the law of conservation of mass relate to balanced equations? *The total mass of reactants and products must be the same, so the number of atoms of each element must be the same for both reactants and products.*

 3-Minute Warm-Up, p. T21

3.3 MOTIVATE

EXPLORE Energy Changes

PURPOSE To observe the energy changes in an endothermic chemical process

TIPS *10 min.* Make sure that students are measuring the temperature of the water as they are adding the Epsom salts.

WHAT DO YOU THINK? *The temperature decreased. Energy was absorbed by the process.*

Ongoing Assessment

Describe how energy changes in a chemical reaction.

Ask: Does breaking a chemical bond require energy or release energy? *requires energy*

KEY CONCEPT

3.3 Chemical reactions involve energy changes.

◀ **BEFORE, you learned**

• Bonds are broken and made during chemical reactions

• Mass is conserved in all chemical reactions

• Chemical reactions are represented by balanced chemical equations

▶ **NOW, you will learn**

• About the energy in chemical bonds between atoms

• Why some chemical reactions release energy

• Why some chemical reactions absorb energy

VOCABULARY

bond energy p. 86
exothermic reaction p. 87
endothermic reaction p. 87
photosynthesis p. 90

EXPLORE Energy Changes

How can you identify a transfer of energy?

PROCEDURE

① Pour 50 ml of hot tap water into the cup and place the thermometer in the cup.

② Wait 30 seconds, then record the temperature of the water.

③ Measure 5 tsp of Epsom salts. Add the Epsom salts to the beaker and immediately record the temperature while stirring the contents of the cup.

④ Continue to record the temperature every 30 seconds for 2 minutes.

MATERIALS

• graduated cylinder
• hot tap water
• plastic cup
• thermometer
• stopwatch
• plastic spoon
• Epsom salts

WHAT DO YOU THINK?

• What happened to the temperature after you added the Epsom salts?

• What do you think caused this change to occur?

COMBINATION NOTES
Use combination notes to organize information on how chemical reactions absorb or release energy.

Chemical reactions release or absorb energy.

Chemical reactions involve breaking bonds in reactants and forming new bonds in products. Breaking bonds requires energy, and forming bonds releases energy. The energy associated with bonds is called **bond energy.** What happens to this energy during a chemical reaction?

Chemists have determined the bond energy for bonds between atoms. Breaking a bond between carbon and hydrogen requires a certain amount of energy. This amount of energy is different from the amount of energy needed to break a bond between carbon and oxygen, or between hydrogen and oxygen.

B **86** Unit: **Chemical Interactions**

RESOURCES FOR DIFFERENTIATED INSTRUCTION

Below Level

UNIT RESOURCE BOOK
• Reading Study Guide A, pp. 154–155
• Decoding Support, p. 177

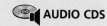 **AUDIO CDS**

Advanced

UNIT RESOURCE BOOK
• Challenge and Extension, p. 160
• Challenge Reading, pp. 173–174

English Learners

UNIT RESOURCE BOOK
Spanish Reading Study Guide, pp. 158–159

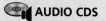 **AUDIO CDS**

• Audio Readings in Spanish
• Audio Readings (English)

Energy is needed to break bonds in reactant molecules. Energy is released when bonds are formed in product molecules. By adding up the bond energies in the reactants and products, you can determine whether energy will be released or absorbed.

If more energy is released when the products form than is needed to break the bonds in the reactants, then energy is released during the reaction. A reaction in which energy is released is called an **exothermic reaction.**

If more energy is required to break the bonds in the reactants than is released when the products form, then energy must be added to the reaction. That is, the reaction absorbs energy. A reaction in which energy is absorbed is called an **endothermic reaction.**

These types of energy changes can also be observed in different physical changes such as dissolving or changing state. The state change from a liquid to a solid, or freezing, releases energy—this is an exothermic process. The state change from a solid to a liquid, or melting, absorbs energy—this is an endothermic process.

 How are exothermic and endothermic reactions different?

Exothermic reactions release energy.

Exothermic chemical reactions often produce an increase in temperature. In exothermic reactions, the bond energies of the reactants are less than the bond energies of the products. As a result, less energy is needed to break the bonds in the reactants than is released during the formation of the products. This energy difference between reactants and products is often released as heat. The release of heat causes a change in the temperature of the reaction mixture.

Even though energy is released by exothermic reactions, some energy must first be added to break bonds in the reactants. In exothermic reactions, the formation of bonds in the products releases more energy. Overall, more energy is released than is added.

Some reactions are highly exothermic. These reactions produce a great deal of heat and significantly raise the temperature of their surroundings. One example is the reaction of powdered aluminum metal with a type of iron oxide, a reaction known as the thermite reaction. The equation for this reaction is

$$2Al + Fe_2O_3 \longrightarrow Al_2O_3 + 2Fe$$

This reaction releases enough heat to melt the iron that is produced. In fact, this reaction is used to weld iron rails together.

 What is evidence for an exothermic chemical reaction?

The white clouds of water vapor are formed by the exothermic reaction between hydrogen and oxygen.

$$2H_2 + O_2 \longrightarrow 2H_2O$$

The thermite reaction releases enough heat to weld pieces of iron together.

DIFFERENTIATE INSTRUCTION

 More Reading Support

A What is a chemical reaction that releases energy? *exothermic reaction*

B What happens during an endothermic reaction? *Energy is absorbed.*

English Learners To help English learners remember the definitions of *exothermic reaction* and *endothermic reaction*, explain the prefixes *endo-* and *exo-*. The prefix *endo-* means "inside" or "within." An endothermic reaction takes in, or absorbs, energy. The prefix *exo-* means "outside" or "external." An exothermic reaction gives off, or releases, energy.

3.3 INSTRUCT

Address Misconceptions

IDENTIFY Ask: Where does the energy that is released by an exothermic reaction come from? If students say it comes from breaking bonds, they may hold the misconception that breaking bonds releases energy.

CORRECT Show students the energy diagram on p. 88. Point out the yellow energy bursts that indicate energy must be added to the reactants before the reaction can take place, and that a larger amount of energy is released when new bonds form.

REASSESS Ask students to describe where energy is absorbed and released in an exothermic reaction. *Energy is always absorbed to break bonds and start a reaction. An exothermic reaction releases energy because, when new bonds form, more energy is released than is required to break the bonds in the reactants.*

Technology Resources

Visit **ClassZone.com** for background on common student misconceptions.

 MISCONCEPTION DATABASE

Teach Difficult Concepts

Explain the difference between temperature and heat. Temperature is a measure of the average amount of kinetic energy of the particles within a substance. Heat is the transfer of energy from an object of higher temperature to an object of lower temperature. Ask: What decreases an object's temperature? *a transfer of energy through heat*

Ongoing Assessment

CHECK YOUR READING *Answer: Exothermic reactions release energy, and endothermic reactions absorb energy.*

CHECK YOUR READING *Answer: change in temperature caused by a release of energy*

To help students interpret the diagrams and graph of energy changes in an exothermic reaction, ask:

• Is more energy added or released during this reaction? *released*

• What does the graph show about reactants and products in an exothermic reaction? *Reactants have a lower bond energy than products, so more energy is released by bond formation than is absorbed to break bonds.*

Real World Example

The production of light by living organisms occurs mainly among marine animals and is the main source of light in the deep ocean. Almost all light produced by marine animals is blue light, because it is visible at the greatest distance under water, and most animals are sensitive only to blue light. The black dragonfish, also called loosejaws, is an exception because it produces red light.

Ongoing Assessment

Identify some chemical reactions that release energy.

Ask: What two forms of energy can be released in exothermic reactions? Identify reactions that release these energy forms. *Energy can be released as heat (all common combustion reactions) or light (the reaction between oxygen and luciferin).*

READING VISUALS *Answer: The size of the yellow energy bursts and the height of the bars in the bar graph indicate that more energy is released than is absorbed.*

CHECK YOUR READING *Answer: as heat or light*

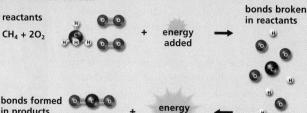

Exothermic Reactions

The products have greater bond energies than the reactants.

Methane Combustion

reactants

$CH_4 + 2O_2$ + energy added → bonds broken in reactants

bonds formed in products

$CO_2 + 2H_2O$ + energy released ←

Difference in Energy

Bond Energy

Reactants (energy added) Products (energy released)

More energy is released than added.

READING VISUALS What information in the diagram shows that methane combustion is exothermic?

C

All common combustion reactions, such as the combustion of methane, are exothermic. To determine how energy changes in this reaction, the bond energies in the reactants—oxygen and methane—and in the products—carbon dioxide and water—can be added and compared. This process is illustrated by the diagram shown above. The difference in energy is released to the surrounding air as heat.

Some chemical reactions release excess energy as light instead of heat. For example, glow sticks work by a chemical reaction that releases energy as light. One of the reactants, a solution of hydrogen peroxide, is contained in a thin glass tube within the plastic stick. The rest of the stick is filled with a second chemical and a brightly colored dye. When you bend the stick, the glass tube inside it breaks and the two solutions mix. The result is a bright glow of light.

These jellyfish glow because of exothermic chemical reactions.

Exothermic chemical reactions also occur in living things. Some of these reactions release energy as heat, and others release energy as light. Fireflies light up due to a reaction that takes place between oxygen and a chemical called luciferin. This type of exothermic reaction is not unique to fireflies. In fact, similar reactions are found in several different species of fish, squid, jellyfish, and shrimp.

CHECK YOUR READING In which ways might an exothermic reaction release energy?

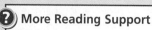
DIFFERENTIATE INSTRUCTION

? More Reading Support

C Are combustion reactions exothermic or endothermic? *exothermic*

Below Level Review the use of bar graphs. Have students identify the x- and y-axes on the bar graph on this page. Ask them to explain what the bar graph is telling them.

Advanced Have students who are interested in learning more about how glow sticks work read the following article:

R • Challenge Reading, pp. 173–174
• Challenge and Extension, p. 160

The bombardier beetle, shown in the photograph on the right, uses natural exothermic reactions to defend itself. Although several chemical reactions are involved, the end result is the production of a hot, toxic spray. The most important reaction in the process is the decomposition of hydrogen peroxide into water and oxygen.

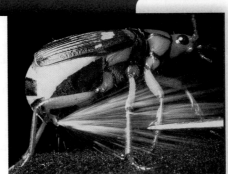

$$2H_2O_2 \longrightarrow 2H_2O + O_2$$

When the hydrogen peroxide rapidly breaks down, the hot, toxic mixture made by the series of reactions is pressurized by the oxygen gas from the reaction in the equation above. After enough pressure builds up, the beetle can spray the mixture.

Endothermic reactions absorb energy.

Endothermic reactions often produce a decrease in temperature. In endothermic reactions, the bond energies of the reactants are greater than the bond energies of the products. As a result, more energy is needed to break the bonds in the reactants than is released during the formation of the products. The difference in energy is usually absorbed from the surroundings as heat. This often causes a decrease in the temperature of the reaction mixture.

All endothermic reactions absorb energy. However, they do not all absorb energy as heat. One example of an endothermic reaction of this type is the decomposition of water by electrolysis. In this case, the energy that is absorbed is in the form of electrical energy. When the electric current is turned off, the reaction stops. The change in energy that occurs in this reaction is shown below.

READING **TiP**

The prefix *endo-* means "inside."

Endothermic Reactions

The products have lower bond energies than the reactants.

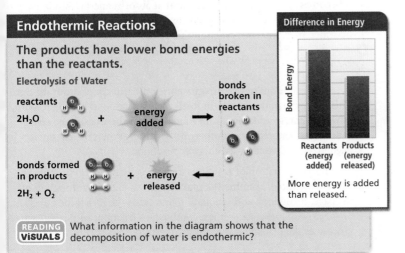

Electrolysis of Water

reactants
2H₂O

+ energy added

bonds broken in reactants

bonds formed in products
2H₂ + O₂

+ energy released

Difference in Energy

Bond Energy

Reactants (energy added)   Products (energy released)

More energy is added than released.

READING **ViSUALS**   What information in the diagram shows that the decomposition of water is endothermic?

Chapter 3: Chemical Reactions **89**   B

Teach from Visuals

To help students interpret the diagrams and graph of energy changes in an endothermic reaction, ask:

• Is more energy added or released during this reaction? *added*

• What does the graph show about reactants and products in an endothermic reaction? *Products have a lower bond energy than reactants, so more energy is absorbed to break bonds than is released by the formation of bonds.*

Integrating the Sciences

The bombardier beetle has a unique method of defending itself from enemies. To prevent toxic chemicals from forming in its body, the beetle stores the reactants in separate compartments. The reactants mix in a reaction chamber when a predator approaches, and the beetle explosively sprays a toxic chemical from a gland at the tip of its abdomen. The gland can revolve, directing the spray in all directions. Bombardier beetles can hit several attackers with a single batch of spray.

Develop Critical Thinking

COMPARE AND CONTRAST Have students compare and contrast exothermic and endothermic reactions by making a chart that lists the characteristics of each. The chart should also present at least one example of each type of reaction.

Ongoing Assessment

READING **ViSUALS**   *Answer: The size of the yellow energy bursts and the height of the bars in the bar graph indicate that more energy is absorbed than is released.*

DIFFERENTIATE INSTRUCTION

More Reading Support

D In endothermic reactions, which has greater bond energies, the reactants or the products? *the reactants*

Alternative Assessment Have students write a paragraph summarizing the diagram on this page. They should describe the reaction that takes place, the energy changes, and the bond energies of the reactants and products.

History of Science

Many scientists were involved in working out how photosynthesis uses light. In 1905, the English plant physiologist F. F. Blackman showed that photosynthesis is a two-step process and that only one of the steps involves light. The role of light was clarified by Cornelis van Niel when he was a graduate student at Stanford University in the 1930s. Ultimately, photosynthesis takes place in the chloroplasts of leaves and chlorophyll absorbs light during the first step, called the light reaction.

Teacher Demo

Purchase or grow two small plants of the same species. Water both plants well. Place one plant in a window or under a bright light. Place the other plant under a foil-wrapped box or in a dark incubator. Have students examine the plants after a week. Ask: Do the results suggest that photosynthesis is an endothermic or exothermic reaction? Why? *Endothermic; the plant exposed to sunlight grew because it absorbed energy.*

Language Arts Connection

Tell students that the word *exothermic* comes from the Greek root words *exo* and *therm,* meaning "heat out." The word *endothermic* comes from the Greek words meaning "heat in." Have students find other *exo* and *endo* words and use them in sentences.

Ongoing Assessment

Identify some chemical reactions that absorb energy.

Ask: What endothermic reaction absorbs light energy rather than heat? *photosynthesis*

 Answer: *The temperature of the reaction mixture decreases.*

 Answer: *It becomes an endothermic reaction that absorbs energy.*

 E

Probably the most important series of endothermic reactions on Earth is photosynthesis. Many steps occur in the process, but the overall chemical reaction is

$$6CO_2 + 6H_2O \longrightarrow C_6H_{12}O_6 + 6O_2$$

Unlike many other endothermic reactions, photosynthesis does not absorb energy as heat. Instead, during **photosynthesis,** plants absorb energy from sunlight to turn carbon dioxide and water into oxygen and glucose, which is a type of sugar molecule. The energy is stored in the glucose molecules, ready to be used when needed.

CHECK YOUR READING How can you determine if a reaction is endothermic?

Exothermic and endothermic reactions work together to supply energy.

When thinking about exothermic and endothermic reactions, it is often useful to consider energy as part of the reaction. An exothermic reaction releases energy, so energy is on the product side of the chemical equation. An endothermic reaction absorbs energy, so energy is on the reactant side of the chemical equation.

VISUALIZATION
CLASSZONE.COM
View examples of endothermic and exothermic reactions.

Exothermic Reaction
Reactants ➞ Products + Energy

Endothermic Reaction
Reactants + Energy ➞ Products

As you can see in the general reactions above, exothermic and endothermic reactions have opposite energy changes. This means that if an exothermic chemical reaction proceeds in the opposite direction, it becomes an endothermic reaction that absorbs energy. Similarly, if an endothermic reaction proceeds in the opposite direction, it becomes an exothermic reaction that releases energy.

CHECK YOUR READING What happens when an exothermic reaction is reversed?

 F

A large amount of the energy we use on Earth comes from the Sun. This energy includes energy in fossil fuels such as coal and petroleum, as well as energy obtained from food. In all of these cases, the energy in sunlight is stored by endothermic reactions. When the energy is needed, it is released by exothermic reactions.

This combination of reactions forms a cycle of energy storage and use. For example, examine the photosynthesis equation at the top of the page. If you look at this equation in reverse—that is, if the direction of the arrow is reversed—it is a combustion reaction, with oxygen and glucose as the reactants, and it is exothermic.

DIFFERENTIATE INSTRUCTION

 More Reading Support

E What is the most important endothermic reaction in nature? *photosynthesis*

F Where does much of the energy on Earth come from? *the Sun*

Advanced Have students list additional examples of exothermic and endothermic reactions and describe what happens when these reactions are reversed, as in the photosynthesis equation on this page. Ask students for examples of other cycles of energy storage and use.

Plants store energy through the endothermic reactions of photosynthesis. Living things can release this energy through a series of exothermic reactions that will be described in the next section.

The energy stored in plants through photosynthesis can also be released in other ways. Consider energy from fossil fuels. Fossil fuels include petroleum, natural gas, and coal. These substances formed from fossilized materials, mainly plants, that had been under high pressures and temperatures for millions of years. When these plants were alive, they used photosynthesis to produce glucose and other molecules from carbon dioxide and water.

The energy stored in the bonds of these molecules remains, even though the molecules have changed over time. The burning of gasoline in a car releases this energy, enabling the car's engine to work. Similarly, the burning of coal in a power plant, or the burning of natural gas in a stove, releases the energy originally stored by the endothermic series of photosynthesis reactions.

Plants such as trees store energy through photosynthesis. Cars and trucks release this energy through combustion.

CHECK YOUR READING How can endothermic and exothermic reactions work together?

3.3 Review

KEY CONCEPTS

1. What are the differences between exothermic and endothermic reactions?

2. Is the combustion of methane an exothermic or endothermic reaction? Explain.

3. Is photosynthesis an exothermic or endothermic reaction? Explain.

CRITICAL THINKING

4. **Synthesize** Describe the connections between the processes of photosynthesis and combustion.

5. **Communicate** Explain how most energy used on Earth can be traced back to the Sun.

CHALLENGE

6. **Synthesize** Electrolysis of water is endothermic. What does this indicate about the bond energy in the reactants and products? What happens when this reaction is reversed?

Chapter 3: Chemical Reactions 91 **B**

ANSWERS

1. Exothermic reactions: products have higher bond energy than reactants and energy is released; endothermic reactions: reactants have higher bond energy than products and energy is absorbed.

2. Exothermic; energy is released as heat and light.

3. Endothermic; energy from light is absorbed in order for the reactions to occur.

4. They are approximately opposite reactions in terms of energy as well as reactants and products.

5. Sample answer: Fossil fuels contain the carbon stored in plants by photosynthesis hundreds of millions of years ago. So, energy in fossil fuels started as energy from the Sun.

6. Bond energy is greater in the reactants; when the reaction is reversed, the bond energy is greater in the products and energy is released.

Ongoing Assessment

CHECK YOUR READING Answer: Energy can be stored in molecules as a result of an endothermic reaction and then released through exothermic reactions.

Reinforce (the BIG idea)

Have students relate the section to the Big Idea.

 Reinforcing Key Concepts, p. 161

3.3 ASSESS & RETEACH

Assess

 Section 3.3 Quiz, p. 43

Reteach

Write the definitions for *endothermic reaction*, *exothermic reaction*, and *photosynthesis* on the board. Ask students whether energy is released or absorbed in each process. Then have students think about the process of baking a cake. Ask:

• Is the process of baking a cake endothermic or exothermic? *endothermic*

• How do you know? *The cake batter will not bake unless it absorbs energy.*

• Where does the energy needed to bake the cake come from? *heat from the oven*

Technology Resources

Have students visit **ClassZone.com** for reteaching of Key Concepts.

 CONTENT REVIEW

 CONTENT REVIEW CD-ROM

Chapter 3 **91** **B**

CHAPTER INVESTIGATION

Focus

PURPOSE To observe energy changes that result from chemical reactions

OVERVIEW Students will measure temperature changes during one exothermic reaction and one endothermic reaction. Students will find the following:

• Adding yeast to hydrogen peroxide produces an increase in temperature due to an exothermic process.

• The reaction between vinegar and baking soda is endothermic, with a decrease in temperature.

Lab Preparation

• Have students bring in stopwatches if you do not have enough for the entire class. A wristwatch with a second hand can substitute for a stopwatch.

• Have students read through the investigation and prepare their data tables. Or you may wish to copy and distribute datasheets and rubrics.

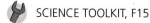 UNIT RESOURCE BOOK, pp. 180–188

SCIENCE TOOLKIT, F15

Lab Management

• Make sure students include necessary time intervals when making data tables.

• Review with students how to use a balance and a graduated cylinder. Remind them to read the bottom of the meniscus in the cylinder.

• Students should measure the yeast on the balance during the 2-minute waiting period. Students should measure the baking soda on the balance while the vinegar is warming.

INCLUSION Ask students if they can observe temperature changes by feeling the beakers.

Teaching with Technology

Temperature probes can be used to measure and record temperatures. Graphing calculators can be used to graph students' data.

Exothermic or Endothermic?

OVERVIEW AND PURPOSE A clue that a chemical reaction has taken place is a transfer of energy, often in the form of heat or light. The chemical reaction used to demolish an old building, as shown in the photograph to the left, is a dramatic example of energy release by a reaction. In this investigation, you will use what you have learned about chemical reactions to

• measure and record temperature changes in two processes
• compare temperature changes during the processes in order to classify them as exothermic or endothermic

▶ Procedure

1. Make a data table like the one shown on the sample notebook page.

2. Work with a partner. One should keep track of time. The other should observe the thermometer and report the temperature.

MATERIALS
• graduated cylinder
• hydrogen peroxide
• 2 beakers
• 2 thermometers
• stopwatch
• measuring spoons
• yeast
• balance
• plastic spoon
• large plastic cup
• hot tap water
• vinegar
• baking soda

PART 1

3. Pour 30 mL of hydrogen peroxide into a beaker. Put a thermometer into the beaker. Wait 2 minutes to allow the thermometer to reach the temperature of the hydrogen peroxide. During the time you are waiting, measure 1 g of yeast with the balance.

4. Record the starting temperature. Add the yeast to the beaker and immediately record the temperature while gently stirring the contents of the beaker. Continue to record the temperature every 30 seconds as you observe the process for 5 minutes.

step 4

INVESTIGATION RESOURCES

CHAPTER INVESTIGATION, Exothermic or Endothermic?
• Level A, pp. 180–183
• Level B, pp. 184–187
• Level C, p. 188

Advanced students should complete Levels B & C.

Writing a Lab Report, pp. D12–13

Technology Resources

Customize this student lab as needed or look for an alternative. Print rubrics to assess student lab reports.

 Lab Generator CD-ROM

PART 2

5. Make a hot water bath by filling a large plastic cup halfway with hot tap water.

6. Measure and pour 30 mL of vinegar into a small beaker. Set this beaker in the hot water bath and place a thermometer in the vinegar. Wait until the temperature of the vinegar rises to between 32 and 38°C (90 to 100°F). While waiting for the vinegar's temperature to increase, measure 1 g of baking soda.

step 6

7. Remove the beaker from the hot water bath. Record the starting temperature.

8. Add the baking soda to the vinegar and immediately record the temperature as you swirl the contents of the beaker. Continue to record the temperature every 30 seconds as you observe the reaction for 5 minutes.

▶ Observe and Analyze   Write It Up

1. **RECORD OBSERVATIONS** Remember to complete your data table.

2. **GRAPH** Use the information from your data table to graph your results. Make a double-line graph, plotting your data in a different color for each part of the investigation. Plot temperature in degrees Celsius on the vertical, or *y*-axis. Plot the time in minutes on the horizontal, or *x*-axis.

3. **ANALYZE DATA** Examine the graph. When did the temperature change the most in each part of the investigation? When did it change the least? Compare the temperature at the start of each process with the temperature after 5 minutes. How do the temperature changes compare?

▶ Conclude   Write It Up

1. **CLASSIFY** Is the mixture of hydrogen peroxide and yeast endothermic or exothermic? Is the reaction between vinegar and baking soda endothermic or exothermic? Provide evidence for your answers.

2. **EVALUATE** Did you have any difficulties obtaining accurate measurements? Describe possible limitations or sources of error.

3. **APPLY** What does the reaction between baking soda and vinegar tell you about their bond energies?

▶ INVESTIGATE Further

CHALLENGE Repeat Part 2, but instead of using the hot water bath, add the hot water directly to the vinegar before pouring in the baking soda. Does this change in procedure change the results of the experiment? Why might your observations have changed? Explain your answers.

Exothermic or Endothermic?
Observe and Analyze
Table 1. Temperature Measurements

Time (min)	Hydrogen Peroxide and Yeast Temperature (°C)	Vinegar and Baking Soda Temperature (°C)
0		
0.5		
1.0		
....		
5.0		

Conclude

▶ Observe and Analyze   Write It Up

1. **SAMPLE DATA** Hydrogen peroxide and yeast: 0 minutes, 21°C; 0.5–2 minutes, 24°C; 2.5–5 minutes, 23°C. Vinegar and baking soda: 0 minutes, 33°C; 0.5–2 minutes, 28°C; 2.5–5 minutes, 27°C.

2. See students' graphs.

3. The reaction involving yeast and hydrogen peroxide showed an immediate increase in temperature. The reaction between vinegar and baking soda showed an immediate decrease in temperature.

▶ Conclude   Write It Up

1. The first reaction is exothermic, as indicated by an increase in temperature; the second reaction is endothermic, as indicated by a decrease in temperature.

2. Answers will vary but could include difficulties with temperature measurements, inconsistent times between measurements, or measurement error during the experimental setup.

3. The bond energies of the reactants are greater than the bond energies of the products.

▶ INVESTIGATE Further

CHALLENGE Answer: Yes; adding water to the reaction mix decreases the concentrations of the reactants, thus slowing the reaction rate. Temperature change should be less after 5 minutes.

Post-Lab Discussion

• Discuss variables in the lab. Ask: What was the independent variable? *time* What was the dependent variable? *temperature*

• Ask: Why was it helpful to heat the vinegar to 32–38°C? *to increase the reaction rate and provide a greater contrast in the change in temperature*

• Ask: Why is it important to record the temperature immediately as the reaction starts? *because the temperature change is relatively large at the beginning of the reaction, but then levels off*

Students will

- Identify the relationship between the reactions of respiration and the reactions of photosynthesis.
- Recognize how chemistry has been used to develop new technology.
- Infer through an experiment how catalysts affect a chemical reaction.

◀ **3-Minute Warm-Up**

Display Transparency 21 or copy this exercise on the board:

Match the definition to the correct term.

Definitions

1. substances formed by a reaction *e*
2. reaction in which energy is released *a*
3. reaction in which energy is absorbed by plants to make glucose and oxygen *c*

Terms

a. exothermic reaction
b. reactants
c. photosynthesis
d. catalysts
e. products

 3-Minute Warm-Up, p. T21

3.4 **MOTIVATE**

THINK ABOUT

PURPOSE To explore examples of chemical reactions in modern life and technology that are adapted from reactions found in nature

DISCUSS Brainstorm possible examples of reactions with students. *Sample answer: reactions involved in the manufacturing of pesticides, fertilizers, medicines, clothing fibers, artificial rubber, dyes*

KEY CONCEPT

3.4 Life and industry depend on chemical reactions.

◀ **BEFORE, you learned**

- Chemical reactions turn reactants into products by rearranging atoms
- Mass is conserved during chemical reactions
- Chemical reactions involve energy changes

▶ **NOW, you will learn**

- About the importance of chemical reactions in living things
- How chemistry has helped the development of new technology

VOCABULARY

respiration p. 94

THINK ABOUT

How is a glow stick like a firefly?

When a firefly glows in the dark, a chemical reaction that emits light is taking place. Similarly, when you activate a glow stick, a chemical reaction that causes the glow stick to emit light occurs. Many reactions in modern life and technology adapt chemical reactions found in nature. Can you think of other examples?

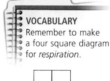 **VOCABULARY**
Remember to make a four square diagram for *respiration*.

Living things require chemical reactions.

In section 3, you saw that photosynthesis stores energy from the Sun in forms that can be used later. These forms of stored energy include fossil fuels and the sugar glucose. The glucose molecules produced by photosynthesis make up the basic food used for energy by almost all living things. For example, animals obtain glucose molecules by eating plants or eating other animals that have eaten plants.

Living cells obtain energy from glucose molecules through the process of **respiration**, which is the "combustion" of glucose to obtain energy. This series of chemical reactions is, in general, the reverse of photosynthesis. It produces carbon dioxide and water from oxygen and glucose. The overall reactions for both photosynthesis and respiration are shown on the top of page 95. From a chemical point of view, respiration is the same as any other combustion reaction.

RESOURCES FOR DIFFERENTIATED INSTRUCTION

Below Level
UNIT RESOURCE BOOK
- Reading Study Guide A, pp. 164–165
- Decoding Support, p. 177

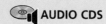 **AUDIO CDS**

Advanced
UNIT RESOURCE BOOK
Challenge and Extension, p. 170

English Learners
UNIT RESOURCE BOOK
Spanish Reading Study Guide, pp. 168–169

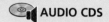 **AUDIO CDS**

- Audio Readings in Spanish
- Audio Readings (English)

Photosynthesis

$$6CO_2 + 6H_2O + energy \longrightarrow C_6H_{12}O_6 + 6O_2$$

Respiration

$$C_6H_{12}O_6 + 6O_2 \longrightarrow 6CO_2 + 6H_2O + energy$$

The energy released by respiration can be used for growth of new cells, movement, or any other life function. Suppose that you are late for school and have to run to get to class on time. Your body needs to activate nerves and muscles right away, without waiting for you to first eat some food as a source of energy. The glucose molecules in food are stored in your body until you need energy. Then, respiration consumes them in a process that includes several steps.

To make these steps go quickly, the body uses catalysts—enzymes— for each step. Some enzymes break the glucose molecules into smaller pieces, while other enzymes break bonds within each piece. Still other enzymes help form the reaction products—carbon dioxide and water. With the help of enzymes, these reactions take place quickly and automatically. You do not have to think about breaking down glucose when you run—you just start to run and the energy is there.

CHECK YOUR READING How are photosynthesis and respiration opposites?

INVESTIGATE Sugar Combustion

How are catalysts important in the combustion of sugar?

PROCEDURE

1. Using the tongs, hold a sugar cube in a candle flame for 30 seconds. Observe what happens.

2. Rub ashes on the second sugar cube.

3. Using the tongs, hold the second sugar cube in the candle flame for 30 seconds. Observe what happens.

WHAT DO YOU THINK?

- What happened to the first sugar cube? What happened to the second sugar cube?

- What may have caused any differences that you observed?

CHALLENGE How might the ashes used in this experiment have a similar function to enzymes in your cells? Explain.

SKILL Inferring

MATERIALS
- candle
- matches
- tongs
- 2 sugar cubes
- ashes

TIME 20 minutes

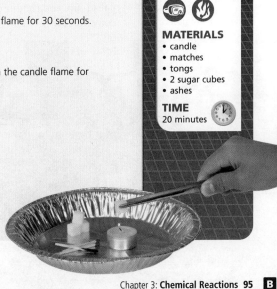

Chapter 3: **Chemical Reactions** 95 **B**

DIFFERENTIATE INSTRUCTION

 More Reading Support

A What series of chemical reactions in the body releases energy that is used for growth and movement? *respiration*

English Learners English learners may not fully understand that respiration is the reverse of photosynthesis. Place the terms and their definitions on the Science Word Wall. Also, point out the chemical equations at the top of this page. The reactants in photosynthesis are the products in respiration. The reactants in respiration are the products in photosynthesis.

Real World Example

Another reaction between chemicals made by humans and natural reactants is the atmospheric reaction between chlorofluorocarbons (CFCs) and ozone (O_3). CFCs, as well as other chemicals, are involved in the decomposition of ozone. A decrease in ozone in the ozone layer high in the atmosphere allows more ultraviolet radiation to reach Earth's surface.

Develop Critical Thinking

COMPARE Ask students to write and compare the chemical equations for the complete and incomplete combustion of methane (CH_4). Tell students that incomplete combustion takes place when insufficient oxygen is present and that it produces carbon monoxide gas (CO) rather than carbon dioxide (CO_2). *complete: $CH_4 + 2O_2 \rightarrow CO_2 + 2H_2O$; incomplete: $2CH_4 + 3O_2 \rightarrow 2CO + 4H_2O$*

Teach Difficult Concepts

The process that occurs inside a catalytic converter takes place in two steps. First, platinum and rhodium are used to change NO and NO_2 into oxygen (O_2) and nitrogen (N_2) by separating the nitrogen and oxygen. Then, platinum and palladium are used in the second step to change CO and unburned hydrocarbons into CO_2 and H_2O.

Ongoing Assessment

CHECK YOUR READING *Answer: to reduce pollution released into the air from vehicles*

Chemical reactions are used in technology.

Every time your cells need energy, they essentially complete respiration—the "combustion" of glucose. The series of chemical reactions in respiration involves enzymes, which are catalysts. Every time someone drives a car, another combustion reaction occurs—the combustion of gasoline. While the combustion of gasoline does not require a catalyst, the chemical reactions that change a car's exhaust gases do use a catalyst.

No chemical reaction is ever completely efficient. It does not matter what the reaction is or how the reaction conditions are set up. There are always some reactants that do not change completely into products. Sometimes a chemical reaction makes unwanted waste products.

In the case of gasoline combustion, some of the original carbon compounds, called hydrocarbons, do not burn completely, and carbon monoxide gas (CO) is produced. Also, nitrogen in the air reacts with oxygen in a car's engine to produce compounds of nitrogen and oxygen, including nitric oxide (NO). The production of these gases lowers the overall efficiency of combustion. More importantly, these gases can react with water vapor in the air to form smog and acid rain.

Sometimes, as you can see with gasoline combustion, chemical technology causes a problem. Then, new chemical technology is designed to treat the problem. For example, it was necessary to reduce carbon monoxide and nitric oxide emissions from car exhaust. As a result, engineers in the 1970s developed a device called a catalytic converter. This device causes chemical reactions that remove the unwanted waste products from the combustion of gasoline.

Many states inspect vehicles to test the pollutants in their exhaust gases.

Catalytic converters contain metal catalysts such as platinum, palladium, and rhodium. The products of the reactions in the catalytic converter are nitrogen (N_2), oxygen (O_2), water (H_2O), and carbon dioxide (CO_2), which are all ordinary parts of Earth's atmosphere.

Even though catalytic converters have been used for many years, scientists and engineers are still trying to improve them. One goal of this research is to use less expensive metals, such as magnesium and zinc, inside catalytic converters, while forming the same exhaust products.

 CHECK YOUR READING Why were catalytic converters developed?

DIFFERENTIATE INSTRUCTION

More Reading Support

B What are waste products of gasoline combustion? *carbon monoxide and nitric oxide*

C What device reduces gasoline's waste products? *catalytic converter*

Alternative Assessment Have students in groups quiz each other on section material. One student can ask a question of the others. The student who correctly answers the question then asks the group another question.

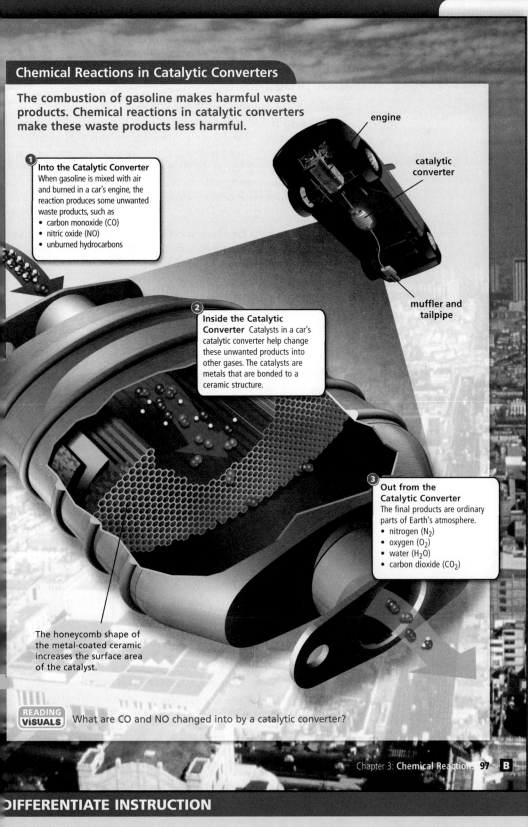

Chemical Reactions in Catalytic Converters

The combustion of gasoline makes harmful waste products. Chemical reactions in catalytic converters make these waste products less harmful.

engine

catalytic converter

muffler and tailpipe

① Into the Catalytic Converter
When gasoline is mixed with air and burned in a car's engine, the reaction produces some unwanted waste products, such as
• carbon monoxide (CO)
• nitric oxide (NO)
• unburned hydrocarbons

② Inside the Catalytic Converter Catalysts in a car's catalytic converter help change these unwanted products into other gases. The catalysts are metals that are bonded to a ceramic structure.

③ Out from the Catalytic Converter
The final products are ordinary parts of Earth's atmosphere.
• nitrogen (N_2)
• oxygen (O_2)
• water (H_2O)
• carbon dioxide (CO_2)

The honeycomb shape of the metal-coated ceramic increases the surface area of the catalyst.

READING VISUALS What are CO and NO changed into by a catalytic converter?

Chapter 3: Chemical Reactions **97** **B**

The development of modern personal computers relied on the invention of the integrated circuit, or microchip. The first silicon chip was invented by Jack St. Claire Kilby at Texas Instruments, and it was produced in September 1958. Computers using chips began to appear in 1962.

Teacher Demo

Discarded silicon chips are usually available from stores that repair computers or other electronic devices. Obtain one, and pass it around the room with a magnifying glass, so that students can examine it.

Develop Number Sense

In order to be used in microchips, silicon must have less than 1 part per billion of impurities. In scientific notation, this value can be expressed as 1×10^{-9}. The larger the absolute value of a negative exponent, the smaller the number. Ask: Is a purity of 1×10^{-11} good enough for microchips? What about a purity of 1×10^{-7}? Why? *Yes; no; 1×10^{-11} is 1 part per hundred billion, which is pure enough, but 1×10^{-7} is 1 part per 10 million, which is not pure enough.*

Ongoing Assessment

Recognize how chemistry has helped develop new technology.

Ask: Why is the element silicon so important to modern technology? *Silicon is a semiconductor. All common electronic devices are based on the electrical properties of semiconductors.*

  Answer: It is a semiconductor.

Industry uses chemical reactions to make useful products.

No area of science and technology has changed today's society as much as the electronics industry has. Just think about all the common electronic products that did not even exist as recently as 30 years ago—from personal computers to CD players to cellular phones. All of these devices are based on the electrical properties of materials called semiconductors. A semiconductor is a material that can precisely control the conduction of electrical signals.

The most common semiconductor material is the element silicon (Si). Silicon is the second most common element in Earth's crust after oxygen, and it is found in most rocks and sand. Pure silicon is obtained from quartz (SiO_2). The quartz is heated with carbon in an electric furnace at 3000°C. The chemical reaction that takes place is

$$SiO_2 + 2C \longrightarrow Si + 2CO$$

This reaction produces silicon that is about 98 percent pure. However, this silicon is still not pure enough to be used in electronics. Several other refining steps must be used to make silicon that is more than 99.999999999 percent pure.

CHECK YOUR READING What property makes silicon useful in electronic devices?

Early electronic devices had to be large enough to fit various types of glass tubes and connecting wires inside. In the 1950s, however, engineers figured out how to replace all of these different tubes and wires with thin layers of material placed on a piece of silicon. The resulting circuits are often called microchips or simply chips.

In order to make these chips, another reaction is used. This reaction involves a material called photoresist (FOH-toh-rih-ZIST), whose properties change when it is exposed to ultraviolet light. Silicon wafers are first coated with photoresist. A stencil is placed over the surface, which allows some areas of the wafer to be exposed to ultraviolet light while other areas are protected. A chemical reaction takes place between the ultraviolet light and the coating of photoresist. The exposed areas of photoresist remain on the silicon surface after the rest of the material is washed away.

The entire process is carried out in special clean rooms to prevent contamination by dust. A typical chip has electrical pathways so small that a single particle of smoke or dust can block the path, stopping the chip from working properly. The process is automated, and no human hand ever touches a chip.

? **D**

? **E**

READING TiP
The prefix *semi-* means "partial," so a semiconductor partially conducts electricity.

Quartz (SiO_2) is the source of silicon for chips.

DIFFERENTIATE INSTRUCTION

? **More Reading Support**

D What property of semiconductors is used in electronic devices? *their electrical properties*

E What is the most common semiconductor material? *silicon*

Advanced Silicon's electrical conductivity can be improved by adding tiny amounts of certain elements, a process called doping. Have students research the process of doping, find out what elements are used, and learn how it improves conductivity of silicon.

R Challenge and Extension, p. 170

From Quartz to Microchips

A chemical reaction makes the tiny circuits that are used to run electronic devices such as cellular phones.

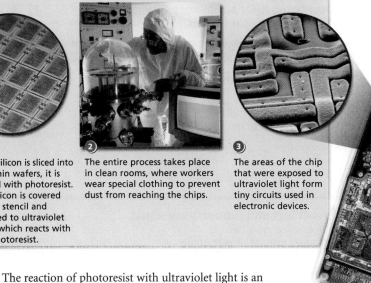

① After silicon is sliced into very thin wafers, it is coated with photoresist. The silicon is covered with a stencil and exposed to ultraviolet light, which reacts with the photoresist.

② The entire process takes place in clean rooms, where workers wear special clothing to prevent dust from reaching the chips.

③ The areas of the chip that were exposed to ultraviolet light form tiny circuits used in electronic devices.

The reaction of photoresist with ultraviolet light is an important chemical reaction. The same type of material is used in the printing of books and newspapers. A similar reaction occurs in photocopiers and laser printers. This is an example of how one type of chemical reaction has helped change industry and society in important ways.

One of the many uses of silicon chips is in cellular phones.

CHECK YOUR READING Describe how chemical reactions are important in industry.

3.4 Review

KEY CONCEPTS

1. Explain how respiration and photosynthesis are chemically opposite from each other.
2. Provide an example of how catalysts are used in technology.
3. Describe two chemical reactions used in making silicon chips.

CRITICAL THINKING

4. **Compare and Contrast** How are respiration and the combustion of gasoline similar? How are they different?
5. **Analyze** In microchip manufacture, what would happen if the clean rooms had outside windows? Explain.

⬤ CHALLENGE

6. **Infer** The gases released from a catalytic converter include N_2, O_2, H_2O, and CO_2. The original reactants must contain atoms of which elements?

Chapter 3: **Chemical Reactions 99** **B**

ANSWERS

1. The reactants of one process are the products of the other process. Respiration is exothermic; photosynthesis is endothermic.

2. Sample answer: Metal catalysts such as platinum and palladium are used in catalytic converters.

3. Silicon dioxide reacts with carbon to purify silicon. Under ultraviolet light, photoresist creates pathways for electronic circuits.

4. Both release energy and require oxygen; both have reactants that depend on photosynthesis. Respiration occurs naturally in living cells and requires catalysts, while gasoline combustion does not.

5. The microchips would likely become contaminated, which would make them unusable.

6. carbon, nitrogen, oxygen, and hydrogen

BACK TO

the **BIG** idea

Have students look at the photograph on pp. 66–67. Ask them to summarize the evidence that a chemical reaction is taking place in the beaker. Ask how this evidence is related to the breaking and forming of bonds. *Color change, gas formation, apparent temperature increase. Bonds are broken in the reactants to form products with a different color and in a different physical state (a gas). Energy is released, and the temperature increases, because more energy is released by the formation of new bonds than was required to break the original bonds.*

◐ KEY CONCEPTS SUMMARY

SECTION 3.1
Ask: What happens when the reactant enters the substance in the beaker? *Bonds of the reactants break. Bonds form to make at least one product, which is visible as a precipitate.*

SECTION 3.2
Ask: How do you know that this equation is balanced? *An equal number of atoms of each element appears on both sides of the equation.*

Ask: How does the equation uphold the law of conservation of mass? *Atoms are neither created nor destroyed.*

SECTION 3.3
Ask: What happens to the chemical energy that is released during an exothermic reaction? *It is converted to other forms of energy, including heat and light.*

SECTION 3.4
Ask: What process in living things releases energy from molecules of glucose? *respiration*

Review Concepts

- Big Idea Flow Chart, p. T17
- Chapter Outline, pp. T23–T24

 # Chapter Review

the **BIG** idea

Chemical reactions form new substances by breaking and making chemical bonds.

 **CONTENT REVIEW**
CLASSZONE.COM

◐ KEY CONCEPTS SUMMARY

3.1 **Chemical reactions alter arrangements of atoms.**
- Chemical changes occur through chemical reactions.
- Evidence of a chemical reaction includes a color change, the formation of a precipitate, the formation of a gas, and a change in temperature.
- Chemical reactions change reactants into products.

VOCABULARY
chemical reaction p. 69
reactant p. 71
product p. 71
precipitate p. 72
catalyst p. 76

3.2 **The masses of reactants and products are equal.**
- Mass is conserved in chemical reactions.
- Chemical equations summarize chemical reactions.
- Balanced chemical equations show the conservation of mass.

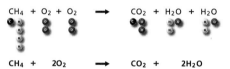

$$CH_4 + O_2 + O_2 \longrightarrow CO_2 + H_2O + H_2O$$

$$CH_4 + 2O_2 \longrightarrow CO_2 + 2H_2O$$

VOCABULARY
law of conservation of mass p. 79
coefficient p. 82

3.3 **Chemical reactions involve energy changes.**
- Different bonds contain different amounts of energy.
- In an exothermic reaction, more energy is released than added.
- In an endothermic reaction, more energy is added than released.

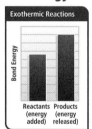
Exothermic Reactions
Bond Energy
Reactants (energy added)   Products (energy released)

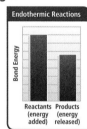
Endothermic Reactions
Bond Energy
Reactants (energy added)   Products (energy released)

VOCABULARY
bond energy p. 86
exothermic reaction p. 87
endothermic reaction p. 87
photosynthesis p. 90

3.4 **Life and industry depend on chemical reactions.**
- Living things rely on chemical reactions that release energy from molecules.
- Different parts of modern society rely on chemical reactions.

VOCABULARY
respiration p. 94

Technology Resources

Have students visit **ClassZone.com** or use the CD-ROM for a cumulative review of concepts.

 CONTENT REVIEW

 CONTENT REVIEW CD-ROM

Engage students in a whole-class interactive review of Key Concepts. Edit content as you wish.

 POWER PRESENTATIONS

Reviewing Vocabulary

Describe how the vocabulary terms in the following pairs are related to each other. Explain the relationship in a one- or two-sentence answer. Underline each vocabulary word or term in your answers.

1. reactant, product

2. conservation of mass, chemical reaction

3. endothermic, exothermic

4. respiration, photosynthesis

Reviewing Key Concepts

Multiple Choice *Choose the letter of the best answer.*

5. During a chemical reaction, reactants always
   a. become more complex
   b. require catalysts
   c. lose mass
   d. form products

6. The splitting of water molecules into hydrogen and oxygen molecules is an example of a
   a. combination reaction
   b. chemical change
   c. synthesis reaction
   d. physical change

7. Combustion reactions
   a. destroy atoms
   b. require glucose
   c. form precipitates
   d. require oxygen

8. Which of the following will increase the rate of a reaction?
   a. breaking the reactants into smaller pieces
   b. removing a catalyst
   c. decreasing the temperature
   d. decreasing the concentration

9. What does a catalyst do in a chemical reaction?
   a. It slows the reaction down.
   b. It speeds the reaction up.
   c. It becomes a product.
   d. It is a reactant.

10. During a chemical reaction, the total amount of mass present
    a. increases
    b. decreases
    c. may increase or decrease
    d. does not change

11. Chemical equations show summaries of
    a. physical changes
    b. changes of state
    c. chemical reactions
    d. changes in temperature

12. A chemical equation must
    a. show energy
    b. be balanced
    c. use subscripts
    d. use coefficients

13. What type of reaction occurs if the reactants have a greater total bond energy than the products?
    a. an endothermic reaction
    b. a synthesis reaction
    c. an exothermic reaction
    d. a decomposition reaction

14. Endothermic reactions always
    a. absorb energy
    b. make more complex products
    c. release energy
    d. make less complex products

Short Answer *Write a short answer to each question.*

15. Describe the differences between physical and chemical changes. How can each be identified?

16. Compare and contrast the overall chemical reactions of photosynthesis and respiration. How can these reactions be described in terms of bond energy in the reactants and products?

17. Describe an example of an advance in technology that makes use of a chemical reaction.

18. When you balance a chemical equation, why can you change coefficients of reactants or products, but not subscripts?

Reviewing Vocabulary

1. The bonds in <u>reactants</u> are broken, and new bonds form to make <u>products</u>.

2. Matter is not created or destroyed during a <u>chemical reaction</u>, so the <u>law of conservation of mass</u> is upheld.

3. <u>Endothermic</u> reactions absorb energy, whereas <u>exothermic</u> reactions release energy.

4. <u>Photosynthesis</u> is an endothermic process. The overall process that occurs during photosynthesis can be considered to be the reverse of the overall process that occurs during respiration, which is exothermic.

Reviewing Key Concepts

5. d	10. d
6. b	11. c
7. d	12. b
8. a	13. a
9. b	14. a

15. A physical change alters the physical characteristics of a substance, but not the substance itself. A chemical change, through a chemical reaction, produces different substances. Chemical changes can be identified by color or temperature change or by formation of a precipitate or a gas.

16. Overall, they are opposite processes. In photosynthesis, the bond energies of the reactants are greater than the bond energies of the products. The process is endothermic. In respiration, the bond energies of the reactants in the overall chemical reaction are less than those of the products of the overall reaction. As a result, the overall process of respiration is exothermic.

17. Answers might cover the reaction that purifies silicon, the reactions in catalytic converters, or the reaction involving photoresist.

18. Changing coefficients changes the amount of the substances. Changing subscripts changes the formula of a substance which makes it a different substance.

ASSESSMENT RESOURCES

UNIT ASSESSMENT BOOK
- Chapter Test A, pp. 45–48
- Chapter Test B, pp. 49–52
- Chapter Test C, pp. 53–56
- Alternative Assessment, pp. 57–58

SPANISH ASSESSMENT BOOK
Spanish Chapter Test, pp. 241–244

Technology Resources

Edit test items and answer choices.

 Test Generator CD-ROM

Visit **ClassZone.com** to extend test practice.

 Test Practice

Thinking Critically

19. *It increased.*

20. *Yes; a gas is still being produced, and the temperature is continuing to change.*

21. *Exothermic; energy is being released, as shown by an increase in temperature.*

22. *The rate of the reaction increased because catalysts increase reaction rate.*

23. *All of it, because catalysts are not consumed during a chemical reaction.*

24. *Bonds are broken in the reactants, and the products that form will be simpler than the reactants.*

Using Math Skills in Science

25. *$2HgO \rightarrow 2Hg + O_2$*

26. *2 to HgO and 2 to Hg*

27. *2*

28. *$4Al + 3O_2 \rightarrow 2Al_2O_3$*

29. *4 to Al, 3 to O_2, and 2 to Al_2O_3*

30. *6*

31. *$S_8 + 12O_2 \rightarrow 8SO_3$*

32. *12 to O_2 and 8 to SO_3*

33. *8*

34. *24*

the BIG idea

35. *Answers will vary but could include life processes and technological applications.*

36. *Answers will vary but should describe how bonds break in reactants and form in products, and how the energy associated with these bonds determines whether energy will be absorbed or released by a chemical reaction.*

UNIT PROJECTS

Collect schedules, materials lists, and questions. Be sure dates and materials are obtainable, and questions are focused.

 Unit Projects, pp. 5–10

Thinking Critically

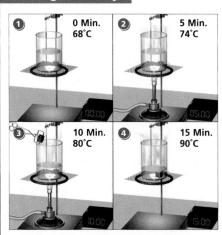

| ① | 0 Min. 68°C | ② | 5 Min. 74°C |
| ③ | 10 Min. 80°C | ④ | 15 Min. 90°C |

The series of illustrations above shows a chemical reaction at five-minute intervals. Use the information in the illustrations to answer the following six questions.

19. OBSERVE What happened to the temperature of the substance in the beaker from the beginning to the end of each five-minute interval?

20. ANALYZE Does the reaction appear to continue in step 4? What evidence tells you?

21. CLASSIFY Is this an endothermic or exothermic reaction? Explain.

22. INFER Suppose the metal cube placed in the beaker in step 3 is a catalyst. What effect did the metal have on the reaction? Why?

23. PREDICT If the metal cube is a catalyst, how much of the metal cube will be left in the beaker when the reaction is completed? Explain.

24. SYNTHESIZE Assume that the reaction shown is a decomposition reaction. Describe what happens to the reactants.

Using Math Skills in Science

Answer the following ten questions based on the equations below.

> Equation 1—$HgO \longrightarrow Hg + O_2$
>
> Equation 2—$Al + O_2 \longrightarrow Al_2O_3$
>
> Equation 3—$S_8 + O_2 \longrightarrow SO_3$

25. Copy and balance equation 1.

26. What coefficients, if any, did you add to equation 1 to balance it?

27. How many Hg atoms take part in the reaction represented by equation 1 when it is balanced?

28. Copy and balance equation 2.

29. What coefficients, if any, did you add to equation 2 to balance it?

30. How many O atoms take part in the reaction represented by equation 2 when it is balanced?

31. Copy and balance equation 3.

32. What coefficients, if any, did you add to equation 3 to balance it?

33. How many S atoms take part in the reaction represented by equation 3 when it is balanced?

34. How many O atoms take part in the reaction represented by equation 3 when it is balanced?

the BIG idea

35. DRAW CONCLUSIONS Describe three ways in which chemical reactions are important in your life.

36. ANALYZE Look back at the photograph and question on pages 66 and 67. Answer the question in terms of the chapter's Big Idea.

UNIT PROJECTS

Check your schedule for your unit project. How are you doing? Be sure that you have placed data or notes from your research in your project folder.

MONITOR AND RETEACH

If students have trouble applying the concept of relative bond energies in reactants and products in items 3, 13, 14, and 16, suggest that they review the diagrams and graphs on pp. 88 and 89.

Students may benefit from summarizing one or more sections of the chapter.

 Summarizing the Chapter, pp. 198–199

Standardized Test Practice

For practice on your state test, go to . . .
TEST PRACTICE
CLASSZONE.COM

Analyzing Theories

Answer the questions based on the information in the following passage.

During the 1700s, scientists thought that matter contained a substance called phlogiston. According to this theory, wood was made of phlogiston and ash. When wood burned, the phlogiston was released and the ash was left behind.

The ash that remained had less mass than the original wood. This decrease in mass was explained by the release of phlogiston. However, when substances such as phosphorus and mercury burned, the material that remained had more mass than the original substances. This increase in mass did not make sense to some scientists.

The scientists who supported the phlogiston theory said that the phlogiston in some substances had negative mass. So, when the substances burned, they released phlogiston and gained mass. Other scientists disagreed, and their research led to the discovery of a scientific law. Antoine Lavoisier carried out several experiments by burning metals in sealed containers. He showed that mass is never lost or gained in a chemical reaction.

1. What did the phlogiston theory successfully explain?
  a. the presence of ash in unburned wood
  b. the apparent gain of mass in some reactions
  c. the chemical makeup of the air
  d. the apparent decrease in mass in some situations

2. Why did some scientists disagree with the phlogiston theory?
  a. Burning a substance always produced an increase in mass.
  b. Burning a substance always produced a decrease in mass.
  c. Burning could produce either an increase or decrease in mass.
  d. Burning wood produced ash and phlogiston.

3. What law did Lavoisier's work establish?
  a. conservation of energy
  b. conservation of mass
  c. conservation of momentum
  d. conservation of resources

4. To carry out his experiments, what kind of equipment did Lavoisier need?
  a. devices to separate the different elements in the air
  b. machines that could separate wood from ash
  c. microscopes that could be used to study rust and ash
  d. balances that could measure mass very accurately

Extended Response

*Answer the following questions in detail.
Include some of the terms from the list on the right.
Underline each term you use in your answers.*

catalyst	coefficient	concentration
temperature	reaction	subscript
surface area		

5. Suppose you wanted to change the rate of a chemical reaction. What might you change in the reaction? Explain each factor.

6. Is the chemical equation shown below balanced? Why or why not? How are balanced chemical equations related to conservation of mass?

$$6CO_2 + 6H_2O \longrightarrow C_6H_{12}O_6 + O_2$$

Chapter 3: **Chemical Reactions** 103 **B**

Analyzing Theories

1. d   *2. c*   *3. b*   *4. d*

Extended Response

5. RUBRIC

4 points for a response that answers the question and uses the following terms accurately:

- temperature
- concentration
- surface area
- catalyst

Sample: Temperature, concentration, or surface area of reactants might be changed. Increases in any of these factors would increase the reaction rate. Decreases in any of these factors would decrease the reaction rate. The addition of a catalyst would also increase reaction rate.

3 points for a response that correctly answers the question and uses three terms correctly
2 points for a response that correctly answers the question and uses two terms correctly
1 point for a response that correctly answers the question and uses one term correctly

6. RUBRIC

4 points for a response that answers the three questions correctly and uses the three following terms accurately:

- subscript
- coefficient
- reaction

The chemical equation is not balanced because oxygen's subscripts on the right side of the equation only give 8 atoms, but the coefficients and subscripts on the left side represent 18 atoms of oxygen. Chemical equations must be balanced because they summarize chemical reactions, which obey the law of conservation of mass.

3 points for a response that answers two of the questions correctly and uses two of the terms correctly
2 points for a response that correctly answers two of the questions
1 point for a response that correctly answers one of the questions

---

METACOGNITIVE ACTIVITY

Have students answer the following questions in their **Science Notebook:**

1. Why do you think it is important for you to learn about chemical reactions?

2. Which topics in this chapter would you like to learn more about?

3. What are the strongest pieces right now in your Unit Project?

⊙ Set Learning Goals

Students will

- Examine how the concept of atomic structure has changed over the years.
- Learn about the tools used to study atoms and subatomic particles.
- Model the discovery of the atomic nucleus.

National Science Education Standards

A.9.a–g Understandings About Scientific Inquiry

E.6.a–c Understandings About Science and Technology

F.5.a–e, F.5.g Science and Technology in Society

G.1.a–b Science as a Human Endeavor

G.2.a Nature of Science

G.3.a–c History of Science

INSTRUCT

Point out that the top half of the timeline depicts scientific progress in describing the atom. The bottom half of the time-line shows technology that has been used to study atoms and their structure and how greater knowledge of atomic structure led to advances in technology.

Teach from Visuals

COLLECTING GASES Discuss the diagram. Tell students that pneumatic troughs are still used today to collect gases from a reaction, so they don't just mix into the air. Ask:

- Where is the gas collected? *over water in a bulb*
- Where does the gas come from? *the fire*

TIMELINES in Science

THE STORY OF
ATOMIC STRUCTURE

About 2500 years ago, certain Greek thinkers proposed that all matter consisted of extremely tiny particles called atoms. The sizes and shapes of different atoms, they reasoned, was what determined the properties of a substance. This early atomic theory, however, was not widely accepted. Many at the time found these tiny, invisible particules difficult to accept.

What everyone could observe was that all substances were liquid, solid, or gas, light or heavy, hot or cold. Everything, they thought, must then be made of only a few basic substances or elements. They reasoned these elements must be water, air, fire, and earth. Different substances contained different amounts of each of these four substances.

The timeline shows a few of the major events that led scientists to accept the idea that matter is made of atoms and agree on the basic structure of atoms. With the revised atomic theory, scientists were able to explain how elements could be basic but different.

1661

Boyle Challenges Concept of the Four Elements
British chemist Robert Boyle proposes that more than four simple substances exist. Boyle also concludes that all matter is made of very tiny particles he calls corpuscles.

EVENTS

| 1600 | 1620 | 1640 | 1660 |

APPLICATIONS AND TECHNOLOGY

TECHNOLOGY

Collecting and Studying Gases
Throughout the 1600s, scientists tried to study gases but had difficulty collecting them. English biologist Stephen Hales designed an apparatus to collect gases. The "pneumatic trough" was a breakthrough in chemistry because it allowed scientists, to collect and study gases for the first time. The pneumatic trough was later used by such chemists as Joseph Black, Henry Cavendish, and Joseph Priestley to study the gases that make up the air we breathe. The work of these scientists showed that air was made of more than a single gas.

DIFFERENTIATE INSTRUCTION

Below Level Point out the structure of the timeline. Tell students that events get closer to the present as you read the line from left to right. Discuss how one discovery often leads to another, farther to the right. Ask: How did electricity, produced by Volta in 1800, lead to Davy's isolation of elements? *Electricity gave investigators a new tool to use to try to break down substances.*

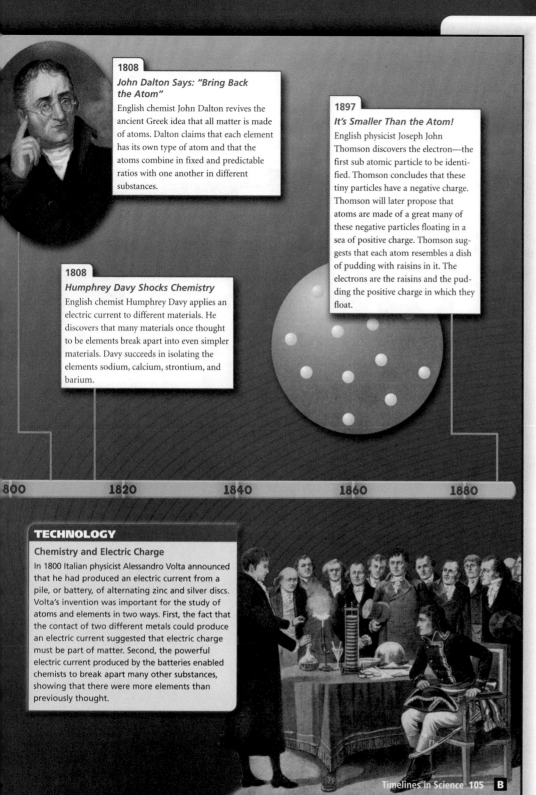

1808

John Dalton Says: "Bring Back the Atom"

English chemist John Dalton revives the ancient Greek idea that all matter is made of atoms. Dalton claims that each element has its own type of atom and that the atoms combine in fixed and predictable ratios with one another in different substances.

1897

It's Smaller Than the Atom!

English physicist Joseph John Thomson discovers the electron—the first sub atomic particle to be identified. Thomson concludes that these tiny particles have a negative charge. Thomson will later propose that atoms are made of a great many of these negative particles floating in a sea of positive charge. Thomson suggests that each atom resembles a dish of pudding with raisins in it. The electrons are the raisins and the pudding the positive charge in which they float.

1808

Humphrey Davy Shocks Chemistry

English chemist Humphrey Davy applies an electric current to different materials. He discovers that many materials once thought to be elements break apart into even simpler materials. Davy succeeds in isolating the elements sodium, calcium, strontium, and barium.

1800   1820   1840   1860   1880

TECHNOLOGY

Chemistry and Electric Charge

In 1800 Italian physicist Alessandro Volta announced that he had produced an electric current from a pile, or battery, of alternating zinc and silver discs. Volta's invention was important for the study of atoms and elements in two ways. First, the fact that the contact of two different metals could produce an electric current suggested that electric charge must be part of matter. Second, the powerful electric current produced by the batteries enabled chemists to break apart many other substances, showing that there were more elements than previously thought.

Timelines in Science 105 **B**

Scientific Process

Ask: What old theory did Thomson's atomic model prove false? *Atoms are the smallest form of matter.*

Technology

1815 Davy and other scientists were using batteries to produce electricity for experiments, but otherwise batteries weren't used by most people. Ask: Why do you think ordinary people weren't using electricity at that time? *Batteries were large, expensive, and difficult to maintain in working order. Also, uses had to be developed for batteries before they could be used.*

Language Arts Connection

Mary Shelley wrote one of the first science fiction novels, *Frankenstein*, about a scientist who created a monster. He used electricity to give it life. Electricity was a new and exciting subject when Shelley wrote her novel.

DIFFERENTIATE INSTRUCTION

Advanced Have students research batteries today. They should chart what types are available, what they are used for, and the chemical reaction that produces electricity.

Scientific Process

Discuss how atomic researchers gathered data in the early 1900s. Point out that they couldn't observe small particles directly; they had to infer what the particles are like by observing what effects they had on other events. Ask: What areas of science can you think of where this process is still essential? *Sample answer: astronomy*

Technology

CHEMISTRY OF COMMUNICATIONS
The audion was the first radio tube to use a control grid as well as a cathode and anode. It became known as a triode. It received radio waves and amplified them.

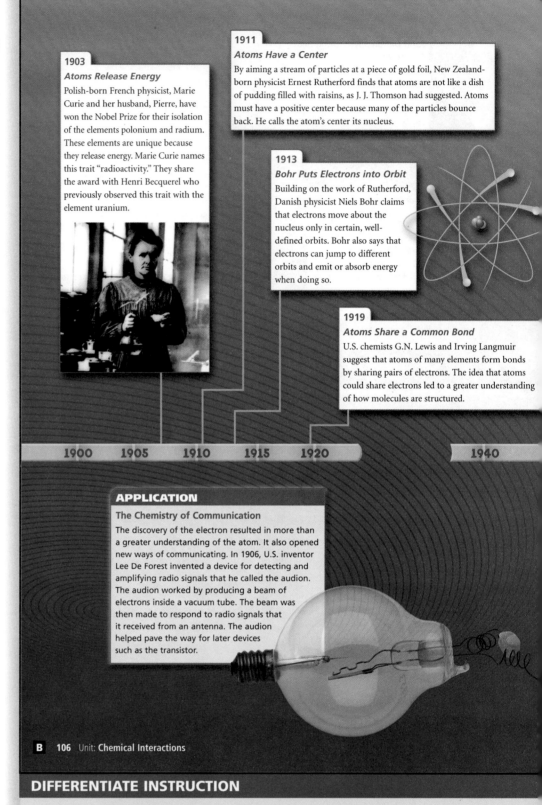

1903
Atoms Release Energy
Polish-born French physicist, Marie Curie and her husband, Pierre, have won the Nobel Prize for their isolation of the elements polonium and radium. These elements are unique because they release energy. Marie Curie names this trait "radioactivity." They share the award with Henri Becquerel who previously observed this trait with the element uranium.

1911
Atoms Have a Center
By aiming a stream of particles at a piece of gold foil, New Zealand-born physicist Ernest Rutherford finds that atoms are not like a dish of pudding filled with raisins, as J. J. Thomson had suggested. Atoms must have a positive center because many of the particles bounce back. He calls the atom's center its nucleus.

1913
Bohr Puts Electrons into Orbit
Building on the work of Rutherford, Danish physicist Niels Bohr claims that electrons move about the nucleus only in certain, well-defined orbits. Bohr also says that electrons can jump to different orbits and emit or absorb energy when doing so.

1919
Atoms Share a Common Bond
U.S. chemists G.N. Lewis and Irving Langmuir suggest that atoms of many elements form bonds by sharing pairs of electrons. The idea that atoms could share electrons led to a greater understanding of how molecules are structured.

1900 1905 1910 1915 1920 1940

APPLICATION
The Chemistry of Communication
The discovery of the electron resulted in more than a greater understanding of the atom. It also opened new ways of communicating. In 1906, U.S. inventor Lee De Forest invented a device for detecting and amplifying radio signals that he called the audion. The audion worked by producing a beam of electrons inside a vacuum tube. The beam was then made to respond to radio signals that it received from an antenna. The audion helped pave the way for later devices such as the transistor.

DIFFERENTIATE INSTRUCTION

English Learners Tell students that the word *audion* was assembled from old words to describe a new technology. *Audio* means "to hear," and *ion* means "to go." List other neologisms on the board, such as *radioactivity* and *electron*. Show how they are made of other words.

...ler Particles Discovered

...nashing atoms into one another,
...ists discover that protons and
...ons are themselves composed
...n smaller particles. In a bit of
...ific humor, these smaller parti-
...re named "quarks," a nonsense
...taken from a novel. Scientists
...these particles by observing the
...s they make in special detectors.

1980s

Tunneling to the Atomic Level

Scanning tunneling microscopes (STMs) allow scientists to interact with matter at the atomic level. Electrons on the tiny tip of an STM "tunnel" through the gap between the tip and target surface. By recording changes in the tunneling current, researchers get an accurate picture.

 RESOURCE CENTER
CLASSZONE.COM
Explore advances in atomic research.

1960   1980   2000

TECHNOLOGY

Particle Accelerators

Particle accelerators speed up charged particles by passing them through an electric field. By smashing subatomic particles into one another, scientists are able to learn what these particles are made of as well as the forces holding them together. The H1 particle detector in Hamburg, Germany, can accelerate protons to 800 billion volts and is used to study the quarks that make up protons.

INTO THE FUTURE

Humans have gone from hypothesizing atoms exist to being able to see and move them. People once considered only four substances to be true elements; today we understand how there are more than a hundred simple substances. Not only have scientists learned atoms contain electric charges, they have also learned how to use these charges.

As scientists learn more and more about the atom, it is difficult to say what they will find next. Is there something smaller than a quark? Is there one type of particle from which all other particles are made? Will we one day be able to move and connect atoms in any way we want? Are there other kinds of atoms to discover? Maybe one day we will find answers to these questions.

ACTIVITIES

Explore a Model Atom

The discovery of the nucleus was one of the most important discoveries in human history. Rutherford's experiment, however, was a simple one that you can model. Take an aluminum pie plate and place a table tennis ball-sized piece of clay at its center. The clay represents a nucleus. Place the end of a grooved ruler at the edge of the plate. Hold the other end up to form a ramp. Roll a marble down the groove toward the clay. Move the ruler to different angles with each roll. Roll the marble 20 times. How many rolls out of 20 hit the clay ball? How do you think the results would be different if the atoms looked like pudding with raisins in it, as Thomson suggested?

Writing About Science

Suppose you are an atom. Choose one of the events on the timeline and describe it from the atom's point of view.

Timelines in Science 107 **B**

Application

1960s People keep inventing new pre-fixes for the basic SI units, to indicate smaller and smaller units. Particle physics has been one reason these terms became necessary. Besides the familiar *nano* (10^{-9}), prefixes from *pico* to *yocto* span factors of 10^{-12} to 10^{-24}.

INTO THE FUTURE

Ask students what they think would be the consequences of being able to take apart and assemble atoms and molecules. Would all consequences be good or would there be some bad results? *Materials such as gold and silver would no longer be valuable because we could make as much as we want. But some people could use this technology to destroy instead of make things.*

ACTIVITIES

Explore a Model Atom

Discuss how the model is like Rutherford's experiment and how it differs. *The plum pudding model would have bounced all the electrons back.*

Writing About Science

Students can have fun with this assignment by imagining themselves as small as an atom. Tell them to think of themselves as the recipient or victim of an attack from whichever historic event they choose. They should write using the first-person noun *I.*

Technology Resources

Students can visit **ClassZone.com** for current news on atomic research.

...IFFERENTIATE INSTRUCTION

...lternative Writing Project Have groups of students research
...ow diagrams of the atom have changed to reflect new ideas. Have
...nem produce labeled diagrams for different atomic theories and write
...rief descriptions of what made the theory change.

CHAPTER 4 Solutions

Physical Science
UNIFYING PRINCIPLES

PRINCIPLE 1

Matter is made of particles too small to see.

PRINCIPLE 2

Matter changes form and moves from place to place.

PRINCIPLE 3

Energy changes from one form to another, but it cannot be created or destroyed.

PRINCIPLE 4

Physical forces affect the movement of all matter on Earth and throughout the universe.

Unit:
Chemical Interactions
BIG IDEAS

CHAPTER 1
Atomic Structure and the Periodic Table
A substance's atomic structure determines its physical and chemical properties.

CHAPTER 2
Chemical Bonds and Compounds
The properties of compounds depend on their atoms and chemical bonds.

CHAPTER 3
Chemical Reactions
Chemical reactions form new substances by breaking and making chemical bonds.

CHAPTER 4
Solutions
When substances dissolve to form a solution, the properties of the mixture change.

CHAPTER 5
Carbon in Life and Materials
Carbon is essential to living things and to modern materials.

CHAPTER 4
KEY CONCEPTS

SECTION (4.1)	SECTION (4.2)	SECTION (4.3)	SECTION (4.4)
A solution is a type of mixture.	**The amount of solute that dissolves can vary.**	**Solutions can be acidic, basic, or neutral.**	**Metal alloys are solid mixtures.**
1. The parts of a solution are mixed evenly.	1. A solution with a high concentration contains a large amount of solute.	1. Acids and bases have distinct properties.	1. Humans have made alloys for thousands of years.
2. Solvent and solute particles interact.	2. The solubility of a solute can be changed.	2. The strengths of acids and bases can be measured.	2. Alloys have many uses in modern life.
3. Properties of solvents change in solutions.	3. Solubility depends on molecular structure.	3. Acids and bases neutralize each other.	

 The Big Idea Flow Chart is available on p. T25 in the **UNIT TRANSPARENCY BOOK.**

Previewing Content

4.1 A solution is a type of mixture.
pp. 111–116

1. The parts of a solution are mixed evenly.
A **solution** is a homogeneous mixture; all portions of a solution have identical properties. The **solute** is the substance that is dissolved. The **solvent** dissolves the solute.
- Solutes, solvents, and solutions can be liquids, solids, or gases.
- The solute and solvent can be in the same or in different physical states.
- A **suspension** is a mixture with large particles. The particles do not dissolve, and the mixture is not a solution.

2. Solvent and solute particles interact.
When a solid dissolves in a liquid, the solute breaks apart. Solute particles are surrounded by solvent particles and are evenly distributed in the solution.
- Ionic compounds break up into individual ions when they dissolve, as the diagram below shows.

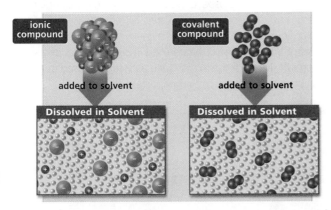

- When a covalent compound dissolves, the molecules separate from each other, but the covalent bonds remain intact and the individual molecules remain whole.

3. Properties of solvents change in solutions.
A solute changes the physical properties of a solvent.
- The freezing point of a solution is lower than the freezing point of the pure solvent.
- The boiling point of a solution is higher than the boiling point of the pure solvent.

4.2 The amount of solute that dissolves can vary. pp. 117–124

1. A solution with a high concentration contains a large amount of solute.
The **concentration** of a solution is the amount of solute dissolved in it at a particular temperature. Solutions can be made more concentrated by adding solute, or more **dilute** by adding solvent.

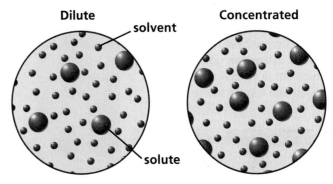

- A **saturated** solution holds as much of a given solute as it can at a given temperature. If a solution contains more solute than can normally dissolve at a given temperature, it is super-saturated. Supersaturated solutions are very unstable, and disturbing the solution will cause the excess solute to come out of the solution as a precipitate.
- Every substance has a characteristic **solubility,** the amount that will dissolve in a certain amount of a certain solvent at a given temperature.

2. The solubility of a solute can be changed.
An increase in temperature increases the solubility of most solid solutes and decreases the solubility of gaseous solutes. An increase in pressure increases the solubility of gaseous solutes. The solubility of solid and liquid solutes are not usually affected by changes in pressure.

3. Solubility depends on molecular structure.
Solubility depends on charges of solute and solvent particles. Molecules with regions of electrical charge (polar molecules) and ions dissolve in polar solvents such as water. Nonpolar molecules (oils) do not have charged regions and do not dissolve in polar solvents, but they will dissolve in nonpolar solvents.

Common Misconceptions

WHY DO MATERIALS DISSOLVE? Students often think that sand does not dissolve in water because it is too dense, too thick, or too hard. Solubility depends instead on the molecular makeup of the solute and solvent and on the polarity of their bonding.

 This misconception is addressed on p. 114.

MISCONCEPTION DATABASE
CLASSZONE.COM Background on student misconceptions

PROPERTIES OF SOLUTIONS Students may think that the mass of the solute is lost when it dissolves in a solute. Although the solute seems to disappear it is still present, and therefore the mass of the solution is equal to the mass of the solute and solvent together.

 This misconception is addressed on p. 115.

Previewing Content

4.3 Solutions can be acidic, basic, or neutral. pp. 125–133

1. Acids and bases have distinct properties.
An **acid** can donate a hydrogen ion to another substance when the acid is dissolved in water.
- HCl is an acid and donates a hydrogen ion in a water solution.
- Acids taste sour, react with carbonates to form carbon dioxide, and react with many metals.

A **base** can accept a hydrogen ion from another substance.
- In water, the base NaOH releases a hydroxide ion, which can accept a hydrogen ion.
- Bases taste bitter and feel slippery or soapy.

Acid   **Base**

$$HCl \xrightarrow{H_2O} H^+ + Cl^- \qquad NaOH \xrightarrow{H_2O} Na^+ + OH^-$$

2. The strengths of acids and bases can be measured.
Strong acids and bases break apart completely into individual ions. No complete molecules of the acid or base remain in the solution. A weak acid or base does not break apart completely into ions, and contains both molecules of the acid or base and its ions.

The acidity of a solution is measured on the **pH** scale.
- Acids produce a high hydrogen ion concentration and have a low pH—from 0 to 7.
- Bases produce a low hydrogen ion concentration and have a high pH—from 7 to 14.
- Solutions of pH 7 are **neutral,** neither acidic nor basic.

3. Acids and bases neutralize each other.
When an acid and a base come in contact with each other, they undergo a neutralization reaction. The hydrogen ion from the acid and the hydroxide ion from the base combine to form water. The negative ion from the acid and the positive ion from the base combine to form a salt. The products of a neutralization reaction—water and a salt—are neutral substances.

Common Misconceptions

ACIDS AND BASES Students often think of a base as a substance that counteracts the dangerous corrosive properties of an acid, and that bases are "good" and acids are "bad." Strong bases have corrosive properties just as do strong acids.

[T E] This misconception is addressed on p. 127.

4.4 Metal alloys are solid mixtures. pp. 134–139

1. Humans have made alloys for thousands of years.
An **alloy** is a solid mixture that has many of the characteristics of a solution. In an alloy, a solid (usually metal) solute is mixed with a solid metallic solvent. Alloys are made by melting the metal components and mixing them in the liquid state. The physical properties of an alloy are different from those of the solvent metal.

There are two general types of alloys. Brass is an example of one type, called a substitutional alloy, in which some of the copper atoms are replaced by zinc atoms. Steel is an example of the other type, called an interstitial alloy, in which carbon atoms occupy gaps between iron atoms.

2. Alloys have many uses in modern life.
New alloys are constantly being developed in response to the development of new technologies. In the transportation industry, medicine, and the aerospace industry, new alloys with unique properties are developed to fulfill specific requirements. For example, aluminum alloys, which have a relatively low density and are lightweight, but very strong, are used in such applications as cars and aircraft. More dense alloys, such as steel, are used in applications in which weight is not an important feature.

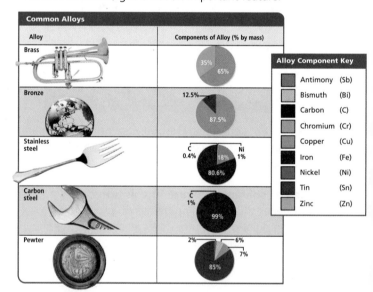

Common Alloys	
Alloy	**Components of Alloy (% by mass)**
Brass	35% / 65%
Bronze	12.5% / 87.5%
Stainless steel	C 0.4% / Ni 1% / 18% / 80.6%
Carbon steel	C 1% / 99%
Pewter	2% / 6% / 7% / 85%

Alloy Component Key
- Antimony (Sb)
- Bismuth (Bi)
- Carbon (C)
- Chromium (Cr)
- Copper (Cu)
- Iron (Fe)
- Nickel (Ni)
- Tin (Sn)
- Zinc (Zn)

 MISCONCEPTION DATABASE
CLASSZONE.COM Background on student misconceptions

Previewing Labs

Lab Generator CD-ROM
Edit these Pupil Edition labs and generate alternative labs.

EXPLORE the BIG idea

Does it Dissolve? p. 109
Students distinguish between solutions and mixtures by observing different substances in water.

TIME 10 minutes
MATERIALS tap water; 4 small clear plastic cups; plastic spoon; 1 teaspoon each of powdered drink mix, vinegar, milk, and sand

---

Acid Test, p. 109
Students use a radish as a pH indicator.

TIME 10 minutes
MATERIALS radish; 3 blank index cards; 3 cotton swabs; a few drops of lemon juice, tap water, and soda water

---

Internet Activity: Alloys, p. 109
Students investigate virtual alloys and change their composition.

TIME 20 minutes
MATERIALS computer with Internet access

SECTION 4.1

EXPLORE Mixtures, p. 111
Students try to dissolve table salt and flour in water to compare the physical properties of a solution and a suspension.

TIME 10 minutes
MATERIALS water, 2 clear plastic cups, plastic spoon, 1 teaspoon each of table salt and flour

INVESTIGATE Solutions, p. 113
Students make a paper chromatogram to observe that the parts of a solution can be separated.

TIME 15 minutes
MATERIALS washable black marking pen, coffee filter, plastic bottle, eyedropper, water, watch

---

SECTION 4.2

EXPLORE Solutions and Temperature, p. 117
Students observe the fizzing of warm and cold soda water to determine how temperature affects a solution of a gas in a liquid.

TIME 10 minutes
MATERIALS 50 mL each of warm and cold soda water, 2 clear plastic cups

INVESTIGATE Solubility, p. 120
Students design an experiment to determine how a change in temperature affects the solubility of table salt in water.

TIME 20 minutes
MATERIALS clear plastic cups, thermometer, water, salt, balance, spoon, hot and cold water baths

---

SECTION 4.3

EXPLORE Acids and Bases, p. 125
Students dissolve an antacid in vinegar to observe the chemical reaction between a base and an acid.

TIME 10 minutes
MATERIALS clear plastic cup, 30 mL of vinegar, 2 antacid tablets containing a carbonate

CHAPTER INVESTIGATION
Acids and Bases, pp. 132–133
Students use pH indicator paper to test and classify common substances as acids or bases.

TIME 40 minutes
MATERIALS 7 plastic cups; 1 tsp. baking soda; salt; 30 mL fruit juice; 30 mL fruit juice; shampoo; soda water; detergent powder; vinegar; masking tape; marking pen; measuring spoons; graduated cylinder; 90 mL distilled water; paper towels; 7 strips universal pH indicator paper

---

SECTION 4.4

INVESTIGATE Alloys, p. 137
Students find the density of alloys and a pure metal to observe that they have different physical properties.

TIME 30 minutes
MATERIALS 3 iron nails, 3 steel nails or 3 stainless steel nails, balance, graduated cylinder, water

---

R **Additional INVESTIGATION,** Rates of Solution, A, B, & C, pp. 259–267; Teacher Instructions, pp. 329–330

Previewing Chapter Resources

INTEGRATED TECHNOLOGY	LABS AND ACTIVITIES

CHAPTER 4
Solutions

 CLASSZONE.COM
- eEdition Plus
- EasyPlanner Plus
- Misconception Database
- Content Review
- Test Practice
- Simulation
- Visualization
- Resource Centers
- Internet Activity: Alloys
- Math Tutorial

 SCILINKS.ORG
SCI LINKS

 CD-ROMS
- eEdition
- EasyPlanner
- Power Presentations
- Content Review
- Lab Generator
- Test Generator

 AUDIO CDS
- Audio Readings
- Audio Readings in Spanish

 EXPLORE the Big Idea, p. 109
- Does It Dissolve?
- Acid Test
- Internet Activity: Alloys

UNIT RESOURCE BOOK
Unit Projects, pp. 5–10

Lab Generator CD-ROM
Generate customized labs.

SECTION
4.1 A solution is a type of mixture.
pp. 111–116

Time: 2 periods (1 block)
 Lesson Plan, pp. 200–201

 UNIT TRANSPARENCY BOOK
- Big Idea Flow Chart, p. T25
- Daily Vocabulary Scaffolding, p. T26
- Note-Taking Model, p. T27
- 3-Minute Warm-Up, p. T28

 • EXPLORE Mixtures, p. 111
- INVESTIGATE Solutions, p. 113

UNIT RESOURCE BOOK
Datasheet, Solutions, p. 209

SECTION
4.2 The amount of solute that dissolves can vary.
pp. 117–124

Time: 2 periods (1 block)
 Lesson Plan, pp. 211–212

 • **VISUALIZATION,** Supersaturated Solutions and Precipitation
- **RESOURCE CENTER,** Aquifers and Purification

 UNIT TRANSPARENCY BOOK
- Daily Vocabulary Scaffolding, p. T26
- 3-Minute Warm-Up, p. T28

 • EXPLORE Solutions and Temperature, p. 117
- INVESTIGATE Solubility, p. 120
- Connecting Sciences, p. 124

UNIT RESOURCE BOOK
- Datasheet, Solubility, p. 220
- Additional INVESTIGATION, Rates of Solution, A, B, & C, pp. 259–267

SECTION
4.3 Solutions can be acidic, basic, or neutral.
pp. 125–133

Time: 3 periods (1.5 blocks)
 Lesson Plan, pp. 222–223

 RESOURCE CENTER, Acids and Bases

 UNIT TRANSPARENCY BOOK
- Daily Vocabulary Scaffolding, p. T26
- 3-Minute Warm-Up, p. T29
- "Common Acids and Bases" Visual, p. T30

 • EXPLORE Acids and Bases, p. 125
- CHAPTER INVESTIGATION, Acids and Bases, pp. 132–133

UNIT RESOURCE BOOK
CHAPTER INVESTIGATION, Acids and Bases, A, B, & C, pp. 250–258

SECTION
4.4 Metal alloys are solid mixtures.
pp. 134–139

Time: 3 periods (1.5 blocks)
 Lesson Plan, pp. 232–233

 • **RESOURCE CENTER,** Alloys
- **MATH TUTORIAL**

 UNIT TRANSPARENCY BOOK
- Big Idea Flow Chart, p. T25
- Daily Vocabulary Scaffolding, p. T26
- 3-Minute Warm-Up, p. T29
- Chapter Outline, pp. T31–T32

 • INVESTIGATE Alloys, p. 137
- Math in Science, p. 139

UNIT RESOURCE BOOK
- Datasheet, Alloys, p. 241
- Math Support, p. 248
- Math Practice, p. 249

READING AND REINFORCEMENT

ASSESSMENT

STANDARDS

- Choose Your Own Strategy, B18–27
- Mind Map, C40–41
- Daily Vocabulary Scaffolding, H1–8

R **UNIT RESOURCE BOOK**
- Vocabulary Practice, pp. 245–246
- Decoding Support, p. 247
- Summarizing the Chapter, pp. 268–269

 Audio Readings CD
Listen to Pupil Edition.

Audio Readings in Spanish CD
Listen to Pupil Edition in Spanish.

PE
- Chapter Review, pp. 141–142
- Standardized Test Practice, p. 143

A **UNIT ASSESSMENT BOOK**
- Diagnostic Test, pp. 59–60
- Chapter Test, A, B, & C, pp. 65–76
- Alternative Assessment, pp. 77–78

SP **A** Spanish Chapter Test, pp. 245–248

Test Generator CD-ROM
Generate customized tests.

Lab Generator CD-ROM
Rubrics for Labs

National Standards
A.1–8, A.9.a–g, B.1.a, B.1.c, E.2–5

See p. 108 for the standards.

R **UNIT RESOURCE BOOK**
- Reading Study Guide, A & B, pp. 202–205
- Spanish Reading Study Guide, pp. 206–207
- Challenge and Extension, p. 208
- Reinforcing Key Concepts, p. 210

TE Ongoing Assessment, pp. 111–112, 114–116

PE Section 4.1 Review, p. 116

A **UNIT ASSESSMENT BOOK**
Section 4.1 Quiz, p. 61

National Standards
A.2–7, A.9.a–b, A.9.e–f, B.1.c

R **UNIT RESOURCE BOOK**
- Reading Study Guide, A & B, pp. 213–216
- Spanish Reading Study Guide, pp. 217–218
- Challenge and Extension, p. 219
- Reinforcing Key Concepts, p. 221

TE Ongoing Assessment, pp. 118–123

PE Section 4.2 Review, p. 123

A **UNIT ASSESSMENT BOOK**
Section 4.2 Quiz, p. 62

National Standards
A.2–7, A.9.a–b, A.9.e–f, B.1.a, E.2–5

R **UNIT RESOURCE BOOK**
- Reading Study Guide, A & B, pp. 224–227
- Spanish Reading Study Guide, pp. 228–229
- Challenge and Extension, p. 230
- Reinforcing Key Concepts, p. 231
- Challenge Reading, pp. 243–244

TE Ongoing Assessment, pp. 126–127, 129–131

PE Section 4.3 Review, p. 131

A **UNIT ASSESSMENT BOOK**
Section 4.3 Quiz, p. 63

National Standards
A.1–7, A.9.a–c, A.9.d–g, B.1.c

R **UNIT RESOURCE BOOK**
- Reading Study Guide, A & B, pp. 234–237
- Spanish Reading Study Guide, pp. 238–239
- Challenge and Extension, p. 240
- Reinforcing Key Concepts, p. 242

TE Ongoing Assessment, pp. 135–138

PE Section 4.3 Review, p. 138

A **UNIT ASSESSMENT BOOK**
Section 4.4 Quiz, p. 64

National Standards
A.2–8, A.9.a–c, A.9.e–f

Previewing Resources for Differentiated Instruction

CHAPTER INVESTIGATION

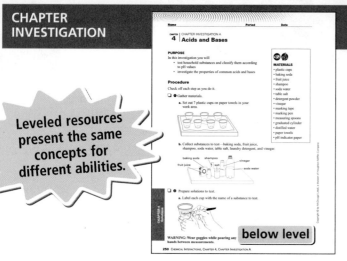

CHAPTER 4 CHAPTER INVESTIGATION A
Acids and Bases

PURPOSE
In this investigation you will
• test household substances and classify them according to pH values
• investigate the properties of common acids and bases

Procedure
Check off each step as you do it.
☐ **①** Gather materials.
a. Set out 7 plastic cups on paper towels in your work area.
b. Collect substances to test—baking soda, fruit juice, shampoo, soda water, table salt, laundry detergent, and vinegar.

☐ **②** Prepare solutions to test.
a. Label each cup with the name of a substance to test.

WARNING: Wear goggles while pouring any hands between measurements.

MATERIALS
• plastic cups
• baking soda
• fruit juice
• shampoo
• soda water
• table salt
• detergent powder
• vinegar
• marking pen
• marking tape
• graduated cylinder
• distilled water
• paper towels
• pH indicator paper

below level

250 CHEMICAL INTERACTIONS, Chapter 4, Chapter Investigation A

CHAPTER 4 CHAPTER INVESTIGATION B
Acids and Bases

OVERVIEW AND PURPOSE
Acids and bases are very common. For example, limestone is made of a substance that is a base when it is dissolved in water. In this investigation you will use what you have learned about solutions, acids, and bases to
• test various household substances and place them in categories according to their pH values
• investigate the properties of common acids and bases

Procedure
① Set out 7 cups in your work area. Collect the substances that you will be testing: baking soda, fruit juice, shampoo, soda water, table salt, laundry detergent, and vinegar.
② Label each cup. Be sure to wear goggles when pouring the substances that you will be testing. Pour 30 mL of each liquid substance into a separate cup. Dissolve 1 tsp of each solid substance in 30 mL of distilled water in a separate cup. To avoid contaminating the test solutions, wash and dry your measuring tools and hands between measurements.

MATERIALS
• plastic cups
• baking soda
• fruit juice
• shampoo
• soda water
• table salt
• detergent powder
• vinegar
• marking pen
• masking tape
• measuring spoons
• graduated cylinder
• distilled water
• paper towels
• pH indicator paper

on level

254 CHEMICAL INTERACTIONS, Chapter 4, Chapter Investigation B

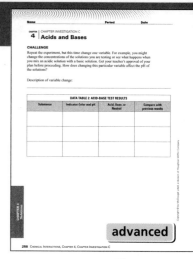

CHAPTER 4 CHAPTER INVESTIGATION C
Acids and Bases

CHALLENGE
Repeat the experiment, but this time change one variable. For example, you might change the concentrations of the solutions you are testing or see what happens when you mix an acidic solution with a basic solution. Get your teacher's approval of your plan before proceeding. How does changing this particular variable affect the pH of the solutions?

Description of variable change:

DATA TABLE 2: ACID-BASE TEST RESULTS

Substance	Indicator Color and pH	Acid, Base, or Neutral	Compare with previous results

advanced

258 CHEMICAL INTERACTIONS, Chapter 4, Chapter Investigation C

R UNIT RESOURCE BOOK, pp. 250–253   **R** pp. 254–257   **R** pp. 254–258

> Leveled resources present the same concepts for different abilities.

READING STUDY GUIDE

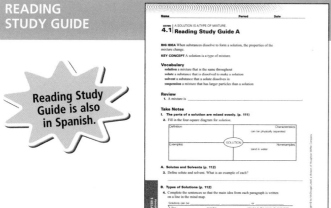

SECTION 4.1 | A SOLUTION IS A TYPE OF MIXTURE.
Reading Study Guide A

BIG IDEA When substances dissolve to form a solution, the properties of the mixture change.
KEY CONCEPT A solution is a type of mixture.

Vocabulary
solution a mixture that is the same throughout
solute a substance that is dissolved to make a solution
solvent a substance that a solute dissolves in
suspension a mixture that has larger particles than a solution

Review
1. A mixture is _____

Take Notes
I. **The parts of a solution are mixed evenly. (p. 111)**
2. Fill in the four-square diagram for *solution*.

Definition	Characteristics
	can be physically separated
SOLUTION	
Examples	Nonexamples
	found in water

A. **Solutes and Solvents (p. 112)**
3. Define solute and solvent. What is an example of each?

B. **Types of Solutions (p. 112)**
4. Complete the sentences so that the main idea from each paragraph is written on a line in the mind map.

Solutions can be _____
The _____ and the _____ can be in the same physical state.
TYPES OF SOLUTIONS
Solid solutions are formed first as _____
Solutions consisting of combinations of _____

below level

202 CHEMICAL INTERACTIONS, Chapter 4, Reading Study Guide A

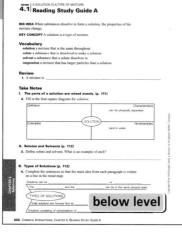

SECTION 4.1 | A SOLUTION IS A TYPE OF MIXTURE.
Reading Study Guide B

BIG IDEA When substances dissolve to form a solution, the properties of the mixture change.
KEY CONCEPT A solution is a type of mixture.

Review
The substances in mixtures do not react chemically.

Take Notes
I. **The parts of a solution are mixed evenly. (p. 111)**
1. Fill in the four-square diagram for *solution*.

Definition	Characteristics
Examples	SOLUTION
	Nonexamples

A. **Solutes and Solvents (p. 112)**
2. Complete the main-idea outline with details about *solutes* and *solvents*.
I. Solutes
   A. _____
   B. _____
II. Solvents
   A. _____
   B. _____

on level

204 CHEMICAL INTERACTIONS, Chapter 4, Reading Study Guide B

SECTION 4.1 | SOLUTIONS
Challenge and Extension

BIG IDEA When substances dissolve to form a solution, the properties of the mixture change.
KEY CONCEPT A solution is a type of mixture.

Liquid Solvents Freezing point is the temperature at which a liquid becomes a solid. Boiling point is the temperature at which a liquid becomes a gas through boiling. When a solute is dissolved in a solvent, the freezing point of the solution is lower than that of the pure solvent, and the boiling point of the solution is higher than that of the pure solvent.

1. Beaker A contains 5 grams of salt dissolved in 500 mL of water. Beaker B contains 500 mL of pure water. Beaker C contains 20 grams of salt dissolved in 500 mL of water. All three beakers are heated. Predict the order in which the contents of each beaker will boil. Give reasons for your predictions.

2. Rosa poured salt on her ice-covered driveway. An hour later, the driveway was still covered with ice. What might have prevented the salt from melting the ice?

3. The greater the concentration of dissolved particles, the greater the change in freezing point and boiling point. Why might calcium chloride ($CaCl_2$) be more useful to melt ice on roads than sodium chloride (NaCl)?

advanced

208 CHEMICAL INTERACTIONS, Chapter 4, Challenge and Extension

R UNIT RESOURCE BOOK, pp. 202–203   **R** pp. 204–205   **R** p. 208

> Reading Study Guide is also in Spanish.

CHAPTER TEST

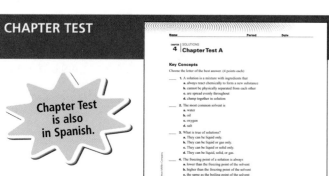

CHAPTER 4 | SOLUTIONS
Chapter Test A

Key Concepts
Choose the letter of the best answer. (4 points each)

_____ 1. A solution is a mixture with ingredients that
a. always react chemically to form a new substance
b. cannot be physically separated from each other
c. are spread evenly throughout
d. clump together in solution

_____ 2. The most common solvent is
a. water
b. oil
c. oxygen
d. salt

_____ 3. What is true of solutions?
a. They can be liquid only.
b. They can be liquid or gas only.
c. They can be liquid or solid only.
d. They can be liquid, solid, or gas.

_____ 4. The freezing point of a solution is always
a. lower than the freezing point of the solvent
b. higher than the freezing point of the solvent
c. the same as the boiling point of the solvent
d. higher than the boiling point of the solvent

_____ 5. A dilute solution always has a
a. low concentration of solute
b. low concentration of solvent
c. high solubility
d. low temperature

below level

CHEMICAL INTERACTIONS, Chapter 4, Chapter Test A 65

CHAPTER 4 | SOLUTIONS
Chapter Test B

Key Concepts
Choose the letter of the best answer. (4 points each)

_____ 1. The components of a solution
a. cannot be separated
b. always chemically react
c. can be separated
d. are always dilute

_____ 2. Which of the following is true about water?
a. It is necessary to make a solution.
b. It is always at least one part of a solution.
c. It is the most common solute in solutions.
d. It is the most common solvent in solutions.

_____ 3. What is true of solvents and solutes?
a. Solutes and solvents are parts of suspensions.
b. Neither solvents nor solutes dissolve in each other.
c. Solvents dissolve in solutes.
d. Solutes dissolve in solvents.

_____ 4. When solid and salt to a glass of pure water,
a. the boiling point of the water is raised
b. the freezing point of the water is raised
c. both the freezing and boiling points of the water are raised
d. the freezing and boiling points of the water remain unchanged

_____ 5. As more solute is added to a solution, the solution becomes more
a. liquid
b. acidic
c. concentrated
d. dilute

on level

CHEMICAL INTERACTIONS, Chapter 4, Chapter Test B 69

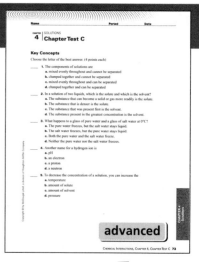

CHAPTER 4 | SOLUTIONS
Chapter Test C

Key Concepts
Choose the letter of the best answer. (4 points each)

_____ 1. The components of solutions are
a. mixed evenly throughout and cannot be separated
b. clumped together and cannot be separated
c. mixed evenly throughout and can be separated
d. clumped together and can be separated

_____ 2. In a solution of two liquids, which is the solute and which is the solvent?
a. The substance that can become a solid or gas more readily is the solvent.
b. The substance that is denser is the solute.
c. The substance that was present first is the solvent.
d. The substance present in the greatest concentration is the solute.

_____ 3. What happens to a glass of pure water and a glass of salt water at 0°C?
a. The pure water freezes, but the salt water stays liquid.
b. The salt water freezes, but the pure water stays liquid.
c. Both the pure water and the salt water freeze.
d. Neither the pure water nor the salt water freezes.

_____ 4. Another name for a hydrogen ion is
a. pH
b. an electron
c. a proton
d. a neutron

_____ 5. To decrease the concentration of a solution, you can increase the
a. temperature
b. amount of solute
c. amount of solvent
d. pressure

advanced

CHEMICAL INTERACTIONS, Chapter 4, Chapter Test C 73

A UNIT ASSESSMENT BOOK, pp. 65–68   **A** pp. 69–72   **A** pp. 73–76

> Chapter Test is also in Spanish.

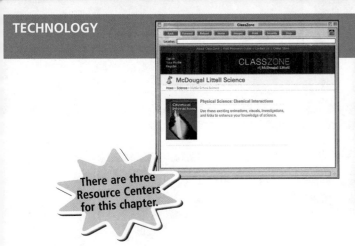

There are three Resource Centers for this chapter.

 CLASSZONE.COM

 CD/CD-ROMS

 CLASSZONE.COM

VISUAL CONTENT

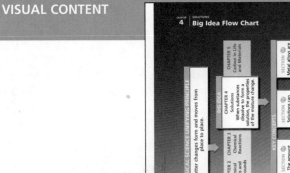

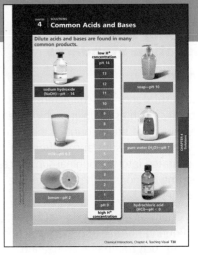

T UNIT TRANSPARENCY BOOK, p. T25

T p. T27

T p. T30

MORE SUPPORT

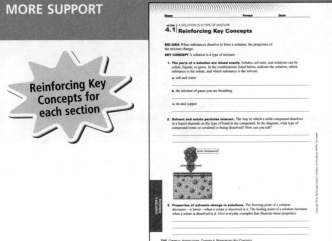

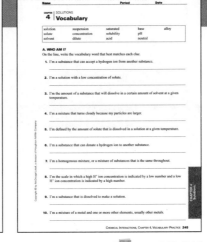

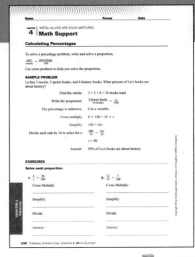

Reinforcing Key Concepts for each section

R UNIT RESOURCE BOOK, p. 210

R pp. 245–246

R p. 248

INTRODUCE

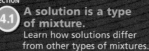
the **BIG** idea

Have students look at the photograph of the seafloor and sunken boat and discuss how the question in the box links to the Big Idea.

- What substances might be dissolved in seawater?

- Why might the various substances and objects on the seafloor not dissolve?

- How might the properties of water change when substances are dissolved in it?

National Science Education Standards

Content

B.1.a A substance has characteristic properties, such as solubility, which are independent of the amount of the sample. A mixture of substances often can be separated into the original substances using one or more of the characteristic properties.

B.1.c Chemical elements do not break down during normal laboratory reactions involving such treatments as heating, exposure to electric current, or reaction with acids.

Process

A.1–8 Identify questions that can be answered through scientific investigations; design and conduct an investigation; use tools; use evidence; think critically between evidence and explanation; recognize different explanations and predictions; communicate procedures and explanations; use mathematics.

A.9.a–g Understand scientific inquiry by using different investigations, methods, mathematics, technology, explanations based on logic, evidence, and skepticism. Data often results in new investigations.

E.1–5 Identify a problem; design, implement, evaluate a solution or product; communicate technological design.

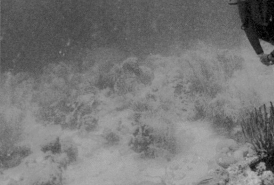

Why might some substances dissolve in the seawater in this photograph, but others do not?

the **BIG** idea

When substances dissolve to form a solution, the properties of the mixture change.

Key Concepts

SECTION
4.1 **A solution is a type of mixture.**
Learn how solutions differ from other types of mixtures.

SECTION
4.2 **The amount of solute that dissolves can vary.**
Learn how solutions can contain different amounts of dissolved substances.

SECTION
4.3 **Solutions can be acidic, basic, or neutral.**
Learn about acids and bases and where they are found.

SECTION
4.4 **Metal alloys are solid mixtures.**
Learn about alloys and how they are used.

Internet Preview

CLASSZONE.COM

Chapter 4 online resources: Content Review, Simulation, Visualization, three Resource Centers, Math Tutorial, Test Practice

---

INTERNET PREVIEW

CLASSZONE.COM For student use with the following pages:

Review and Practice
- Content Review, pp. 110, 140
- Math Tutorial: Understanding Percents, p. 139
- Test Practice, p. 143

Activities and Resources
- Internet Activity, p. 109
- Visualization: Supersaturated Solutions and Precipitation, p. 118
- Resource Centers: Aquifers and Purification, p. 124; Acids and Bases, p. 126; Alloys, p. 136

NSTA *SCiLINKS*
scilinks.org

Solutions **Code: MDL025**

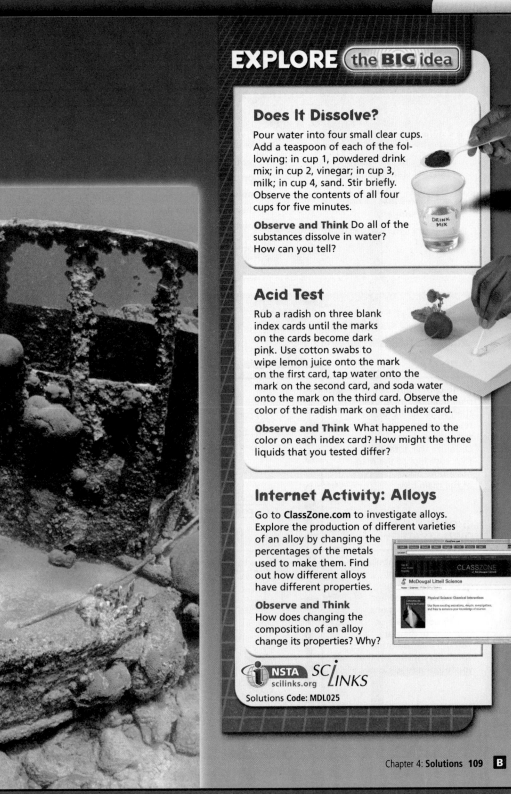

Does It Dissolve?

Pour water into four small clear cups. Add a teaspoon of each of the following: in cup 1, powdered drink mix; in cup 2, vinegar; in cup 3, milk; in cup 4, sand. Stir briefly. Observe the contents of all four cups for five minutes.

Observe and Think Do all of the substances dissolve in water? How can you tell?

Acid Test

Rub a radish on three blank index cards until the marks on the cards become dark pink. Use cotton swabs to wipe lemon juice onto the mark on the first card, tap water onto the mark on the second card, and soda water onto the mark on the third card. Observe the color of the radish mark on each index card.

Observe and Think What happened to the color on each index card? How might the three liquids that you tested differ?

Internet Activity: Alloys

Go to **ClassZone.com** to investigate alloys. Explore the production of different varieties of an alloy by changing the percentages of the metals used to make them. Find out how different alloys have different properties.

Observe and Think How does changing the composition of an alloy change its properties? Why?

NSTA
scilinks.org
SC**LINKS**

Solutions Code: MDL025

These inquiry-based activities are appropriate for use at home or as a supplement to classroom instruction.

Does It Dissolve?

PURPOSE To observe differences between solutions and mixtures. Students observe different substances in water.

TIP *10 min.* Glass jars can be used instead of plastic cups.

Answer: The drink mix and vinegar dissolve; the milk and sand do not. The drink mix and vinegar are not distinct from the water; the milk and sand are.

REVISIT after p. 112.

Acid Test

PURPOSE To observe a pH indicator test. Students use a radish as an indicator.

TIP *10 min.* Cut cotton swabs in half. Vinegar can be used instead of lemon juice.

Answer: The pink color changes on the index card when exposed to an acid. The change is most evident with lemon juice because it is the most acidic.

REVISIT after p. 129.

Internet Activity: Alloys

PURPOSE To investigate and change virtual alloys.

TIP *20 min.* Emphasize that an alloy changes when the types and amounts of metals in it are changed.

Answer: Changing the materials or percentages of materials in an alloy can make the alloy stronger, lighter, or otherwise physically different, because the mixture is altered.

REVISIT after p. 135.

EACHING WITH TECHNOLOGY

BL and Probeware Students can use a H probe in the Chapter Investigation, p. 132–133.

Spreadsheet Have students use a spreadsheet to record their data and do the calculations in "Investigate Alloys" on p. 137.

CONCEPT REVIEW

Activate Prior Knowledge

Add table salt to a small amount of water to make a saltwater solution. Tell students what is in the solution.

- Heat the saltwater solution until the water evaporates. Ask: Have physical changes or chemical changes taken place? *physical*

- Ask: Why do you think the changes are physical? *The salt and water have not changed into other substances. Only their physical properties changed.*

TAKING NOTES

Mind Map

Students can include as many lines of information as they want with their diagrams. If students have trouble remembering what they have written, they may be including too much detail.

Choose Your Own Strategy

Allow students to choose the vocabulary strategy that works best for them.

Vocabulary and Note-Taking Resources

- Vocabulary Practice, pp. 245–246
- Decoding Support, p. 247

- Daily Vocabulary Scaffolding, p. T26
- Note-Taking Model, p. T27

- Choose Your Own Strategy, B18–27
- Mind Map, C40–41
- Daily Vocabulary Scaffolding, H1–8

CONCEPT REVIEW

- Matter can change from one physical state to another.
- A mixture is a blend of substances that do not react chemically.
- Particles can have electrical charges.

VOCABULARY REVIEW

proton p. 11

ion p. 14

molecule p. 51

chemical reaction p. 69

mixture *See Glossary.*

CONTENT REVIEW
CLASSZONE.COM
Review concepts and vocabulary.

TAKING NOTES

MIND MAP

Write each main idea, or blue heading, in an oval; then write details that relate to each other and to the main idea. Organize the details so that each line of the map has a note about one part of the main idea.

CHOOSE YOUR OWN STRATEGY

For each new vocabulary term, take notes by choosing one of the strategies from earlier chapters—**frame game**, **description wheel**, or **four square** diagram. You can also use other vocabulary strategies that you might already know.

See the Note-Taking Handbook on pages R45–R51.

SCIENCE NOTEBOOK

parts not easily separated or differentiated
substances dissolved in a solvent

A solution is a type of mixture.

can be solid, liquid, or gas
physical properties differ from solvent

Frame Game

example
example | TERM | example
example

Description Wheel

feature
feature | TERM | feature
feature

Four Square

definition	characteristics
TERM	
examples	nonexamples

CHECK READINESS

Administer the Diagnostic Test to determine students' readiness for new science content and their mastery of requisite math skills.

 Diagnostic Test, pp. 59–60

Technology Resources

Students needing content and math skills should visit **ClassZone.com**.

- **CONTENT REVIEW**
- **MATH TUTORIAL**

 CONTENT REVIEW CD-ROM

KEY CONCEPT

4.1 A solution is a type of mixture.

◀ **BEFORE, you learned**

- Ionic or covalent bonds hold a compound together
- Chemical reactions produce chemical changes
- Chemical reactions alter the arrangements of atoms

▶ **NOW, you will learn**

- How a solution differs from other types of mixtures
- About the parts of a solution
- How properties of solutions differ from properties of their separate components

VOCABULARY

solution p. 111
solute p. 112
solvent p. 112
suspension p. 113

VOCABULARY
Remember to use the strategy of your choice. You might use a four square diagram for *solution*.

EXPLORE Mixtures

Which substances dissolve in water?

PROCEDURE

① Pour equal amounts of water into each cup.

② Pour one spoonful of table salt into one of the cups. Stir.

③ Pour one spoonful of flour into the other cup. Stir.

④ Record your observations.

WHAT DO YOU THINK?
- Did the salt dissolve? Did the flour dissolve?
- How can you tell?

MATERIALS
- tap water
- 2 clear plastic cups
- plastic spoon
- table salt
- flour

The parts of a solution are mixed evenly.

A mixture is a combination of substances, such as a fruit salad. The ingredients of any mixture can be physically separated from each other because they are not chemically changed—they are still the same substances. Sometimes, however, a mixture is so completely blended that its ingredients cannot be identified as different substances. A **solution** is a type of mixture, called a homogeneous mixture, that is the same throughout. A solution can be physically separated, but all portions of a solution have the same properties.

If you stir sand into a glass of water, you can identify the sand as a separate substance that falls to the bottom of the glass. Sand in water is a mixture that is not a solution. If you stir sugar into a glass of water, you cannot identify the sugar as a separate substance. Sugar in water is a common solution, as are examples such as seawater, gasoline, and the liquid part of your blood.

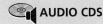

Chapter 4: **Solutions** 111 **B**

▶ **Set Learning Goals**

Students will

- Recognize how a solution differs from other types of mixtures.
- Name the different parts of a solution.
- Distinguish how properties of solutions differ from properties of their original components.
- Observe in an experiment how to separate the components of a solution.

○ **3-Minute Warm-Up**

Display Transparency 28 or copy this exercise on the board:

Draw diagrams of an ionic compound such as sodium chloride (NaCl) and a covalent compound such as carbon dioxide (CO_2). What is the difference in how these compounds are held together? *An ionic compound is held together by the attraction of oppositely charged ions; a covalent compound is held together by the sharing of pairs of electrons between two atoms.*

T 3-Minute Warm-Up, p. T28

4.1 MOTIVATE

EXPLORE Mixtures

PURPOSE To compare the physical appearance of solutions and suspensions

TIP *10 min. setup* Have students observe the two cups after 30 minutes to check for settling.

WHAT DO YOU THINK? *The salt dissolved; the flour did not. Salt disappeared into the water. Flour remained visible, and the water was cloudy.*

Ongoing Assessment

Recognize how a solution differs from other types of mixtures.

Ask: Why is a solution called a homogeneous mixture? *It is the same throughout.*

RESOURCES FOR DIFFERENTIATED INSTRUCTION

Below Level
UNIT RESOURCE BOOK
Reading Study Guide A, pp. 202–203
Decoding Support, p. 247

🎧 **AUDIO CDS**

Advanced
UNIT RESOURCE BOOK
Challenge and Extension, p. 208

English Learners
UNIT RESOURCE BOOK
Spanish Reading Study Guide, pp. 206–207

💿 **AUDIO CDS**

- Audio Readings in Spanish
- Audio Readings (English)

Teach from Visuals

To help students interpret the visual showing types of solutions, have them read each block of text in the photograph. Ask: What do all three solutions have in common? *Each contains substances dissolved in another substance.*

EXPLORE (the BIG idea)

Revisit "Does It Dissolve?" on p. 109. Have students explain their results.

Develop Critical Thinking

ANALYZE Have students analyze the atmosphere as a solution. Ask:

- Is Earth's atmosphere a solution? Why? *No; because Earth's atmosphere is constantly in motion and changing, every portion of it will not be the same as all other portions. Thus, the atmosphere as a whole is not a solution.*

- If you took a sample of Earth's atmosphere and placed it in a container, would the sample be a solution? Why? *Yes; all parts of the sample are the same, so the sample is a solution.*

Ongoing Assessment

Name the different parts of a solution.

Ask: What are the two components of all solutions? *solute and solvent*

CHECK YOUR READING *Answer: A solute is the substance that is dissolved; a solvent dissolves a solute.*

CHECK YOUR READING *Answer: The solvent is the substance that is present in the greatest amount.*

Solutes and Solvents

Like other mixtures, a solution has definite components. A **solute** (SAHL-yoot) is a substance that is dissolved to make a solution. When a solute dissolves, it separates into individual particles. A **solvent** is a substance that dissolves a solute. Because a solute dissolves into individual particles in a solvent, it is not possible to identify the solute and solvent as different substances when they form a solution.

In a solution of table salt and water, the salt is the solute and the water is the solvent. In the cells of your body, substances such as calcium ions and sugar are solutes, and water is the solvent. Water is the most common and important solvent, but other substances can also be solvents. For example, if you have ever used an oil-based paint you know that water will not clean the paintbrushes. Instead, a solvent like turpentine must be used.

CHECK YOUR READING What is the difference between a solute and a solvent?

Types of Solutions

Many solutions are made of solids dissolved in liquids. However, solutes, solvents, and solutions can be gases, liquids, or solids. For example, oxygen, a gas, is dissolved in seawater. The bubbles in carbonated drinks come from the release of carbon dioxide gas that was dissolved in the drink.

In some solutions, both the solute and the solvent are in the same physical state. Vinegar, for example, is a solution of acetic acid in water. In a solution of different liquids, it may be difficult to say which substance is the solute and which is the solvent. In general, the substance present in the greater amount is the solvent. Since there is more water than acetic acid in vinegar, water is the solvent and acetic acid is the solute.

Although you may usually think of a solution as a liquid, solid solutions also exist. For example, bronze is a solid solution in which tin is the solute and copper is the solvent. Solid solutions are not formed as solids. Instead, the solvent metal is heated until it melts and becomes a liquid. Then the solute is added, and the substances are thoroughly mixed together. When the mixture cools, it is a solid solution.

Solutions made of combinations of gases are also common. The air you breathe is a solution. Because nitrogen makes up the largest portion of air, it is the solvent. Other gases present, such as oxygen and carbon dioxide, are solutes.

CHECK YOUR READING When substances in a solution are in the same physical state, which is the solvent?

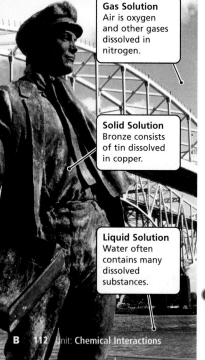

Gas Solution
Air is oxygen and other gases dissolved in nitrogen.

Solid Solution
Bronze consists of tin dissolved in copper.

Liquid Solution
Water often contains many dissolved substances.

DIFFERENTIATE INSTRUCTION

 More Reading Support

A What is the most common solvent? *water*

B What are many solutions made of? *solids dissolved in liquids*

English Learners Words with multiple meanings may confuse English learners. For example, the term *solution*, introduced on p. 111, has another common use that English learners may be more familiar with (as in a solution to a problem).

Below Level The terms *solution*, *solute*, and *solvent* look and sound similar. Have students write the definition for each term on an index card. They should keep the cards available for quick reference.

INVESTIGATE Solutions

How can you separate the parts of a solution?

PROCEDURE

① Draw a solid black circular region 6 cm in diameter around the point of the filter.

② Place the filter, point up, over the top of the bottle.

③ Squeeze several drops of water onto the point of the filter.

④ Observe the filter once every minute for 10 minutes. Record your observations.

WHAT DO YOU THINK?

• What happened to the ink on the filter?

• Identify, in general, the solutes and the solution in this investigation.

CHALLENGE Relate your observations of the ink and water on the coffee filter to the properties of solutions.

Suspensions

When you add flour to water, the mixture turns cloudy, and you cannot see through it. This mixture is not a solution but a suspension. In a **suspension,** the particles are larger than those found in a solution. Instead of dissolving, these larger particles turn the liquid cloudy. Sometimes you can separate the components of a suspension by filtering the mixture.

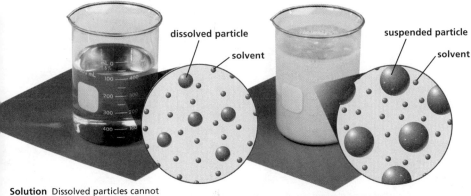

dissolved particle

solvent

suspended particle

solvent

Solution Dissolved particles cannot be identified as a substance different from the solvent.

Suspension Particles that do not dissolve make a suspension look cloudy.

Chapter 4: Solutions 113 **B**

INVESTIGATE Solutions

PURPOSE To separate the parts of a solution

TIP *15 min.* Use a cone-shaped coffee filter for best results. The marker must have water-soluble ink.

WHAT DO YOU THINK? *The black ink separated into several colors. The solutes are the different-colored substances, and the solution is the black ink.*

CHALLENGE *Substances of different colors are dissolved in the ink, and these substances can be separated.*

Develop Critical Thinking

COMPARE A solution is a homogeneous mixture; a suspension is a heterogeneous mixture. Ask: How do the terms *homogeneous* and *heterogeneous* describe the different types of mixtures? *A homogeneous mixture is the same throughout; a heterogeneous mixture is different throughout.*

Teach from Visuals

To help students interpret the illustration comparing solutions and suspensions, ask: How are the appearances of the mixtures in the photographs related to the blow-up illustrations? *The larger particle size in the suspension makes the mixture appear cloudy.*

DIFFERENTIATE INSTRUCTION

? More Reading Support

C How does the particle size compare in a solution and a suspension? *It is larger in a suspension.*

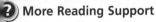

Alternative Assessment Have students answer the questions in "Investigate Solutions" by drawing a diagram.

Teach from Visuals

To help students interpret the diagrams showing how solutes dissolve, ask:

- How do the solvent molecules interact with the solute particles. *The solvent molecules surround each solute particle, dissolving and evenly distributing the solute.*

- Identify the difference between the dissolved ionic compound and the dissolved covalent compound. *The ions separate from one another; each molecule of the covalent compound separates from each other, but the covalent bonds within the molecules remain intact.*

Address Misconceptions

IDENTIFY Ask: Why doesn't sand dissolve in water? If students suggest that it doesn't dissolve because of physical properties such as density, hardness, or roughness, they may hold a misconception about why some substances dissolve but others do not.

CORRECT Have students observe the physical characteristics of salt or granulated sugar, sand, and flour, but do not tell the students what the substances are. Have students predict, based on their observations, which of the substances will dissolve. Add the substances to water and have the students observe which substance dissolves. Explain that substances dissolve when the interaction between solute and solvent particles is greater than the interaction between the solute or solvent particles alone.

REASSESS Ask students again why sand does not dissolve in water. *The forces holding the particles of sand together are stronger than the force of water trying to pull them apart.*

Ongoing Assessment

READING VISUALS *Answer: An ionic compound dissociates into individual ions; a covalent compound dissociates into individual molecules.*

Solvent and solute particles interact.

The parts of a solution—that is, the solute and the solvent—can be physically separated because they are not changed into new substances. However, individual particles of solute and solvent do interact. When a solid dissolves in a liquid, the particles of the solute are surrounded by particles of the liquid. The solute particles become evenly distributed through the solvent.

The way in which a solid compound dissolves in a liquid depends on the type of bonds in the compound. Ionic compounds such as table salt (NaCl) split apart into individual ions. When table salt dissolves in water, the sodium and chloride ions separate, and each ion is surrounded by water molecules. When a covalent compound such as table sugar ($C_{12}H_{22}O_{11}$) dissolves, each molecule stays together and is surrounded by solvent molecules. The general processes that take place when ionic compounds dissolve and when covalent compounds dissolve are shown below.

How Solutes Dissolve

Ionic compounds separate into ions. Covalent compounds separate into individual molecules.

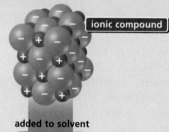

ionic compound

covalent compound

added to solvent

added to solvent

Ionic Compound Dissolved in Solvent

Covalent Compound Dissolved in Solvent

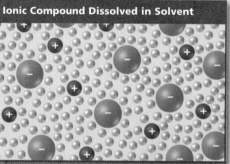

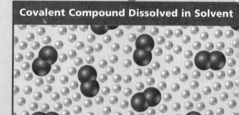

READING VISUALS What difference between the two illustrations tells you whether a compound is ionic or covalent?

DIFFERENTIATE INSTRUCTION

? More Reading Support

D What determines the way a solid dissolves in a liquid? *the type of bonds in the solid*

English Learners To help English learners recognize cause-and-effect relationships, have them make a two-column Cause-and-Effect chart and put each part of a sentence where it belongs, as in the example. Tell students to look for introductory phrases that set up cause-and-effect sentences.

Cause	Effect
When you add flour to water,	the mixture turns cloudy, and you cannot see through it.

Properties of solvents change in solutions.

In every solution—solid, liquid, and gas—solutes change the physical properties of a solvent. Therefore, a solution's physical properties differ from the physical properties of the pure solvent. The amount of solute in the solution determines how much the physical properties of the solvent are changed.

Lowering the Freezing Point

Recall that the freezing point is the temperature at which a liquid becomes a solid. The freezing point of a liquid solvent decreases—becomes lower—when a solute is dissolved in it. For example, pure water freezes at 0°C (32°F) under normal conditions. When a solute is dissolved in water, the resulting solution has a freezing point below 0°C.

Lowering the freezing point of water can be very useful in winter. Road crews spread salt on streets and highways during snowstorms because salt lowers the freezing point of water. When snow mixes with salt on the roads, a saltwater solution that does not freeze at 0°C is formed. The more salt that is used, the lower the freezing point of the solution.

Since salt dissolves in the small amount of water usually present on the surface of ice, it helps to melt any ice already present on the roads. However, there is a limit to salt's effectiveness because there is a limit to how much will dissolve. No matter how much salt is used, once the temperature goes below –21°C (–6°F), the melted ice will freeze again.

REMINDER
In temperature measurements, *C* stands for "Celsius" and *F* stands for "Fahrenheit."

CHECK YOUR READING How does the freezing point of a solvent change when a solute is dissolved in it?

Making ice cream also depends on lowering the freezing point of a solvent. Most hand-cranked ice cream makers hold the liquid ice cream ingredients in a canister surrounded by a mixture of salt and ice. The salt added to the ice lowers the freezing point of this mixture. This causes the ice to melt—absorbing heat from its surroundings, including the ice cream ingredients. The ice cream mix is chilled while its ingredients are constantly stirred. As a result, tiny ice crystals form all at once in the ice cream mixture instead of a few crystals forming and growing larger as the mix freezes. This whole process helps to make ice cream that is smooth and creamy.

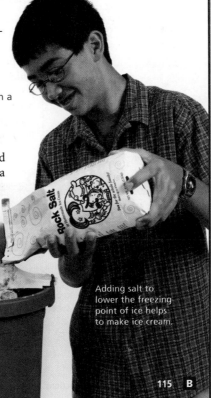

Adding salt to lower the freezing point of ice helps to make ice cream.

115 B

DIFFERENTIATE INSTRUCTION

? More Reading Support

E What changes the physical properties of a solvent when it is in a solution? *the solute*

Advanced Not only does the melting point (same as freezing point) of a solvent decrease after a solute is added, but in some cases it stops being an actual point. Copper melts at 1083.4°C (1982.12°F). However, brass (zinc dissolved in copper) melts somewhere between 900°C and 1000°C. Brass may be a either a solid or a liquid in that temperature range.

R Challenge and Extension, p. 208

Real World Example

Calcium chloride ($CaCl_2$) is better than sodium chloride (NaCl) for melting ice and snow on roads. When calcium chloride dissolves in water, it separates into three ions (one calcium ion and two chloride ions), while sodium chloride separates into two ions. The greater number of dissolved ions causes a greater decrease in the freezing point of water.

Teacher Demo

To demonstrate freezing-point depression, perform this demonstration with the help of one or two students. Fill two beakers with crushed ice and add 50 milliliters of cold water to each. Stir each beaker with a separate stirring rod until the temperature stabilizes. Read and record the temperature in each beaker. Add about 75 grams of rock salt or kosher salt to one of the beakers, and continue stirring both beakers. When the temperature in each beaker is stabilized, read and record the temperatures. The freezing point in the beaker containing salt will be depressed by about 5°C.

Address Misconceptions

IDENTIFY Ask students how much mass a liter of a saturated sugar-water solution has. If they answer that it would have a mass of one kilogram, they may hold the misconception that when the sugar dissolves to make a solution, it does not contribute to the solution's mass.

CORRECT Using a balance, find the mass of a liter of water. Next, find the mass of 400 to 500 mL of sugar. Place both on the balance at the same time, if possible. Pour the sugar into the water and stir. Draw attention to the mass of the solution.

REASSESS Ask students how much mass a liter of water with 300 mL of salt dissolved in it would have. *1.3 kilograms*

Technology Resources

Visit **ClassZone.com** for background on common student misconceptions.

 MISCONCEPTION DATABASE

Ongoing Assessment

CHECK YOUR READING *Answer: It decreases.*

Distinguish how properties of solutions differ from properties of their original components.

Ask: What determines how much the physical properties of a solvent change when you add a solute? *the amount of the solute in the solution*

PHOTO CAPTION Answer: The antifreeze would increase the boiling point of water, thus keeping it liquid at higher temperatures.

 CHECK YOUR READING *Answer: The more solute that is added, the greater the increase in the boiling point.*

Reinforce (the BIG idea)

Have students relate the section to the Big Idea.

 Reinforcing Key Concepts, p. 210

4.1 ASSESS & RETEACH

Assess

 Section 4.1 Quiz, p. 61

Reteach

Have students write a short paragraph describing the dissolving of an ionic compound in water, from the point of view of an ion in the compound. *Answers should describe how the solvent molecules are attracted to the ion, pull it out of the compound, and surround it.*

Technology Resources

Have students visit **ClassZone.com** for reteaching of Key Concepts.

 CONTENT REVIEW

CONTENT REVIEW CD-ROM

Raising the Boiling Point

The boiling point of a liquid is the temperature at which the liquid forms bubbles in its interior and becomes a gas. Under normal conditions, a substance cannot exist as a liquid at a temperature greater than its boiling point. However, the boiling point of a solution is higher than the boiling point of the pure solvent. Therefore, a solution can remain a liquid at a higher temperature than its pure solvent.

For example, the boiling point of pure water is 100°C (212°F) under normal conditions. Saltwater, however, can be a liquid at temperatures above 100°C because salt raises the boiling point of water. The amount of salt in the water determines how much the boiling point is increased. The more solute that is dissolved in a solution, the greater the increase in boiling point.

APPLY Why might the addition of antifreeze to the water in this car's radiator have prevented the car from overheating?

 CHECK YOUR READING How does the boiling point of a solution depend on the amount of solute in it?

A solute lowers the freezing point and raises the boiling point of the solvent in the solution. The result is that the solute extends the temperature range in which the solvent remains a liquid. One way in which both a decrease in freezing point and an increase in boiling point can be useful in the same solution involves a car's radiator. Antifreeze, which is mostly a chemical called ethylene glycol, is often added to the water in the radiator. This solution prevents the water from freezing in the winter and also keeps it from boiling in the summer.

4.1 Review

KEY CONCEPTS

1. How is a solution different from other mixtures?
2. Describe the two parts of a solution. How can you tell them apart?
3. How does the boiling point of a solvent change when a solute is dissolved in it? How does the freezing point change?

CRITICAL THINKING

4. **Compare** Contrast the way in which an ionic compound, such as table salt, dissolves with the way in which a covalent compound, such as sugar, dissolves.
5. **Infer** Pure water freezes at 0°C and boils at 100°C. Would tap water likely freeze and boil at those exact temperatures? Why or why not?

CHALLENGE

6. **Synthesize** People often sprinkle salt on icy driveways and sidewalks. Would a substance like flour have a similar effect on the ice? Explain.

ANSWERS

1. All of a solution appears the same, and the components of a solution are evenly distributed, unlike a typical mixture.

2. The solute is the substance that dissolves. The solvent dissolves the solute. The substance present in the greatest amount is the solvent.

3. Boiling point increases; freezing point decreases.

4. Ionic compounds separate into individual ions, and covalent compounds split into individual molecules.

5. No; tap water contains dissolved substances.

6. No; the salt lowers the freezing point of water because it makes a salt solution. Flour does not make a solution, so the freezing point will not decrease and the ice will not melt.

4.2 The amount of solute that dissolves can vary.

BEFORE, you learned

- Solutions are a type of mixture
- A solution is made when a solute is dissolved in a solvent
- Solutes change the properties of solvents

NOW, you will learn

- About the concentration of a solution
- How a solute's solubility can be changed
- How solubility depends on molecular structure

VOCABULARY

concentration p. 117
dilute p. 118
saturated p. 118
solubility p. 119

EXPLORE Solutions and Temperature

How does temperature affect a solution?

PROCEDURE

1. Pour cold soda water into one cup and warm soda water into another cup. Record your observations.

2. After 5 minutes, observe both cups of soda water. Record your observations.

MATERIALS
- soda water
- 2 clear plastic cups

WHAT DO YOU THINK?
- Which solution bubbled more at first?
- Which solution bubbled for a longer period of time?

A solution with a high concentration contains a large amount of solute.

MIND MAP
Remember to use a mind map to take notes on the concentration of a solution.

Think of water from the ocean and drinking water from a well. Water from the ocean tastes salty, but water from a well does not. The well water does contain salt, but in a concentration so low that you cannot taste it. A solution's **concentration** depends on the amount of solute dissolved in a solvent at a particular temperature. A solution with only a small amount of dissolved solute, such as the salt dissolved in well water, is said to have a low concentration. As more solute is dissolved, the concentration gets higher.

If you have ever used a powdered mix to make lemonade, you probably know that you can change the concentration of the drink by varying the amount of mix you put into a certain amount of water. Two scoops of mix in a pitcher of water makes the lemonade stronger than just one scoop. The lemonade with two scoops of mix has a higher concentration of the mix than the lemonade made with one scoop.

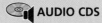

4.2 FOCUS

○ Set Learning Goals

Students will
- Explain how the concentration of a solution varies.
- Describe how a solute's solubility can be changed.
- Recognize that solubility depends on molecular structure.
- Design an experiment to demonstrate the effect of temperature on solubility.

◁ 3-Minute Warm-Up

Display Transparency 28 or copy this exercise on the board:

Decide if these statements are true. If not, correct them.

1. Dissolving a solute in a solvent is a chemical change. *physical change*

2. Adding a solute to a solvent raises the solvent's boiling point. *true*

3. The particles in a solution are larger than the particles in a suspension. *smaller*

[T] 3-Minute Warm-Up, p. T28

4.2 MOTIVATE

EXPLORE Solutions and Temperature

PURPOSE To observe how temperature affects a solution of a gas in a liquid

TIP *10 min.* Refrigerate one bottle of soda water and leave the other one at room temperature.

WHAT DO YOU THINK? *The warm solution bubbled more at first. The cold solution contains more bubbles after five minutes.*

Teach from Visuals

To help students interpret the graphic of dilute and concentrated solutions, ask:

- How can you make a solution more concentrated? *by adding more solute*

- What is the difference between the dilute solution and the concentrated solution, the amount of solute particles or the amount of solvent particles? *the amount of solute particles in relation to the amount of solvent*

Integrate the Sciences

The concentration of salt among bodies of water varies. Fresh water is usually defined as water containing less than 1% total dissolved solute concentration and less than 0.2% salt. The average concentration of salt across all of Earth's oceans is approximately 3.5%. The Great Salt Lake in Utah has a much higher concentration of salt. The southern end of the lake has a salt concentration of approximately 15% and the northern end has a salt concentration of approximately 27%.

Teacher Demo

To demonstrate how solutions can become supersaturated, dissolve 2 cups of sugar in 1 cup of boiling water. Carefully pour it into a clean jar. Dip a piece of string into the solution, stretch it out on a piece of wax paper, and let it dry overnight. Place the dried string in the jar and let the jar stand for 7 days. Eventually, rock candy will form on the string. The solution becomes more concentrated as the water evaporates and then becomes supersaturated.

Ongoing Assessment

CHECK YOUR READING   *Answer: A dilute solution contains a low concentration of solute. A saturated solution contains the highest concentration of solute that can dissolve in that solvent at that temperature.*

Degrees of Concentration

READING TIP

The word *dilute* can be used as either an adjective or a verb. A dilute solution has a low concentration of solute. To dilute a solution is to add more solvent to it, thus lowering the concentration of the solution.

A A solution that has a low concentration of solute is called a **dilute** solution. Salt dissolved in the drinking water from a well is a dilute solution. The concentration of a solution can be even further reduced, or diluted, by adding more solvent. On the other hand, as more solute is added to a solution, the solution becomes more concentrated. A concentrated solution has a large amount of solute.

Dilute

solvent

Concentrated

solute

Less solute is dissolved in a dilute solution.

More solute is dissolved in a concentrated solution.

B Have you ever wondered how much sugar can be dissolved in a glass of iced tea? If you keep adding sugar to the tea, eventually no more sugar will dissolve. The tea will contain as much dissolved sugar as it can hold at that temperature. Such a solution is called a **saturated** solution because it contains the maximum amount of solute that can be dissolved in the solvent at a given temperature. If a solution contains less solute than this maximum amount, it is an unsaturated solution.

CHECK YOUR READING   How are the terms *dilute* and *saturated* related to the concept of concentration?

Supersaturated Solutions

VISUALIZATION
CLASSZONE.COM
Explore supersaturated solutions and precipitation.

Sometimes, a solution contains more dissolved solute than is normally possible. This type of solution is said to be supersaturated. A saturated solution can become supersaturated if more solute is added while the temperature is raised. Then if this solution is slowly cooled, the solute can remain dissolved. This type of solution is very unstable, though. If the solution is disturbed, or more solute is added in the form of a crystal, the excess solute will quickly solidify and form a precipitate. This process is shown in the photographs on the top of page 119.

DIFFERENTIATE INSTRUCTION

More Reading Support

A What do we call a solution with a low concentration of solute? *dilute*

B What is a solution with the maximum amount of solute? *saturated*

English Learners Tell English learners that *given* (p. 119) is often used as an adjective meaning "specified" or "certain." The phrase "said to be" is used to introduce unfamiliar expressions or a description. Have students make note cards that list definitions of *solute, solvent,* and *solubility.* If students have a hard time comprehending *soluble,* the adjective form of *solubility,* tell them it means "able to dissolve."

(1) A supersaturated solution contains more dissolved solute than is normally possible.

(2) After a crystal of solute is added, or the solution is disturbed, a precipitate forms.

REMINDER
A precipitate is a solid substance that comes out of a solution.

One example of a supersaturated solution is a chemical heat pack that contains sodium acetate and water. The pack contains more sodium acetate than can normally dissolve at room temperature, but when the pack is heated in a microwave oven, all of the sodium acetate dissolves. The solution inside the pack is supersaturated. The heat pack is activated by bending it. This disturbs the solution, solidifying the sodium acetate and releasing a large amount of heat over a long period of time.

Solubility

The **solubility** (SAHL-yuh-BIHL-ih-tee) of a substance is the amount of that substance that will dissolve in a certain amount of solvent at a given temperature. For example, consider household ammonia used for cleaning. This ammonia is not pure ammonia—it is a solution of ammonia in water.

Because a large amount of ammonia can dissolve in water, ammonia is said to have a high solubility in water. However, other substances do not dissolve in such large amounts in water. Only a small amount of carbon dioxide will dissolve in water, so carbon dioxide has a low solubility in water. Oils do not dissolve at all in water, so oils are said to be insoluble in water.

The amount of solute needed to make a saturated solution depends on the solubility of a solute in a particular solvent.

- If the solute is highly soluble, a saturated solution will be very concentrated.
- If the solute has a low solubility, the saturated solution will be dilute.

In other words, a saturated solution can be either dilute or concentrated, depending on the solubility of a solute in a particular solvent.

READING TIP
The word *solubility* is related to the words *solute* and *solvent*, and means "ability to be dissolved." A substance that is insoluble will not dissolve.

CHECK YOUR READING How does solubility affect a solution?

Teach Difficult Concepts

Students often confuse a concentrated solution with a saturated solution. Ask students how these two concepts differ. *A concentrated solution holds a large amount of solute per unit volume of solvent. A saturated solution holds as much solute as possible at a given temperature. A concentrated solution may or may not be saturated.* To help students understand solubility, you might try the following demonstration.

Teacher Demo

To demonstrate the solubility of Styrofoam in acetone (nail polish remover) in a dramatic way, pour about 400 mL of acetone into a 1 L beaker. Add foam packing peanuts, one at a time at first, then several at a time. Stir with a glass stirring rod. The peanuts will dissolve so quickly that they will appear to melt. Although the resulting solution is concentrated, it is not saturated. Caution: Acetone is flammable, and you should avoid prolonged exposure to its fumes. Let most of the solution evaporate outdoors before properly disposing of the residue in a chemical waste container.

Ongoing Assessment

Explain how the concentration of a solution varies.

Ask: Why is a supersaturated solution unstable? *It holds more dissolved solute than is normally possible at that temperature. Any disturbance will cause the excess solute to precipitate out of the solution.*

CHECK YOUR READING *Answer: The more soluble a substance is, the greater the amount that will dissolve and the greater its concentration before the solution becomes saturated.*

DIFFERENTIATE INSTRUCTION

More Reading Support

C What do you call the amount of a substance that will dissolve in a certain amount of solvent at a certain temperature? *solubility*

Below Level To teach the concept of concentration, provide students with marbles (or another object) in a box. A dilute solution has few marbles in the box; a concentrated solution has many. A saturated solution has a completely filled layer of marbles in the bottom of the box with no room to add more. A supersaturated solution has a filled layer of marbles with additional marbles resting on top of the layer.

INVESTIGATE Solubility

PURPOSE To design an experiment to demonstrate the effect of temperature on solubility

TIPS *20 min.* Students can bring many of the materials—salt, cups, spoon—from home. For best results, an ice bath could be used instead of a cold-water bath.

WHAT DO YOU THINK? *The variable being changed should involve heating or cooling water. The amount of water should stay the same. An increase in water temperature increases the solubility of salt; a decrease in water temperature decreases the solubility of salt.*

 Datasheet, Solubility, p. 220

Technology Resources

Customize this student lab as needed or look for an alternative. Print rubrics to assess student lab reports.

 Lab Generator CD-ROM

Metacognitive Strategy

Ask students to write a short paragraph discussing whether they find it difficult to design an experiment. What parts of the process do they find most challenging? What parts are easy for them?

Ongoing Assessment

CHECK YOUR READING *Answer: An increase in temperature increases the solubility of most solid solutes. An increase in temperature decreases the solubility of gases.*

The solubility of a solute can be changed.

The solubility of a solute can be changed in two ways. Raising the temperature is one way to change the solubility of the solute, because most solids are more soluble at higher temperatures. Another way to change solubility when the solute is a gas is to change the pressure. The solubility of gases in a liquid solvent increases at high pressure.

D

Temperature and Solubility

REMINDER
An increase in temperature means an increase in particle movement.

An increase in temperature has two effects on most solid solutes—they dissolve more quickly, and a greater amount of the solid dissolves in a given amount of solvent. In general, solids are more soluble at higher temperatures, and they dissolve faster.

The opposite is true of all gases—an increase in temperature makes a gas less soluble in water. You can see this by warming tap water in a pan. As the water approaches its boiling point, any air that is dissolved in the water comes out of solution. The air forms tiny bubbles that rise to the surface.

E

CHECK YOUR READING What effect does temperature have on most solid solutes? on gaseous solutes?

INVESTIGATE Solubility

How can you change solubility?

Use what you know about solubility to design an experiment that shows how a change in temperature can change the amount of table salt that will dissolve in water.

DESIGN — YOUR OWN — EXPERIMENT

PROCEDURE

1. Use the materials in the list to identify the relationship between temperature and solubility.

2. Write your procedure, identifying the constants and variables.

3. Perform your experiment and record your results.

WHAT DO YOU THINK?

- Which variables did you change? What were your constants? Why?

- How do your results demonstrate the effect of temperature on solubility?

SKILL FOCUS
Designing experiments

MATERIALS
- clear plastic cups
- thermometer
- tap water
- table salt
- balance
- plastic spoon
- hot-water bath
- cold-water bath

TIME
20 minutes

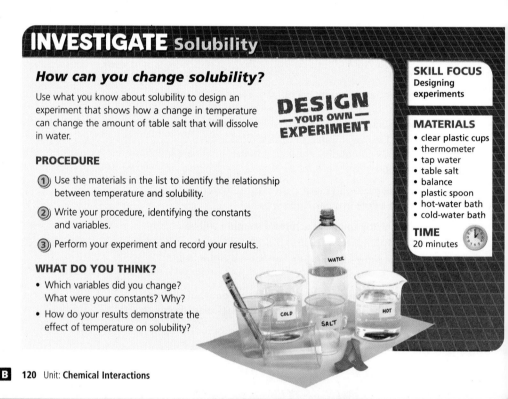

DIFFERENTIATE INSTRUCTION

More Reading Support

D Are most solid solutes more soluble in warm or cool solvents? *in warm solvent*

E Are gases more soluble in warm or in cool solvents? *in cool solvent*

Additional Investigation To reinforce Section 4.2 learning goals, use the following full-period investigation:

 Additional INVESTIGATION, Rates of Solution, A, B, & C, pp. 259–267, 329–330
(Advanced students should complete Levels B and C.)

Alternative Assessment Have each group give an oral report to the class describing their experimental procedures in the investigation. The class should then critique each group's procedure. Students should discuss what was good about each experimental design as well as what was incorrect or unclear.

Think back to the earlier discussion of supersaturated solutions. One way in which a solution can become supersaturated is through a change in temperature. For example, suppose that a solution is saturated at 50°C (122°F), and is then allowed to cool slowly. The solid is less soluble in the cooler solution, but the excess solute may not form a precipitate. As a result, the solution contains more of the dissolved solute than would be possible under normal conditions because of the change in temperature.

Temperature and Solubility		
Solute	Increased Temperature	Decreased Temperature
Solid	increase in solubility	decrease in solubility
Gas	decrease in solubility	increase in solubility

A change in temperature can produce changes in solutions in the environment. For example, a factory located on the shore of a lake may use the lake water as a coolant and then return heated water to the lake. This increase in temperature decreases the solubility of oxygen in the lake water. As a result, less oxygen will remain dissolved in the water. A decrease in the oxygen concentration can harm plant and animal life in the lake.

Changing Temperature Changes Solubility

More sugar dissolves in hot water than in cold water.

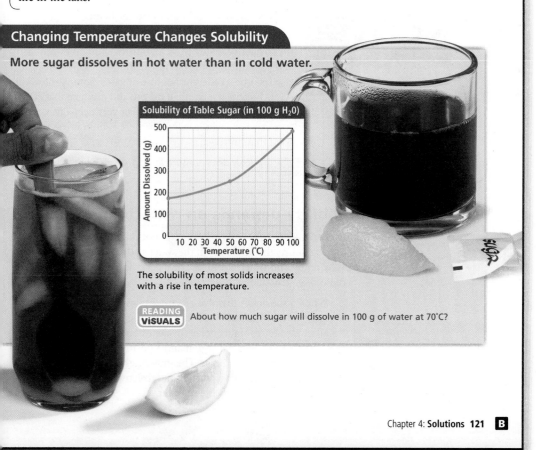

Solubility of Table Sugar (in 100 g H_2O)

The solubility of most solids increases with a rise in temperature.

READING VISUALS About how much sugar will dissolve in 100 g of water at 70°C?

Chapter 4: **Solutions** 121 **B**

Teach from Visuals

To help students interpret the graph of solubility of table sugar (sucrose) versus temperature, ask:

- Can you dissolve any sugar in iced tea? Explain. *Yes; even at almost-freezing temperatures, sugar is soluble in water.*

- What would the temperature of 100 g of tea have to be for 200 g of sugar to dissolve in it? *at least 20°C*

Integrate the Sciences

Thermal pollution occurs when warm wastewater is dumped into a body of water. This causes the temperature of the water to increase above its normal level. When the water temperature raises, it can harm animals and plants that live in the water. Fish can die from the quick change in temperature, and algae in the water grow more rapidly, blocking the sunlight from other plant life in the water. One of the main sources of thermal pollution is factories or power plants that use water to cool equipment or to produce steam and then discharge the water into a nearby river, lake, or ocean. The Environmental Protection Agency (EPA) has regulations for discharging wastewater. In the United States, many factories try to reduce thermal pollution by cooling the water before it is released so that the heat escapes through the air.

Ongoing Assessment

Describe how a solute's solubility can be changed.

Ask: How could a soft-drink manufacturer increase the amount of dissolved carbon dioxide to make the drink fizzier? *Cool the drink before adding carbon dioxide.*

READING VISUALS *Answer: about 340 g*

History Connection

Three workers involved in the building of the Brooklyn Bridge, which opened in 1883, died from the bends. The disorder was called caisson disease because workers were susceptible to it while working underwater in a caisson, or airlock, on the riverbed.

Teacher Demo

Remind students of the Teacher Demo on p. 119. Repeat the demonstration using 100 milliliters of acetone and starch-based packing peanuts. These peanuts will not dissolve in acetone. Dissolve the peanuts in water. Point out that the solubility of the two kinds of peanuts is different because the molecular structure of the peanuts is different. Both starch and water are polar. Polystyrene and acetone are nonpolar.

Integrate the Sciences

Many vitamins are fat soluble but not water soluble. Because vitamin molecules are nonpolar they are stored in a person's body fat. Vitamin C and the B-complex vitamins are exceptions—they are water soluble, and so are not stored in the body for a long period of time.

Ongoing Assessment

INFER If these divers are breathing regular air, why might they be looking at their depth gauges?

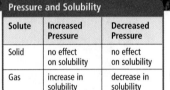

Pressure and Solubility

Solute	Increased Pressure	Decreased Pressure
Solid	no effect on solubility	no effect on solubility
Gas	increase in solubility	decrease in solubility

Pressure and Solubility

 F

A change in pressure does not usually change the solubility of solid or liquid solutes. However, the solubility of any gas increases at higher pressures and decreases at lower pressures.

When manufacturers make carbonated beverages, such as soda, they add carbon dioxide gas at a pressure slightly greater than normal air pressure. When you open the can or bottle, the pressure decreases and the carbon dioxide bubbles out of solution with a fizz.

Another example is shown in the photograph on the left. When a diver's tank contains regular air, about 79 percent of the air is nitrogen. People breathe air like this all the time without any problem, but the pressure underwater is much greater than on Earth's surface. The higher pressure increases the solubility of nitrogen in the diver's blood.

When a diver heads up to the surface too fast, the pressure decreases, and so does the solubility of the nitrogen. The nitrogen comes out of solution, forming bubbles in the diver's blood vessels. These bubbles can cause a painful and sometimes fatal condition called the bends.

Divers can avoid the bends in two ways. They can rise to the surface very slowly, so that nitrogen bubbles stay small and pass through the bloodstream more easily. They can also breathe a different mixture of gases. Some professional divers breathe a mixture of oxygen and nitrogen that contains only about 66 percent nitrogen. For very deep dives, the mixture can also include helium because helium is less soluble in blood than nitrogen.

 CHECK YOUR READING How does pressure affect the solubility of solids? of gases?

Solubility depends on molecular structure.

Everyone knows that oil and water do not mix. When a tanker spills oil near shore, the oil floats on the water and pollutes the beaches. Why do oil and water not mix? The answer involves their different molecular structures.

When a substance dissolves, its molecules or ions separate from one another and become evenly mixed with molecules of the solvent. Recall that water contains polar covalent bonds. As a result, water molecules have a negative region and a positive region. Water molecules are said to be polar. The molecules of an oil are nonpolar—the molecules do not have positive and negative regions. This difference makes oil insoluble in water.

 G

DIFFERENTIATE INSTRUCTION

? More Reading Support

F What type of solute is more soluble at higher pressures? *gas*

G What type of bonds does a water molecule contain? *polar covalent bonds*

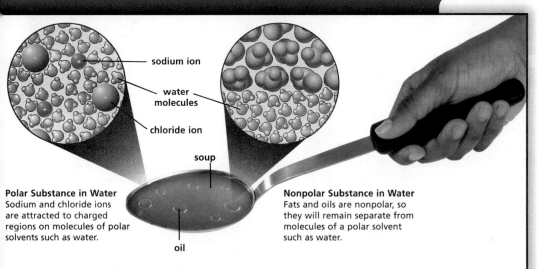

sodium ion

water molecules

chloride ion

soup

oil

Polar Substance in Water
Sodium and chloride ions are attracted to charged regions on molecules of polar solvents such as water.

Nonpolar Substance in Water
Fats and oils are nonpolar, so they will remain separate from molecules of a polar solvent such as water.

Because water is polar and oil is nonpolar, their molecules are not attracted to each other. The molecules of a polar solvent like water are attracted to other polar molecules, such as those of sugar. This explains why sugar has such a high solubility in water. Ionic compounds, such as sodium chloride, are also highly soluble in water. Because water molecules are polar, they interact with the sodium and chloride ions. In general, polar solvents dissolve polar solutes, and nonpolar solvents dissolve nonpolar solutes. This concept is often expressed as "Like dissolves like."

So many substances dissolve in water that it is sometimes called the universal solvent. Water is considered to be essential for life because it can carry so many different ions and molecules—just about anything the body needs or needs to get rid of—through the body.

CHECK YOUR READING Why will a nonpolar substance not dissolve in a polar substance?

4.2 Review

KEY CONCEPTS

1. How can a solution be made more concentrated? less concentrated?

2. What two factors can change the solubility of a gas?

3. Are nonpolar compounds highly soluble in water? Why or why not?

CRITICAL THINKING

4. **Predict** Suppose you stir sugar into ice water. Some sugar remains on the bottom of the glass. After the glass sits out for an hour, you stir it again. What will happen? Why?

5. **Infer** A powder dissolves easily in water but not in oil. Are the molecules in the powder probably polar or nonpolar? Explain.

◑ CHALLENGE

6. **Synthesize** If mixing a substance with water forms a suspension, does the substance have a high or a low solubility in water? Explain.

Chapter 4: **Solutions 123** **B**

ANSWERS

1. Add more solute, keeping the solvent constant; add more solvent, keeping the solute constant.

2. temperature and pressure

3. No; water is polar, and nonpolar substances are not highly soluble in polar solvents.

4. More of the sugar will dissolve as the water's temperature increases.

5. They are probably polar. They dissolve well in water, which is polar, but not in oil, which is nonpolar.

6. It has a low solubility because little if any of the substance dissolves.

Ongoing Assessment

Recognize that solubility depends on molecular structure.

Ask: What must happen between molecules for a solute to dissolve? *Solute and solvent particles must attract each other.*

CHECK YOUR READING *Answer: Nonpolar solutes do not have charged regions that would be attracted to charged regions of polar solvents.*

Reinforce (the **BIG** idea)

Have students relate the section to the Big Idea.

 R Reinforcing Key Concepts, p. 221

4.2 ASSESS & RETEACH

Assess

 A Section 4.2 Quiz, p. 62

Reteach

Discuss with students the types of solutions that have high solubility and low solubility. For example, ammonia has high solubility because a large amount of it is dissolved in water. Remind students that the amount of solute to make a saturated solution depends on its solubility. Ask students what happens to a saturated solution when a solute is highly soluble. What if it has low solubility? *When a solute is highly soluble, a saturated solution will be very concentrated; with low solubility, the solution will be dilute.*

Technology Resources

Have students visit **ClassZone.com** for reteaching of Key Concepts.

 CONTENT REVIEW

 CONTENT REVIEW CD-ROM

Set Learning Goal

To understand that drinking water and groundwater are solutions

Present the Science

About two-thirds of the water drawn from aquifers in the United States is used for irrigation. Many aquifers are being drawn down below their natural recharge rates, resulting in subsidence and changes in water quality.

Some minerals that dissolve in groundwater are annoying and others are harmful. For example, well water that contains a large amount of sulfur smells like rotten eggs. Also, some minerals contain arsenic, which is a poison.

Discussion Questions

- Ask: Where do solutes in groundwater come from? *soil, rocks, rain, wastewater*

- Ask: How do the solutes in water change underground? *They may be trapped on clay or removed in chemical reactions with the soil and rocks.*

- Ask: What might be some sources of water pollution in our area? *Answers might include pesticide and fertilizer from lawns and fields, underground tanks, byproducts of manufacturing.*

DIFFERENTIATION TIP Demonstrate the difference between hard water and soft water. Pour equal amounts of hard and soft water into two jars. Add the same amount of liquid soap to each jar, shake them, and have students observe the different amounts of suds produced and the formation of scum or insoluble particulates.

Close

Ask: How could you soften hard water? *Sample answer: By adding a chemical that would cause solutes to precipitate out of the solution; physical filtering can't do it alone, because the minerals are dissolved.*

CONNECTING SCIENCES

PHYSICAL SCIENCE AND EARTH SCIENCE

Water that looks clear, clean, and pure may not be. The water you drink every day contains many dissolved substances.

Cool, Clear Water

The drinking water that comes out of a tap is a solution. Many minerals, chemicals, and even gases are dissolved in it. Some drinking water comes from rivers, lakes, or reservoirs, but about half of the drinking water in the United States is pumped from wells. Well water comes from underground aquifers. The water in aquifers flows through gaps in broken or porous rocks.

Filtering Impurities

The water in a puddle is not pure water. It contains suspended dirt and dissolved chemicals. This water can be cleaned underground. As the solution flows through soil and rocks, the soil and rocks filter and trap particles. Some chemicals in the water are removed from the solution by clay particles in the soil. Other chemicals, such as acids from acid rain, are neutralized by limestone and other rocks.

Adding Minerals

The rocks surrounding aquifers do not just remove chemicals from water. As water flows underground, minerals dissolve in the water. The solutes include compounds of calcium, magnesium, and iron. These compounds do not harm the quality of drinking water because they are necessary parts of your diet. Water with a high concentration of dissolved minerals is called hard water.

Copying Earth

Water that has been used by people must be cleaned before it is returned to the environment. Waste treatment plants copy some of the natural cleansing processes of Earth. Wastewater solutions may contain many dissolved impurities and harmful chemicals, but the water can be filtered through beds of sand and gravel. Water must be treated after it is used because so many substances dissolve in it.

Aquifer Layer

EXPLORE

1. **INFER** A white solid often forms around a tiny leak in a water pipe. Where does the white solid come from?

2. **CHALLENGE** Use the Internet or call your local water company to find out the source of your drinking water. Find out whether you have hard or soft water and what dissolved chemicals are in your drinking water.

RESOURCE CENTER
CLASSZONE.COM
Learn more about aquifers and water purification.

EXPLORE

1. *INFER Leaking water evaporates, leaving the solutes.*

2. *CHALLENGE Hard water usually contains relatively high concentrations of calcium, magnesium, or iron. Soft water usually has small quantities of these (and other) elements.*

KEY CONCEPT

Solutions can be acidic, basic, or neutral.

BEFORE, you learned

- Substances dissolved in solutions can break apart into ions
- Concentration is the amount of a substance dissolved in a solution
- Water is a common solvent

NOW, you will learn

- What acids and bases are
- How to determine if a solution is acidic or basic
- How acids and bases react with each other

VOCABULARY

acid p. 126
base p. 126
pH p. 129
neutral p. 129

EXPLORE Acids and Bases

What happens when an antacid mixes with an acid?

PROCEDURE

1. Fill the cup halfway with vinegar.
2. Observe the vinegar in the cup. Record your observations.
3. Crush two antacid tablets and place them in the vinegar.
4. Observe the contents of the cup for 5 minutes. Record your observations.

MATERIALS
- clear plastic cup
- vinegar
- 2 antacid tablets

WHAT DO YOU THINK?
- What did you observe before adding the antacid tablets?
- What happened after you added the tablets?

Acids and bases have distinct properties.

Many solutions have certain properties that make us call them acids or bases. Acids are found in many foods, such as orange juice, tomatoes, and vinegar. They taste slightly sour when dissolved in water and produce a burning or itchy feeling on the skin. Strong acids should never be tasted or touched—these solutions are used in manufacturing and are dangerous chemicals.

Bases are the chemical opposite of acids. They tend to taste bitter rather than sour and often feel slippery to the touch. Bases are also found in common products around the home, including soap, ammonia, and antacids. Strong bases, like the lye used for unclogging drains, are also dangerous chemicals.

READING TiP

The prefix *ant-* means "against," so an antacid is a substance that works against an acid.

RESOURCES FOR DIFFERENTIATED INSTRUCTION

Below Level
UNIT RESOURCE BOOK
- Reading Study Guide A, pp. 224–225
- Decoding Support, p. 247

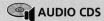 **AUDIO CDS**

Advanced
UNIT RESOURCE BOOK
- Challenge and Extension, p. 230
- Challenge Reading, pp. 243–244

English Learners
UNIT RESOURCE BOOK
Spanish Reading Study Guide, pp. 228–229

AUDIO CDS
- Audio Readings in Spanish
- Audio Readings (English)

Set Learning Goals
Students will
- Explain what acids and bases are.
- Determine if a solution is acidic or basic.
- Describe how acids and bases react with each other.

3-Minute Warm-Up

Display Transparency 29 or copy this exercise on the board:

Look at the illustration of the dilute solution on p. 118. Draw diagrams that show this solution as more dilute and as more concentrated. *A more dilute solution should have fewer solute particles and the same number of solvent particles, or the same number of solute particles and more solvent particles. A more concentrated solution should have more solute particles and the same number of solvent particles, or the same number of solute particles and fewer solvent particles.*

T 3-Minute Warm-Up, p. T29

4.3 MOTIVATE

EXPLORE Acids and Bases

PURPOSE To observe the chemical reaction between an acid and a base

TIP *10 min.* The antacid used must contain a carbonate. Check the label.

WHAT DO YOU THINK? *The vinegar looked like water but had a strong, sour smell. White foamy bubbles appeared and made a fizzing sound. The vinegar odor was not as strong.*

Teach from Visuals

To help students interpret atomic diagrams of an acid and a base dissolving in water, ask:

- What in the diagrams indicates that the acid and base are dissolved in water? *the H₂O shown above the arrows*

- How does the size of a hydrogen ion compare with the size of other ions? *It is the smallest.*

- Why do you think this is so? *Hydrogen has only one electron to begin with. When hydrogen loses its only electron to become a hydrogen ion, it is a bare hydrogen nucleus, or a proton.*

History of Science

The definitions of acids and bases have changed over time. The first generally accepted theory was the Arrhenius model of acids and bases, which states that an acid contains hydrogen and produces hydrogen ions in solution. A base contains a hydroxide group and produces hydroxide ions in solution. Currently, acids and bases are defined by the Brønsted-Lowry theory. This model states that acids are hydrogen-ion donors and bases are hydrogen-ion acceptors. The definitions used in this book are based on the Brønsted-Lowry theory.

Ongoing Assessment

Explain what acids and bases are.

Ask: What happens to the hydrogen and hydroxide ions that are released from acids and bases? *They are free to combine with other substances in the solution.*

CHECK YOUR READING *Answer: Acids donate a proton; bases accept a proton.*

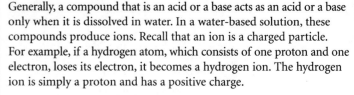
RESOURCE CENTER
CLASSZONE.COM
Find out more about acids and bases.

 **A**

 B

Acids, Bases, and Ions

Generally, a compound that is an acid or a base acts as an acid or a base only when it is dissolved in water. In a water-based solution, these compounds produce ions. Recall that an ion is a charged particle. For example, if a hydrogen atom, which consists of one proton and one electron, loses its electron, it becomes a hydrogen ion. The hydrogen ion is simply a proton and has a positive charge.

An **acid** can be defined as a substance that can donate a hydrogen ion—that is, a proton—to another substance. The diagram below shows what happens when the compound hydrogen chloride (HCl) is dissolved in water. The compound separates into hydrogen ions (H^+) and chloride ions (Cl^-). Hydrogen ions are free to react with other substances, so the solution is an acid. When hydrogen chloride is dissolved in water, the solution is called hydrochloric acid.

Acid

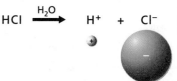

$$HCl \xrightarrow{H_2O} H^+ + Cl^-$$

In water, acids release a proton (H^+) into the solution.

A **base** can be defined as a substance that can accept a hydrogen ion from another substance. The diagram below shows what happens when the compound sodium hydroxide (NaOH) is dissolved in water. The compound separates into sodium ions (Na^+) and hydroxide ions (OH^-). The hydroxide ions are free to accept protons from other substances, so the solution is a base. The solution that results when NaOH is dissolved in water is called sodium hydroxide.

Base

$$NaOH \xrightarrow{H_2O} Na^+ + OH^-$$

In water, many bases release a hydroxide ion (OH^-), which can accept a proton.

On the atomic level, the difference between acids and bases is that acids donate protons and bases accept protons. When a proton—a hydrogen ion—from an acid is accepted by a hydroxide ion from a base, the two ions join together and form a molecule of water. This simple transfer of protons between substances is involved in a great many useful and important chemical reactions.

 CHECK YOUR READING How are protons related to acids and bases?

DIFFERENTIATE INSTRUCTION

 **More Reading Support**

A What type of substance can donate a hydrogen ion? *an acid*

B What type of substance can accept a hydrogen ion? *a base*

English Learners The many word forms of *acid* and *base* in this section may confuse English learners. *Acid* and *acidity* are nouns, while *acidic* is an adjective. *Base* and *basicity* are nouns, while *basic* is an adjective. Help English learners use the correct word form when they are writing.

Characteristics of Acids

As you read earlier, acids in foods taste sour and produce a burning or prickling feeling on the skin. However, since tasting or touching an unknown chemical is extremely dangerous, other methods are needed to tell whether a solution is an acid.

One safe way to test for an acid is to place a few drops of a solution on a compound that contains a carbonate (CO_3). For example, limestone is a rock that contains calcium carbonate ($CaCO_3$). When an acid touches a piece of limestone, a reaction occurs that produces carbon dioxide gas.

Acids also react with most metals. The reaction produces hydrogen gas, which you can see as bubbles in the photograph on the right. Such a reaction is characteristic of acids.

The feature of acids most often used to identify them is their ability to change the colors of certain compounds known as acid-base indicators. One common indicator is litmus, which is often prepared on slips of paper. When a drop of an acid is placed on litmus paper, the paper turns red.

Acids react with some metals, such as zinc, and release hydrogen gas.
$2HCl + Zn \longrightarrow H_2 + ZnCl_2$

CHECK YOUR READING What are three safe methods to test for an acid?

Characteristics of Bases

Bases also have certain common characteristics. Mild bases in foods taste bitter and feel slippery, but as with acids, tasting and touching are not safe ways of testing whether a solution is a base. In fact, some strong bases can burn the skin as badly as strong acids.

Bases feel soapy or slippery because they react with acidic molecules in your skin called fatty acids. In fact, this is exactly how soap is made. Mixing a base—usually sodium hydroxide—with fatty acids produces soap. So, when a base touches your skin, the combination of the base with your own fatty acids actually makes a small amount of soap.

Like acids, bases change the colors of acid-base indicators, but the colors they produce are different. Bases turn litmus paper blue. A base will counteract the effect that an acid has on an acid-base indicator. You might put a few drops of acid on litmus paper to make it turn red. If you put a few drops of a base on the red litmus paper, the litmus paper will change colors again.

CHECK YOUR READING How do the characteristics of bases differ from those of acids?

Bases are found in many cleaning agents, including soap.

Chapter 4: **Solutions** 127 **B**

More Reading Support

C What happens when an acid reacts with a carbonate or a metal? *It releases a gas.*

D How does a base feel? *slippery or soapy*

Alternative Assessment Have students make a two-column checklist that they can use to identify unknown substances for characteristics of acids and bases.

Address Misconceptions

IDENTIFY Ask students what actions they would take if they were to spill a strong base on exposed skin. If they reply that they would need to take no actions or that bases are not harmful, they may hold the misconception that acids are bad and bases are good.

CORRECT Ask students if they think drain or oven cleaners are safe chemicals. Present labels from such products and draw attention to words such as *danger* and *caustic* on them. Explain that *caustic* usually refers to a chemical that is a strong base.

REASSESS Ask students again if a base can be as dangerous as an acid. *It can.* Ask them what information they would put on the label of a product that contained a strong base. *Keep out of reach of children. Warning: can irritate skin.*

Technology Resources

Visit **ClassZone.com** for background on common student misconceptions.

MISCONCEPTION DATABASE

Real World Example

Saponification is the process that makes soap. Fats, which contain fatty acids, react with a strong base such as sodium hydroxide to form glycerol and a soap. Soaps are the sodium salts of fatty acids. They have a polar end and a nonpolar end. The nonpolar end can bond to nonpolar dirt and oils, and the polar end is soluble in water. Dirt can then be rinsed away.

Ongoing Assessment

Determine if a solution is acidic or basic.

Ask: What colors do acids and bases turn litmus paper? *Acids turn litmus paper red; bases turn it blue.*

CHECK YOUR READING *Answer: Look for a reaction with a carbonate, a metal, or an indicator such as litmus.*

CHECK YOUR READING *Answer: Bases taste bitter (not sour, like acids) and feel slippery or soapy. Bases turn litmus blue, rather than red.*

Teach from Visuals

To help students interpret the molecular depiction of strong and weak acids, ask:

- Are more ions present in the solution of a strong acid or a weak acid? *strong acid*

- Which solution contains intact acid molecules? *the weak acid*

- In which solution are more hydrogen ions available to be donated to other substances? *the strong acid*

Health Connection

Stomach acid, which is mainly HCl, is a strong acid that can seriously burn skin. Why doesn't stomach acid burn the lining of the stomach? The inner surface of the stomach is covered with a protective coating of mucus. If a break develops in this protective coating, the acid begins to burn the cells of the stomach's inner surface, and an ulcer forms. Medical research suggests that acid-resistant bacteria can make holes in the mucus and contribute to ulcer formation.

Develop Critical Thinking

PREDICT Tell students that when ions are present in a solution, the solution can conduct electricity. The more ions present, the stronger the electric current that will be conducted. Ask: What can you predict about the conductivity of a solution containing a weak acid? *The solution does not conduct electricity very well, if at all.*

Art Connection

Hydrofluoric acid (HF) is a weak but extremely dangerous acid that is used to decorate and etch glass. Hydrofluoric acid may be used to give glass a frosted appearance. Designs can be etched in glass by coating the glass with wax and then cutting the pattern to be etched through the wax layer to the glass. The acid will etch the uncovered glass, but will not harm the wax-covered areas.

MIND MAP Remember to use a mind map to take notes about acid and base strength.

The strengths of acids and bases can be measured.

Battery fluid and many juices contain acids. Many people drink some type of juice every morning, but you would not want to drink, or even touch, the liquid in a car battery. Similarly, you probably wash your hands with soap several times a day, but you would not want to touch the liquid used to unclog drains. Both soap and drain cleaners are bases. Clearly, some acids and bases are stronger than others.

Acid and Base Strength

Strong acids break apart completely into ions. For example, when hydrogen chloride (HCl) dissolves in water to form hydrochloric acid, it breaks down into hydrogen ions and chloride ions. No hydrogen chloride remains in the solution. Because all of the hydrogen chloride forms separate ions, hydrochloric acid is a strong acid.

A weak acid does not form many ions in solution. When acetic acid ($HC_2H_3O_2$), which is the acid in vinegar, dissolves in water, only about 1 percent of the acetic acid breaks up into hydrogen ions and acetate ions. The other 99 percent of the acetic acid remains unchanged. As a result, acetic acid is a weak acid.

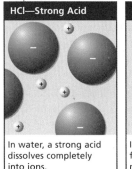

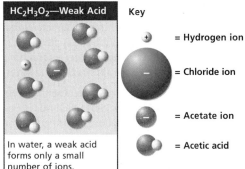

HCl—Strong Acid

In water, a strong acid dissolves completely into ions.

$HC_2H_3O_2$—Weak Acid

In water, a weak acid forms only a small number of ions.

Key

+ = Hydrogen ion

− = Chloride ion

= Acetate ion

= Acetic acid

Bases also can be strong or weak. When sodium hydroxide (NaOH) dissolves in water, it forms sodium ions (Na^+) and hydroxide ions (OH^-). None of the original NaOH remains in the solution, so sodium hydroxide is a strong base. However, when ammonia (NH_3) dissolves in water, only about 1 percent of the ammonia reacts with water to form OH^- ions.

$$NH_3 + H_2O \longrightarrow NH_4^+ + OH^-$$

The other 99 percent of the ammonia remains unchanged, so ammonia is a weak base. The ions formed when NaOH or NH_3 is dissolved in water are shown on the top of page 129.

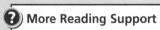

DIFFERENTIATE INSTRUCTION

? More Reading Support

E What happens to strong acids and bases when they dissolve? *They break apart completely into ions.*

Advanced Students may notice that acid and base molecules, such as HCl or NH_3, are covalent compounds rather than ionic compounds, yet dissociate into ions in solution. These substances are exceptions to the information presented earlier in the chapter. The covalent bond that holds HCl together is very polar, and is intermediate between a covalent bond and an ionic bond.

 Challenge Reading, pp. 243–244

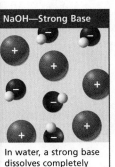

NaOH—Strong Base	NH₃—Weak Base	Key

In water, a strong base dissolves completely into ions.

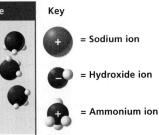

In water, a weak base forms only a small number of ions.

+ = Sodium ion

− = Hydroxide ion

+ = Ammonium ion

= Ammonia

READING TiP

Look at the reaction on the bottom of page 128 for help with the illustration of NH_3 in water.

Note that the strength of an acid or base is not the same as its concentration. Dilute hydrochloric acid is still strong and can burn holes in your clothing, whereas acetic acid cannot. The strengths of acids and bases depend on the percentage of the substance that forms ions.

 CHECK YOUR READING   What determines acid and base strength?

Measuring Acidity

The acidity of a solution depends on the concentration of H^+ ions in the solution. This concentration is often measured on the **pH** scale. In this scale, a high H^+ concentration is indicated by a low number, and a low H^+ concentration is indicated by a high number. The numbers of the pH scale usually range from 0 to 14, but numbers outside this range are possible. The middle number, 7, represents a neutral solution. A **neutral** substance is neither an acid nor a base. Pure water has a pH of 7.

Numbers below 7 indicate acidic solutions. A concentrated strong acid has a low pH value—the pH of concentrated hydrochloric acid, for example, is less than 0. Numbers above 7 indicate a basic solution. A concentrated strong base has a high pH value—the pH of concentrated sodium hydroxide, for example, is greater than 14. The illustration on page 130 shows the pH values of some common acids and bases.

Today, electronic pH meters are commonly used to measure pH. A probe is placed in a solution, and the pH value is indicated by the meter. An older method of measuring pH is to use an acid-base indicator. You read earlier that acids turn litmus paper red and bases turn litmus paper blue. Other acid-base indicators, such as a universal pH indicator, show a variety of colors at different pH values.

The strip of universal indicator paper in the bottom front of the photograph shows a nearly neutral pH.

 CHECK YOUR READING   Is the pH of a base higher or lower than the pH of an acid?

DIFFERENTIATE INSTRUCTION

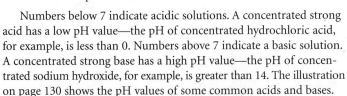 **More Reading Support**

F What does the pH scale measure? *concentration of hydrogen ions*

G What does a pH below 7 mean? *The solution is acidic.*

Inclusion Print these formulas on cardboard: HCl, NaOH, NH₃, and HC₂H₃O₂. Cut apart the H and Cl and the Na and OH, like jigsaw puzzle pieces. Leave the other two formulas intact. Students with cognitive disabilities will benefit by manipulating the "molecules" and separating the strong acid and base into ions. They will see that the weak acid and base cannot be separated.

Teach from Visuals

To help students interpret the molecular depiction of strong and weak bases, ask:

- How does the number of sodium ions compare with the number of hydroxide ions in a solution of sodium hydroxide? *They are equal.*
- Where does the hydroxide ion come from when ammonia is dissolved in water? *from a water molecule*

Teach Difficult Concepts

If students have trouble differentiating between the strength and the concentration of an acid or base, use this analogy. Two boxes of the same size represent two solutions. One box contains 100 pennies. The other box contains 100 quarters. Ask:

- How does the concentration of coins in the boxes compare? *They have the same concentration.*
- How does the buying "strength" of the two boxes compare? *The box containing quarters has much more buying strength.*

Integrate the Sciences

The pH of human blood has a very narrow range, around 7.4. This is slightly basic. Even a small deviation in blood pH can lead to, or be an indication of, serious health problems.

EXPLORE the **BIG** idea

Revisit "Acid Test" on p. 109. Have students explain their results.

Ongoing Assessment

CHECK YOUR READING *Answer: The greater their dissociation into ions in water, the stronger the acid or base will be.*

CHECK YOUR READING *Answer: higher*

To help students interpret the pH chart, ask:

- Would you expect fatty acids to be weak or strong acids? Explain. *Weak; they make milk only slightly acidic.*

- Why is the hydrogen ion concentration equal to the hydroxide ion concentration in pure water? *A water molecule dissociates into one hydrogen ion and one hydroxide ion.*

 This transparency is also available as T30 in the Unit Transparency Book.

Mathematics Connection

The pH scale is based on a negative logarithmic scale that measures the concentration of hydrogen ions. Therefore, a pH of 0 is 1×10^0, or 1.0, while a pH of 14 represents a hydrogen ion concentration of 1×10^{-14}, or 0.00000000000001. Ask students how many times greater the hydrogen ion concentration is if the pH of a solution decreases by one unit, as from a pH of 4 to a pH of 3. *ten times*

Teacher Demo

To make an indicator solution that will identify the pH of substances, place three or four chopped-up, fresh, red cabbage leaves in a plastic bottle and fill the bottle halfway with hot tap water. Screw the top on tightly, and shake for several minutes until the water turns deep purple. Cool to room temperature. Strain the indicator solution, and add water to make a final volume of one liter. Place a small amount of each liquid to be tested in a plastic cup. Fill the cup halfway with indicator solution, and stir. The indicator solution will turn color ranging from cherry red for very acidic solutions through purple (neutral), blue, green, and yellow (strongly basic).

Ongoing Assessment

READING VISUALS *Answer: The strong acids are at the low end of the pH scale; the strong bases are at the high end. The hydrogen ion concentration decreases as pH increases.*

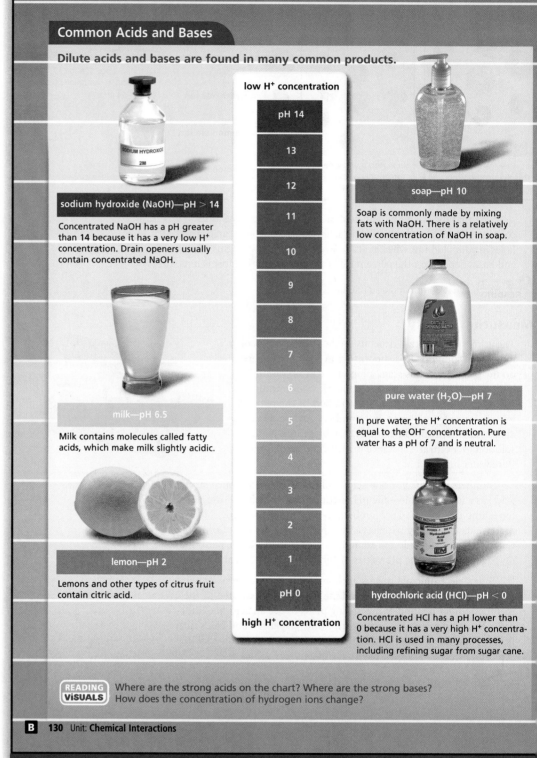

Common Acids and Bases

Dilute acids and bases are found in many common products.

low H⁺ concentration

| pH 14 |
| 13 |
| 12 |
| 11 |
| 10 |
| 9 |
| 8 |
| 7 |
| 6 |
| 5 |
| 4 |
| 3 |
| 2 |
| 1 |
| pH 0 |

high H⁺ concentration

sodium hydroxide (NaOH)—pH > 14

Concentrated NaOH has a pH greater than 14 because it has a very low H⁺ concentration. Drain openers usually contain concentrated NaOH.

milk—pH 6.5

Milk contains molecules called fatty acids, which make milk slightly acidic.

lemon—pH 2

Lemons and other types of citrus fruit contain citric acid.

soap—pH 10

Soap is commonly made by mixing fats with NaOH. There is a relatively low concentration of NaOH in soap.

pure water (H₂O)—pH 7

In pure water, the H⁺ concentration is equal to the OH⁻ concentration. Pure water has a pH of 7 and is neutral.

hydrochloric acid (HCl)—pH < 0

Concentrated HCl has a pH lower than 0 because it has a very high H⁺ concentration. HCl is used in many processes, including refining sugar from sugar cane.

READING VISUALS Where are the strong acids on the chart? Where are the strong bases? How does the concentration of hydrogen ions change?

DIFFERENTIATE INSTRUCTION

Below Level Students who have difficulty reading can give an oral report summarizing the main science concepts represented in the chart. Reports should include the meaning of pH and identification of the nature of substances at various pH points on the chart.

Advanced

 Challenge and Extension, p. 230

Acids and bases neutralize each other.

Acids donate hydrogen ions, and bases accept hydrogen ions. Therefore, it is not surprising that acids and bases react when they come into contact with each other. Recall that when a hydrogen ion (H^+) from an acid collides with a hydroxide ion (OH^-) from a base, the two ions join to form a molecule of water (H_2O).

The negative ion of an acid (Cl^-) joins with the positive ion of a base (Na^+) to form a substance called a salt. Since both the salt and water are neutral, an acid-base reaction is called a neutralization (NOO-truh-lih-ZAY-shuhn) reaction. The reactants are an acid and a base, and the products are a salt and water.

READING TIP

The salt produced by a neutralization reaction is not necessarily table salt.

A common example of a neutralization reaction occurs when you swallow an antacid tablet to relieve an upset stomach. The acid in your stomach has a pH of about 1.5, due mostly to hydrochloric acid produced by the stomach lining. If your stomach produces more acid than is needed, you may feel a burning sensation. An antacid tablet contains a base, such as sodium bicarbonate, magnesium hydroxide, or calcium carbonate. The base reacts with the stomach acid and produces a salt and water. This reaction lowers the acidity—and raises the pH—to its normal value.

Acid rain forms when certain gases in the atmosphere dissolve in water vapor, forming acidic solutions. During rainstorms these acids fall to Earth. They can harm forests by making soil acidic and harm aquatic life by making lakes acidic. Acid rain can also dissolve marble and limestone in buildings and statues, because both marble and limestone contain calcium carbonate, which is a base.

 CHECK YOUR READING How is neutralization an example of a chemical reaction?

4.3 Review

KEY CONCEPTS

1. Use the concept of ions to explain the difference between an acid and a base.

2. How do the properties of an acid differ from the properties of a base?

3. What happens when an acid and a base react with each other?

CRITICAL THINKING

4. **Infer** When an acid reacts with a metal, such as zinc, what is released? Where does that product come from?

5. **Infer** Suppose that you have 1 L of an acid solution with a pH of 2. You add 1 L of pure water. What happens to the pH of the solution? Explain.

◔ CHALLENGE

6. **Synthesize** Suppose that equal amounts of solutions of HCl and NaOH with the same concentration are mixed together. What will the pH of the new solution be? What are the products of this reaction?

Chapter 4: **Solutions** 131 **B**

ANSWERS

. An acid is a substance that an give up a proton (a ydrogen ion). A base is a ubstance that can accept a roton.

. taste, feel, reactions with netals and carbonates, color hange in indicators

3. A neutralization reaction occurs, forming a salt and water.

4. The original acid releases hydrogen gas.

5. Adding water dilutes the solution. The concentration of hydrogen ions decreases, so the pH goes up.

6. The pH will move toward 7 (neutral). The products are water and table salt (NaCl).

Focus

PURPOSE To test and classify common substances as acidic or basic

OVERVIEW Students will test various household substances to determine whether they are acidic or basic. Students should find that:

- fruit juice, soda water, and vinegar are acidic
- baking soda, shampoo, and detergent powder are basic
- table salt is neutral

Lab Preparation

- Remind students to measure to the bottom of the meniscus in the graduated cylinder.
- Use a shampoo that is not deeply colored. The color could interfere with the testing. Not all shampoos are basic.
- Prior to the investigation, have students read through the investigation and prepare their data tables. Or you may wish to copy and distribute datasheets and rubrics.

 UNIT RESOURCE BOOK, pp. 250–258

 SCIENCE TOOLKIT, F15

Lab Management

- This lab cannot use litmus paper; it requires universal indicator paper that detects a wide pH range.
- Clear film canisters can be used in place of plastic cups.
- You could divide the class into seven groups, with each group testing a different substance.

Teaching with Technology

Students can use a probeware system with a pH probe to measure the pH of the substances.

CHAPTER INVESTIGATION

Acids and Bases

OVERVIEW AND PURPOSE Acids and bases are very common. For example, the limestone formations in the cave shown on the left are made of a substance that is a base when it is dissolved in water. In this activity you will use what you have learned about solutions, acids, and bases to

- test various household substances and place them in categories according to their pH values
- investigate the properties of common acids and bases

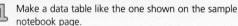 Procedure

1. Make a data table like the one shown on the sample notebook page.

2. Set out 7 cups in your work area. Collect the substances that you will be testing: baking soda, fruit juice, shampoo, soda water, table salt, laundry detergent, and vinegar.

3. Label each cup. Be sure to wear goggles when pouring the substances that you will be testing. Pour 30 mL of each liquid substance into a separate cup. Dissolve 1 tsp of each solid substance in 30 mL of distilled water in a separate cup. To avoid contaminating the test solutions, wash and dry your measuring tools and hands between measurements.

MATERIALS
- plastic cups
- baking soda
- fruit juice
- shampoo
- soda water
- table salt
- detergent powder
- vinegar
- masking tape
- marking pen
- measuring spoons
- graduated cylinder
- distilled water
- paper towels
- pH indicator paper

INVESTIGATION RESOURCES

 CHAPTER INVESTIGATION, Acids and Bases
- Level A, pp. 250–253
- Level B, pp. 254–257
- Level C, p. 258

Advanced students should complete Levels B & C.

 Writing a Lab Report, D12–13

Technology Resources

Customize this student lab as needed or look for an alternative. Print rubrics to assess student lab reports.

 Lab Generator CD-ROM

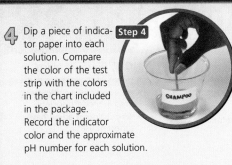

4. Dip a piece of indicator paper into each solution. Compare the color of the test strip with the colors in the chart included in the package. Record the indicator color and the approximate pH number for each solution. **Step 4**

5. After you have tested all of the solutions, arrange the cups in order of their pH values.

▶ Observe and Analyze [Write It Up]

1. **RECORD DATA** Check to be sure that your data table is complete.

2. **ANALYZE DATA** What color range did the substances show when tested with the indicator paper? What do your results tell you about the pH of each substance you tested?

3. **CLASSIFY** Look for patterns in the pH values. Use your test results to place each household substance in one of three groups—acids, bases, or neutral.

4. **MODEL** Draw a diagram of the pH scale from 0 to 14. Use arrows and labels to show where the substances you tested fall on this scale.

▶ Conclude [Write It Up]

1. **GENERALIZE** What general conclusions can you draw about the hydrogen ion concentration in many acids and bases found in the home? Are the hydrogen ion concentrations very high or very low? How do you know?

2. **EVALUATE** What limitations or difficulties did you experience in interpreting the results of your tests or other observations?

3. **APPLY** Antacid tablets react with stomach acid containing hydrochloric acid. What is this type of reaction called? What are the products of this type of reaction?

▶ INVESTIGATE Further

CHALLENGE Repeat the experiment, changing one variable. You might change the concentrations of the solutions you are testing or see what happens when you mix an acidic solution with a basic solution. Get your teacher's approval of your plan before proceeding. How does changing one particular variable affect the pH of the solutions?

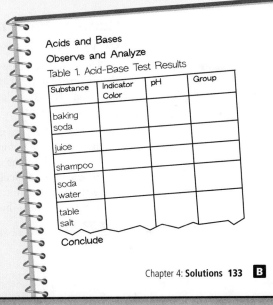

Acids and Bases
Observe and Analyze
Table 1. Acid-Base Test Results

Substance	Indicator Color	pH	Group
baking soda			
juice			
shampoo			
soda water			
table salt			

Conclude

Chapter 4: **Solutions** 133 **B**

▶ Observe and Analyze [Write It Up]

1. *Sample data: baking soda, pH 8; detergent powder, pH 10; fruit juice, pH 4; shampoo, pH 5; soda water, pH 5; table salt, pH 7; vinegar, pH 2*

2. *The range of indicator colors and pH should vary between relatively weak acids and relatively weak bases.*

3. *The placement of substances on the pH scale will vary with observations and the particular products tested. Baking soda, detergents, and some shampoos are basic. Soda water, vinegar, fruit juices, and some shampoos are acidic. Table salt (dissolved in distilled water) should be neutral.*

4. *See students' diagrams, which should properly place their test results on a pH scale.*

▶ Conclude [Write It Up]

1. *In general, the hydrogen ion concentrations of many household products is neither very high nor very low. They are often in a middle range, between a pH of 2 and a pH of 12 as indicated by the sample of substances tested.*

2. *Answers will vary, but may indicate pH tests that do not correspond exactly to the color standards.*

3. *A neutralization reaction; water and a salt*

▶ INVESTIGATE Further

CHALLENGE Changing the concentrations changes the pH slightly but does not make an acidic substance basic, or vice versa.

Post-Lab Discussion

Ask: Why might it be important to test the pH of the distilled water used to dissolve the solids? *If the pH of the water is not neutral, it will affect the pH of the solution being tested.*

Discuss why it is important for manufacturers and water-treatment plants to regulate the pH of their products. Lead the discussion to include the dangers of extreme pH to plants and animals.

▶ Set Learning Goals

Students will

- Describe how metal alloys are made.
- Identify how a variety of alloys are used in modern society.
- Explain why different alloys have different uses.
- Differentiate between a pure metal and its alloy by observing their properties in an experiment.

◑ 3-Minute Warm-Up

Display Transparency 29 or copy this exercise on the board:

Match the definition to the correct term.

Definitions

1. amount of a substance that will dissolve in a certain amount of solvent at a certain temperature *e*

2. describes a solution that can hold no more of a solute at a particular temperature *b*

3. amount of dissolved solute in a solution *a*

Terms

a. concentration   d. solution
b. saturated   e. solubility
c. dilute

📘 3-Minute Warm-Up, p. T29

◢◢ MOTIVATE

THINK ABOUT

PURPOSE To examine why mixing metals might be desirable

DISCUSS Brainstorm possible reasons why goldsmiths might want to add other metals to pure gold. *to make the gold harder and more durable, to change its color*

CHECK YOUR READING *Answer: by melting and mixing the metals, then waiting for the alloy to cool and solidify*

4.4 Metal alloys are solid mixtures.

◁ **BEFORE, you learned**

- A solution can be a solid
- Solutes change the properties of solvents
- The concentration of a solution can vary

▷ **NOW, you will learn**

- How metal alloys are made
- How a variety of alloys are used in modern society
- Why different alloys have different uses

VOCABULARY

alloy p. 134

 VOCABULARY
Remember to use the strategy of your choice. You might use a description wheel diagram for *alloy*.

THINK ABOUT

If gold jewelry is not pure gold, what is it?

People have prized gold since ancient times—archaeologists have found gold jewelry that was made thousands of years ago. Gold is a very soft metal, and jewelry made of pure gold bends very easily. Today, most gold jewelry is about 75 percent gold and 25 percent other metals. Why might these metals be mixed in?

Humans have made alloys for thousands of years.

The gold used in jewelry is an example of an alloy. An **alloy** is a mixture of a metal and one or more other elements, usually metals as well. The gold alloys used in jewelry contain silver and copper in various amounts.

Many alloys are made by melting the metals and mixing them in the liquid state to form a solution. For example, bronze is made by melting and mixing copper and tin and then letting the solution cool. Bronze is not difficult to make, because both copper and tin melt at relatively low temperatures. Bronze was probably the first alloy made in ancient times—historians say it was discovered about 3800 B.C.

 CHECK YOUR READING How is an alloy usually made?

RESOURCES FOR DIFFERENTIATED INSTRUCTION

Below Level
UNIT RESOURCE BOOK
- Reading Study Guide A, pp. 234–235
- Decoding Support, p. 247

💿 **AUDIO CDS**

Advanced
UNIT RESOURCE BOOK
Challenge and Extension, p. 240

English Learners
UNIT RESOURCE BOOK
Spanish Reading Study Guide, pp. 238–239

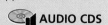 **AUDIO CDS**

- Audio Readings in Spanish
- Audio Readings (English)

Recall that the addition of a solute changes the properties of a solvent. The alloy bronze is harder than either copper or tin alone. This hardness made bronze a better material than stones or animal bones for making tools. The transition from the Stone Age to the Bronze Age, when humans first began to use metals, was an important period in human history.

Even though alloys have been made for thousands of years, new alloys with new properties are still being developed. One alloy with a very interesting property is Nitinol, which is made of nickel and titanium. Nitinol is called a memory alloy because it can be given a particular shape and then reshaped. What makes Nitinol unusual is that it will return to its original shape after being heated. Because of this property, Nitinol is used in several common products, including eyeglass frames.

A short list of useful alloys is given in the table below. The percentages shown in the table are those for only one type of each alloy.

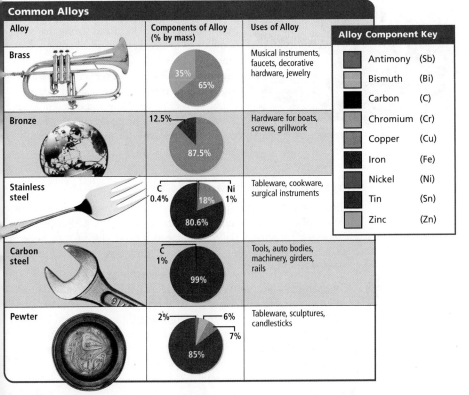

Common Alloys

Alloy	Components of Alloy (% by mass)	Uses of Alloy	Alloy Component Key
Brass	35% / 65%	Musical instruments, faucets, decorative hardware, jewelry	Antimony (Sb), Bismuth (Bi), Carbon (C), Chromium (Cr), Copper (Cu), Iron (Fe), Nickel (Ni), Tin (Sn), Zinc (Zn)
Bronze	12.5% / 87.5%	Hardware for boats, screws, grillwork	
Stainless steel	C 0.4% / 18% / Ni 1% / 80.6%	Tableware, cookware, surgical instruments	
Carbon steel	C 1% / 99%	Tools, auto bodies, machinery, girders, rails	
Pewter	2% / 6% / 7% / 85%	Tableware, sculptures, candlesticks	

READING VISUALS How are brass and bronze different from each other? How are stainless steel and carbon steel different from each other?

DIFFERENTIATE INSTRUCTION

More Reading Support

A What was the Bronze Age? *the time when humans began to use metals*

Inclusion For students with visual impairments, make an enlarged copy of the chart of common alloys and label the percentages with the corresponding elements. Then have students reinterpret the data by making a scaled bar graph for bronze and for brass.

Social Studies Connection

Humans in the Stone Age made their tools and weapons out of stone. It was difficult to make sharp edges on these crude tools. Around 3000 B.C., people in the Middle East discovered that copper could be made harder by melting it with tin to form bronze. Bronze tools and weapons were less likely to break than stone tools, and they could be sharpened. It took more than 1000 years for the Bronze Age to spread to the rest of the Old World.

EXPLORE (the **BIG** idea)

Revisit "Internet Activity: Alloys" on p. 109. Have students explain how changing the composition of an alloy changes its properties.

Teach from Visuals

To help students interpret the table of alloys, ask:

• What is the solvent in carbon steel? *iron*

• What is the solute in carbon steel? *carbon*

• What is the solute in brass? *zinc*

Ongoing Assessment

Describe how alloys are made.

Ask: What property of metals makes them relatively easy to be mixed? *relatively low melting point*

Identify how a variety of alloys are used in modern society.

Ask: Why are new alloys still being developed? *New technology often requires new alloys that have different properties from existing alloys.*

READING VISUALS *Answer: Brass and bronze have the same solvent (copper) but different solutes (zinc and tin, respectively). Carbon steel contains more carbon, and stainless steel contains nickel and chromium in addition to carbon.*

History of Science

Sir Henry Bessemer developed the method of making steel by blasting compressed air through molten iron. The oxygen in the air combines with excess carbon and other impurities to form oxides. The heat generated by this process raises the temperature and keeps the mass of iron molten. Large batches of steel can be made in this way. The Bessemer process has been replaced by even more efficient methods in most of the industrialized world.

Develop Critical Thinking

ANALYZE Discuss qualities of an alloy that would be useful in an airplane engine. *lightweight, heat resistance, freedom from rust, high strength*

Ask: Would steel be the best alloy to use in an airplane engine? *No; it is not lightweight, and it will rust.* What alloys would be better to use? *aluminum, titanium, because they are lighter*

Ongoing Assessment

Explain why different alloys have different uses.

Ask: Why would an alloy used for a racing bicycle be unsuitable for the framework of a building? *Racing bikes must be lightweight for maximum speed. The steel used in the framework of buildings must be strong; weight is not important.*

CHECK YOUR READING *Answer: Steel is used in ships and railroads. Aluminum alloys are used in aircraft, high-speed ferries, and cars. Titanium alloys are also used in aircraft.*

RESOURCE CENTER
CLASSZONE.COM
Explore alloys and their uses.

B

Alloys have many uses in modern life.

The advances in materials science that began with bronze almost 6000 years ago continue today. Modern industry uses many different alloys. Some alloys, based on lightweight metals such as aluminum and titanium, are relatively recent developments. However, the most important alloy used today—steel—has been around for many years.

A major advance in technology occurred in the 1850s with the development of the Bessemer process. This process made it possible to manufacture large amounts of steel in a short time. Until then, steel could be made only in batches of less than 100 pounds. The Bessemer process made it possible to produce up to 30 tons of steel in about 20 minutes. Since it began to be mass-produced, steel has been used in everything from bridges to cars to spoons.

Most steel used in construction is an alloy of iron and carbon. Iron is too soft to be a good building material by itself, but adding only a small amount of carbon—about 1 percent by mass—makes a very hard and strong material. Some types of steel contain small amounts of other metals as well, which give the alloys different properties. As you can see on the chart on page 135, one type of stainless steel contains only 1 percent nickel. However, different types of stainless steel can be made, and they have different uses. For instance, stainless steel used in appliances has 8 to 10 percent nickel and 18 percent chromium in it.

Steel is the main material used in the structure of this sphere at Epcot Center in Florida.

Alloys in Transportation

Different forms of transportation rely on steel. Wooden sailing ships were replaced by steel ships in the late 1800s. Today, steel cargo ships carry steel containers. Railroads depended on steel from their very beginning. Today's high-speed trains still run on steel wheels and tracks.

C

Modern vehicles use more recently developed alloys as well. For example, aluminum and titanium are lightweight metals that are relatively soft, like iron. However, their alloys are strong, like steel, and light. Airplane engines are made from aluminum alloys, and both aluminum and titanium alloys are used in aircraft bodies. Aluminum alloys are also commonly used in high-speed passenger ferries and in the bodies of cars. Because the alloys are light, they help to improve the fuel efficiency of these vehicles.

CHECK YOUR READING How are alloys used in transportation?

DIFFERENTIATE INSTRUCTION

 More Reading Support

B What is the most important alloy used today? *steel*

C What makes aluminum alloys useful? *They are strong and light.*

Advanced Have students research the economic advantages of using aluminum alloys in automobiles. Have them make a table comparing the properties of the structural materials used in cars.

 Challenge and Extension, p. 240

Alloys in Medicine

You may have noticed that most medical equipment is shiny and silver-colored. This equipment is made of stainless steel, which contains nickel and chromium in addition to iron and carbon. Surgical instruments are often made of stainless steel because it can be honed to a very sharp edge and is also rust resistant.

Cobalt and titanium alloys are also widely used in medicine because they do not easily react with substances in the body, such as blood and digestive juices. These alloys can be surgically placed inside the body with a minimum of harm to either the body or the metal. The photographs on the right show one use of alloys—making artificial joints.

Memory alloys similar to the Nitinol alloy described earlier also have a wide range of medical uses. These alloys are used in braces for teeth, and as implants that hold open blocked arteries or correct a curve in the spine. Medical devices made of memory alloys can be made in a particular shape and then reshaped for implantation. After the device is in place, the person's body heat causes it to return to its original shape.

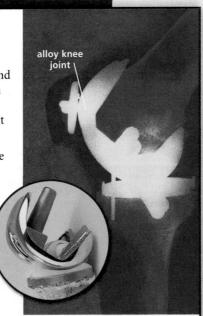

alloy knee joint

 CHECK YOUR READING What properties make alloys useful in medicine?

Artificial knee joints are often made of a titanium alloy. The x-ray image shows the device in place.

INVESTIGATE Alloys

How is a pure metal different from its alloy?

PROCEDURE

1. Examine the iron nails and the alloy (steel or stainless steel) nails. Record your observations.

2. Find and record the mass of the three iron nails. Repeat with the three alloy nails.

3. Find the volume of the nails by displacement, as follows: Into the empty graduated cylinder, pour water to a height that is higher than the nails are long. Note the water level. Add the iron nails and record the change in water level. Repeat this step with the alloy nails.

4. Calculate the density of each type of nail. $\text{Density} = \dfrac{\text{mass}}{\text{Volume}}$

WHAT DO YOU THINK?
- Compare your observations of the metals contained in the nails.
- Which metal has the greater density? How might a metal's density be important in how it is used?

CHALLENGE How can you identify different alloys of a metal?

SKILL FOCUS
Observing

MATERIALS
- 3 iron nails
- 3 steel nails or 3 stainless steel nails
- balance
- graduated cylinder
- water

TIME
30 minutes

Chapter 4: **Solutions** 137 **B**

DIFFERENTIATE INSTRUCTION

More Reading Support

D What are surgical instruments usually made of? *stainless steel*

Below Level Students may have a difficult time understanding why volume can be determined by measuring water displacement. Remind them that when a solid object is immersed in water, it displaces the same amount of water that it occupies as a solid object, its volume.

Chapter 4 **137** **B**

Reinforce (the **BIG** idea)

Have students relate the section to the Big Idea.

 R Reinforcing Key Concepts, p. 242

 4.4 ASSESS & RETEACH

Assess

 A Section 4.4 Quiz, p. 64

Reteach

Ask students to write a short paragraph explaining why this lesson on alloys is included in a chapter on solutions.
Alloys are solid mixtures that have many of the properties of solutions. One metal is dissolved in another metal, and the properties of the alloy differ from the properties of the components in the alloy.

Technology Resources

Have students visit **ClassZone.com** for reteaching of Key Concepts.

 CONTENT REVIEW

CONTENT REVIEW CD-ROM

Alloys in Space Flight

The aerospace industry develops and uses some of the newest and most advanced alloys. The same qualities that make titanium and aluminum alloys useful in airplanes—lightness, strength, and heat resistance—also make them useful in spacecraft. Titanium alloys were used in the Gemini space program of the 1960s. Large portions of the wings of today's space shuttle are made of aluminum alloys.

Research on the International Space Station may lead to the development of new alloys.

For more than 20 years, the heat shield on the shuttle's belly has been made from ceramic tiles. However, engineers have experimented with a titanium heat shield as well.

Construction of the International Space Station, which is shown in the photograph on the left, began in 1998. Alloys are a major part of the space station's structure. More important, research on the space station may lead to the development of new alloys.

One of the goals of research on the space station is to make alloys in a microgravity environment, which cannot be done on Earth. For example, astronauts have experimented with thick liquids, made with iron, that harden or change shape when a magnet is placed nearby, and then return to their previous shapes when the magnet is removed. These liquid alloys may be useful in robots or in artificial organs for humans.

 CHECK YOUR READING Why is research into new alloys on the International Space Station important?

 **4.4 Review**

KEY CONCEPTS

1. How can one metal be made to dissolve in another metal?

2. Name three metal alloys and a use of each one.

3. Why are alloys of cobalt or titanium, instead of pure iron, used for medical devices that are implanted inside people?

CRITICAL THINKING

4. **Infer** In industry, all titanium alloys are simply called titanium. What might this tell you about the use of pure titanium?

5. **Compare and Contrast** How are modern alloys similar to alloys made hundreds or thousands of years ago? How are they different?

⬥ CHALLENGE

6. **Synthesize** The melting point of copper is 1083°C. Tin is dissolved in copper to make bronze. Will bronze have a melting point of 1083°C? Why or why not?

 B 138 Unit: **Chemical Interactions**

ANSWERS

1. by melting both, mixing the two together, and cooling the solution into a solid

2. Sample answer: Stainless steel: medical equipment; Aluminum alloys: airplane engines; Titanium alloys: artificial joints

3. The alloys do not react with substances within the body, and they are light but strong.

4. It is not used.

5. Making an alloy still involves melting and mixing different substances. Modern

alloys use more complex mixtures, with carefully measured amounts of materials for specific purposes.

6. No; adding a solute changes the properties of a solvent, so copper and bronze will not have the same melting point.

MATH in SCIENCE

The steel in girders like these contains iron and 0.6 percent carbon by mass.

SKILL: CALCULATING PERCENTAGES

The Mixtures in Alloys

An alloy is a mixture of a metal with other substances. Because even a small change in the percentages of materials in an alloy can change its properties, alloys are made according to strict specifications. For example, steel is an alloy of iron and carbon. Steel that contains 0.6 percent carbon by mass is used in steel beams, whereas steel that contains 1.0 percent carbon by mass, which makes the steel harder, is used to make tools and springs. How can the percentages of materials in an alloy be calculated?

Example

Calculate the percentage of nickel in an alloy if a small portion of the alloy has 10 atoms, 3 of which are nickel.

(1) Convert the number of atoms into a fraction.

3 of 10 atoms in the alloy are nickel $= \dfrac{3}{10}$

(2) To calculate a percentage, first find an equivalent fraction that has a denominator of 100. Use x as the numerator.

$$\dfrac{3}{10} = \dfrac{x}{100}$$

(3) Convert the fraction into a percentage by using cross products

$$3 \cdot 100 = 10 \cdot x$$
$$300 = 10x$$
$$30 = x$$

ANSWER The percentage of nickel atoms in the alloy is 30%.

Answer the following questions.

1. A sample of an alloy contains 4 iron atoms, 3 zinc atoms, 2 aluminum atoms, and 1 copper atom.

   a. What percentage of the alloy is aluminum by number of atoms?

   b. What percentage is zinc by number of atoms?

2. A sample of an alloy contains 12 titanium atoms, 4 niobium atoms, and 4 aluminum atoms.

   a. What percentage of the alloy is titanium by number of atoms?

   b. What percentage is niobium by number of atoms?

CHALLENGE Suppose there is an alloy in which 2 of every 3 atoms are silver atoms, 1 of every 4 atoms is a copper atom, and 1 of every 12 atoms is a tin atom. What are the percentages of each metal in the alloy by number of atoms?

Chapter 4: **Solutions** 139 **B**

ANSWERS

1a. *2/10 = x/100; 10x = 200; x = 20%*

1b. *3/10 = x/100; 10x = 300; x = 30%*

2a. *12/20 = x/100; 20x = 1200; x = 60%*

2b. *4/20 = x/100; 20x = 400; x = 20%*

CHALLENGE *2/3 = x/100; 3x = 200; x = 66.67; 67% silver,*
1/4 = x/100; 4x = 100; x = 25; 25% copper,
1/12 = x/100; 12x = 100; x = 8.33; 8% tin

Set Learning Goal

To calculate the percentage of a metal in an alloy

Present the Science

Steel contains materials other than carbon and iron. The steel-making process introduces traces of various elements. Some metals are deliberately added. For example, adding more than 10.5% of chromium creates stainless steel. Typically, when percentages of components of an alloy are given, the percentage is in terms of mass, not the number of atoms. So, if an alloy contains 10.5% chromium, it means that 10.5% of the alloy's mass is from chromium, not that 10.5% of the atoms are chromium atoms.

Develop Algebra Skills

- Point out that whenever you look for a percentage, one side of the cross-products equation will be x over 100.

- Ask: If you knew that 4 percent of the atoms in the sample of an alloy were zinc and wanted to find out how many atoms of zinc would be in 50 atoms of alloy, what would your cross product look like? *x/50 = 4/100*

DIFFERENTIATION TIP Students may find it helpful to represent atoms in different-colored disks or other distinctive small objects to model the word problems.

Close

Ask students to think of other situations where cross products would be useful. *everyday situations such as scaling a recipe up or down, calculating percentages of other types of mixtures*

 • Math Support, p. 248
• Math Practice, p. 249

Technology Resources

Students can visit **ClassZone.com** for practice in understanding percents.

 MATH TUTORIAL

the **BIG** idea

Give students the following list of properties and ask them to pick out any properties of a solvent that do not change when a solute dissolves in the solvent: density, boiling point, freezing point, chemical reactivity, physical state. *Dissolving is a physical process that changes only the physical properties of a solvent. All properties listed are physical properties except chemical reactivity, which does not change.*

○ KEY CONCEPTS SUMMARY

SECTION 4.1
Ask: What is happening in the diagram? *An ionic solute breaks up in a solvent to form individual ions. Solvent molecules surround each ion.*

SECTION 4.2
Ask: Explain why some solutes cannot form a concentrated solution. *If a solute is only slightly soluble in a solvent, not enough of the solute will dissolve to form a concentrated solution.*

SECTION 4.3
Ask: What is the relationship between hydrogen ions and whether a substance is an acid or a base? *An acid donates hydrogen ions; a base (through, for example, a hydroxide ion) can accept hydrogen ions.*

SECTION 4.4
Ask: What structures in the photograph are most likely made of alloys? *the sphere, the monorail cars, and the tracks*

Review Concepts

- Big Idea Flow Chart, p. T25
- Chapter Outline, pp. T31–T32

---

4 Chapter Review

the **BIG** idea

When substances dissolve to form a solution, the properties of the mixture change.

◀ KEY CONCEPTS SUMMARY

4.1 A solution is a type of mixture.

- A solution is a mixture in which one or more solutes are dissolved in a solvent.
- A solution is a homogeneous mixture.

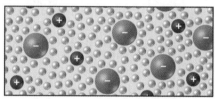

Ionic compound dissolved in solvent

VOCABULARY
solution p. 111
solute p. 112
solvent p. 112
suspension p. 113

4.2 The amount of solute that dissolves can vary.

- The amount of dissolved solute determines a solution's concentration.
- The more soluble a substance is, the more of it will dissolve in a solution.

Dilute   Concentrated

VOCABULARY
concentration p. 117
dilute p. 118
saturated p. 118
solubility p. 119

4.3 Solutions can be acidic, basic, or neutral.

- Acids donate protons (H+) in solutions, and bases accept protons in solutions.
- Acidity is measured by the H+ concentration on the pH scale.

Acid $HCl \xrightarrow{H_2O} H^+ + Cl^-$

Base $NaOH \xrightarrow{H_2O} Na^+ + OH^-$

VOCABULARY
acid p. 126
base p. 126
pH p. 129
neutral p. 129

4.4 Metal alloys are solid mixtures.

- Many of the metals used in modern transportation and medicine are alloys.
- The properties of a metal can be changed by adding one or more substances to produce a more useful material.

VOCABULARY
alloy p. 134

Technology Resources

Have students visit **ClassZone.com** or use the CD-ROM for a cumulative review of concepts.

 CONTENT REVIEW

 CONTENT REVIEW CD-ROM

Engage students in a whole-class interactive review of Key Concepts. Edit content as you wish.

 POWER PRESENTATIONS

Reviewing Vocabulary

Draw a diagram similar to the example shown below to connect and organize the concepts of related vocabulary terms. After you have completed your diagram, explain in two or three sentences why you organized the terms in that way. Underline each of the terms in your explanation.

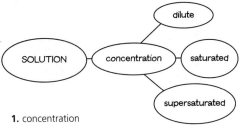

1. concentration
2. acid
3. base
4. neutral
5. pH

Latin Roots *Several of the vocabulary terms in this chapter come from the Latin word* solvere, *which means "to loosen." Describe how each of the following terms is related to the Latin word.*

6. solution
7. solute
8. solvent
9. solubility

Reviewing Key Concepts

Multiple Choice *Choose the letter of the best answer.*

10. What makes a solution different from other types of mixtures?
    a. Its parts can be separated.
    b. It is the same throughout.
    c. Its parts can be seen.
    d. It is a liquid.

11. When a solute is dissolved in a solvent, the solvent's
    a. boiling point decreases
    b. boiling point decreases and its freezing point increases
    c. freezing point increases
    d. freezing point decreases and its boiling point increases

12. When a compound held together by ionic bonds dissolves, the compound
    a. releases molecules into the solution
    b. forms a suspension
    c. releases ions into the solution
    d. becomes nonpolar

13. Water is called the universal solvent because it
    a. dissolves many substances
    b. dissolves very dense substances
    c. has no charged regions
    d. is nonpolar

14. How does an increase in temperature affect the solubility of solids and gases?
    a. It increases solubility of most solids and decreases the solubility of gases.
    b. It decreases solubility of most solids and gases.
    c. It increases solubility of gases and decreases the solubility of most solids.
    d. It increases solubility of both solids and gases.

15. A solution with a very high H^+ concentration has a
    a. very high pH   c. pH close to 5
    b. very low pH   d. pH close to 7

16. Why are oils insoluble in water?
    a. They are acids.   c. They are bases.
    b. They are polar.   d. They are nonpolar.

Short Answer *Write a short answer to each question.*

17. Describe the reaction that occurs when a strong acid reacts with a strong base.

18. How might an alloy be changed for different uses? Explain.

Reviewing Vocabulary

1–5. *Answers should indicate that concentration is the amount of solute dissolved. Answers should indicate differences between acids, bases, and neutral solutions in pH, and that the three are closely related due to hydrogen-ion concentrations in the different solutions.*

6. *A solution is a mixture of substances whose bonds have been "loosened."*

7. *A solute is the substance that is "loosened," or dispersed, in the solvent.*

8. *A solvent is the substance that "loosens," or disperses, the solute.*

9. *Solubility is the degree to which a solute can be "loosened," or dissolved, by a given amount of a particular solvent.*

Reviewing Key Concepts

10. *b*
11. *d*
12. *c*
13. *a*
14. *a*
15. *b*
16. *d*

17. *The hydrogen ion of the acid combines with the hydroxide ion of the base to form water. The negative ion of the acid combines with the positive ion of the base to form a neutral salt. The resulting solution's pH is close to 7 if equal amounts of acid and base react.*

18. *Different substances, or differing concentrations of substances, can be mixed with the primary metal in the alloy to make a new alloy with different properties. Examples include differing carbon concentrations in steel for different applications.*

Thinking Critically

19. *Acids: strips A & D; Bases: strips B & C*

20. *strip D; because the litmus paper is dark red, which indicates the lowest pH*

21. *The acid and the base would neutralize each other, and form a salt and water.*

22. *It will be orange; the base on strip C is not concentrated, so it will not neutralize all of the strong acid on strip D.*

23. *As the solution cooled, the solubility of the solute decreased. Some of it fell out of the solution onto the bottom of the beaker.*

24. *No; iron is relatively soft. It would not have the strength to support a bridge.*

25. *The properties depend on the amount of the solute. Bronze has different melting points, depending on the amount of tin in the alloy. Gold alloys differ in hardness depending on the amount of other metals dissolved in the gold.*

Using Math Skills in Science

26. *It increases; it decreases*

27. *23 g; 63 g*

28. *A solid; solubility of a gas decreases as the temperature increases.*

the **BIG** idea

29. *Answers will vary.*

30. *Both dissolve in water. Table salt forms separate sodium and chloride ions, while sugar forms individual sugar molecules. Both will decrease freezing point and increase boiling point because the change in these properties depends on the amount of the substance that is dissolved rather than the identity of the substance that is dissolved.*

UNIT PROJECTS

Students should have begun designing their models or presentations by this time. Encourage them to try different solutions to the problems they encounter.

 Unit Projects, pp. 5–10

Thinking Critically

The illustration below shows the results of pH tests of four different solutions. Assume the solutions are made with strong acids or strong bases. Use the diagram to answer the next four questions.

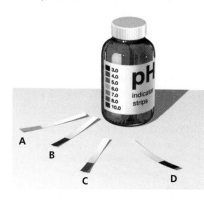

19. OBSERVE Which of the indicator strips show an acidic solution? Which show a basic solution?

20. INFER Which strip of indicator paper detected the highest concentration of H^+ ions? How do you know?

21. PREDICT What would happen if you mixed together equal amounts of the solutions that produced the results of strip B and strip D?

22. INFER Suppose you mix together equal amounts of the solutions that produced the results of strip C and strip D, then test the pH of this new solution. What color will the indicator paper be? Explain.

23. CAUSE AND EFFECT Suppose that you place a beaker containing a solution in a refrigerator. An hour later there is a white solid on the bottom of the beaker. What happened? Why?

24. INFER Do you think iron by itself would be a good material to use in the frame of a bridge? Why or why not?

25. SYNTHESIZE How might the concentration of a solute in an alloy be related to the properties of the alloy? Explain.

Using Math Skills in Science

Use the graph below to answer the next three questions.

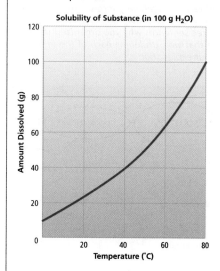

Solubility of Substance (in 100 g H₂O)

26. What happens to the solubility of the substance as the temperature increases? decreases?

27. Approximately how many grams of the substance dissolve at 20°C? 60°C?

28. Is the substance a solid or a gas? Explain.

the **BIG** idea

29. APPLY Look back at pages 108–109. Think about the answer you gave to the question about the photograph. How has your understanding of solutions and their properties changed?

30. COMPARE Describe the similarities and differences between solutions of table salt (NaCl) in water and sugar in water. Do both solutes have similar effects on the properties of the solvent? Explain.

UNIT PROJECTS

Check your schedule for your unit project. How are you doing? Be sure that you have placed data or notes from your research in your project folder.

MONITOR AND RETEACH

If students have trouble connecting the properties of an alloy to its composition in items 24 and 25, suggest that they review the material in Section 4.4. Students should make a table summarizing the use and properties of various alloys in everyday life, transportation, medicine, and space flight.

Students may benefit from summarizing one or more sections of the chapter.

 Summarizing the Chapter, pp. 268–269

Standardized Test Practice

For practice on your state test, go to . . .

TEST PRACTICE
CLASSZONE.COM

Interpreting Graphs

Use the information in the paragraph and the graph to answer the questions.

Acid rain is an environmental concern in the United States and in other countries. Acid rain is produced when the burning of fuels releases certain chemicals into the air. These chemicals can react with water vapor in Earth's atmosphere to form acids. The acids then fall back to the ground in either rain or snow. The acids can damage plants, animals, and buildings. Normally, rain has a pH of about 5.6, which is slightly acidic. But rain in some areas of the United States has a pH that is lower than 4.0. The graph shows the pH of water in several lakes.

Lake Water pH Values

1. Which lake is the most acidic?
   - **a.** Lake A
   - **b.** Lake B
   - **c.** Lake C
   - **d.** Lake D

2. Which lake is the least acidic?
   - **a.** Lake A
   - **b.** Lake B
   - **c.** Lake C
   - **d.** Lake D

3. Which lake has water the closest to neutral?
   - **a.** Lake A
   - **b.** Lake B
   - **c.** Lake E
   - **d.** Lake G

4. Lakes that form on a bed of limestone are less likely to suffer from high acidity. The limestone reacts with acids to neutralize them. Which of the following lakes is most likely to have a limestone bed?
   - **a.** Lake C
   - **b.** Lake D
   - **c.** Lake F
   - **d.** Lake G

5. Lake trout are fish that live in many freshwater lakes. When the pH of the water in a lake drops below 5.5, this species of fish can no longer reproduce, because its eggs cannot hatch. Which of the following statements is most likely true?
   - **a.** Lake trout have probably stopped reproducing in all the lakes.
   - **b.** In terms of reproducing, lake trout are not in danger in any of the lakes.
   - **c.** Lake trout will probably be able to reproduce in lakes A, B, and G but not in the others.
   - **d.** Lake trout have probably stopped reproducing only in lakes C, D, and F.

Extended Response

Answer the following two questions in detail. Include some of the terms from the list in the box. Underline each term you use in your answers.

concentration	solute	solubility
polar	solution	solvent

6. Suppose you are trying to make two solutions. One contains water and salt. The other contains water and oil. What do you think will happen in both cases? How might charges on particles affect your results?

7. Explain why some substances dissolve more easily than others. How can this characteristic of a solute be changed by changing the temperature or pressure of a solution?

Chapter 4: **Solutions** 143 **B**

Extended Response

6. RUBRIC

4 points for a response that correctly uses each of the following terms to answer the question:

- polar
- solution
- solute
- solvent
- solubility

Sample answer: Salt has high <u>solubility</u> in water because sodium and chloride ions are charged and water molecules are <u>polar</u>. The <u>solute</u> salt and the <u>solvent</u> water form a <u>solution</u> as the sodium and chloride ions separate and become surrounded by water molecules. Oil is insoluble in water because its molecules are nonpolar and water is polar. Oil and water do not form a solution.

3 points for a response that correctly uses four terms

2 points for a response that correctly uses three terms

1 point for a response that correctly uses one or two terms

7. RUBRIC

4 points for a response that correctly describes the dissolving process using the following terms:

- solute
- solvent
- solubility
- concentration

Sample: Each <u>solute</u> has a unique <u>solubility</u> that depends on the characteristics of the solute and the <u>solvent</u>. The higher a solute's solubility, the higher the <u>concentration</u> can be. The solubility of most solids can be increased by increasing the temperature; the solubility of gases can be increased by lowering the temperature or by increasing the pressure.

3 points for a response that correctly uses three terms

2 points for a response that correctly uses two terms

1 point for a response that correctly uses one term

METACOGNITIVE ACTIVITY

Have students answer the following questions in their **Science Notebook:**

1. Why do you think it is important for you to learn about acids and bases?

2. Which topics in this chapter would you like to learn more about?

3. What have you learned from your research on your Unit Project?

Chapter 4 **143** **B**

5 Carbon in Life and Materials

Physical Science
UNIFYING PRINCIPLES

PRINCIPLE 1

Matter is made of particles too small to see.

PRINCIPLE 2

Matter changes form and moves from place to place.

PRINCIPLE 3

Energy changes from one form to another, but it cannot be created or destroyed.

PRINCIPLE 4

Physical forces affect the movement of all matter on Earth and throughout the universe.

Unit: Chemical Interactions
BIG IDEAS

CHAPTER 1
Atomic Structure and the Periodic Table

A substance's atomic structure determines its physical and chemical properties.

CHAPTER 2
Chemical Bonds and Compounds

The properties of compounds depend on their atoms and chemical bonds.

CHAPTER 3
Chemical Reactions

Chemical reactions form new substances by breaking and making chemical bonds.

CHAPTER 4
Solutions

When substances dissolve to form a solution, the properties of the mixture change.

CHAPTER 5
Carbon in Life and Materials

Carbon is essential to living things and to modern materials.

CHAPTER 5
KEY CONCEPTS

SECTION 5.1

Carbon-based molecules have many structures.

1. Living and nonliving things contain carbon.

2. Carbon forms many different compounds.

SECTION 5.2

Carbon-based molecules are life's building blocks.

1. Carbon-based molecules have many functions in living things.

2. Living things contain four major types of carbon-based molecules.

SECTION 5.3

Carbon-based molecules are in many materials.

1. Carbon-based compounds from ancient organisms are used to make new materials.

2. Polymers contain repeating carbon-based units.

The Big Idea Flow Chart is available on p. T33 in the **UNIT TRANSPARENCY BOOK.**

Previewing Content

SECTION

5.1 **Carbon-based molecules have many structures.** pp. 147–153

1. Living and nonliving things contain carbon.

All **organic compounds** contain carbon and are the basis of all living things on Earth. There are different definitions of *organic* because it is an arbitrary classification system, and not all carbon compounds are considered to be organic. Compounds such as carbon dioxide, as well as those that contain cyanides (a CN^- group) or carbonates (a CO_3^{2-} group), are not considered to be organic and are called **inorganic compounds.** Also, all compounds that do not contain carbon are considered to be inorganic.

2. Carbon forms many different compounds.

Carbon can form millions of different compounds because a carbon atom forms four covalent bonds with other atoms. In addition, two carbon atoms can form one, two, or three bonds with each other, as the diagram below shows.

Single Bond	Double Bond	Triple Bond
$CH_3 — CH_3$	$CH_2 = CH_2$	$CH \equiv CH$

- Carbon atoms can bond together to form long chains. The chains can be straight or branched.
- Carbon atoms also form rings. Benzene is an important carbon ring. It has six carbon atoms with alternating double and single bonds.
- Compounds that contain the same types and numbers of atoms are called **isomers.** Isomers can have very different properties because their structures are different.

Isobutane

$CH_3 — CH_2 — CH_2 — CH_3$

$$CH_3 — \overset{\displaystyle CH_3}{\overset{\displaystyle |}{CH}} — CH_3$$

SECTION

5.2 **Carbon-based molecules are life's building blocks.** pp. 154–162

1. Carbon-based molecules have many functions in living things.

The many carbon-based molecules found in living things have different structures and different functions but work together to fulfill all life processes.

2. Living things contain four major types of carbon-based molecules.

Organic molecules in living things comprise four groups.

- **Carbohydrates** include sugars, starches, and cellulose and are used primarily for energy and for cell structure. Glucose ($C_6H_{12}O_6$) is a sugar used as a source of energy by cells of most living things. Starch is a macromolecule that stores glucose; it is made of many glucose molecules linked together. Cellulose is similar to starch but is used by plants to build cell walls. Cellulose is made of linked glucose molecules, but the bonds between each glucose subunit give cellulose a different structure than starch.
- **Lipids** include fats and oils that are used primarily for energy and for cell structure. Lipids typically contain up to three carbon chains called fatty acids. When all of the carbon-carbon bonds in the fatty acids are single bonds, the lipid is a saturated fat. If one or more of the bonds are double bonds, it is an unsaturated fat. Phospholipids are structural lipids that contain two fatty acid chains and a phosphate group, and are a major part of cell membranes.
- **Proteins** are macromolecules made of linked units called amino acids. The order of amino acids gives a protein its structure and its function. If the order of amino acids changes, the structure and function of the protein changes. Proteins have many roles in living things, including cell and tissue structure, transport, and immunity. **Enzymes** are proteins that are catalysts for chemical reactions within cells.
- **Nucleic acids** carry an organism's genetic code and assemble proteins from that code. DNA (deoxyribonucleic acid) carries the genetic code. The backbone of DNA is made of deoxyribose sugar molecules and phosphate groups. The genetic information of DNA is carried by four molecules called bases. The sequence of DNA bases directs the production of proteins. Several types of RNA (ribonucleic acid) read DNA and assemble proteins.

Common Misconceptions

PROTEINS FOR ENERGY Students may think that protein cannot be used by the body to supply energy. In fact, the body breaks down proteins and uses them as an energy source if carbohydrates or lipids are not available. The body does this only as a last resort.

 MISCONCEPTION DATABASE
CLASSZONE.COM Background on student misconceptions

 This misconception is addressed on p. 159.

Previewing Content

5.3 Carbon-based molecules are in many materials. pp. 163–171

1. Carbon-based compounds from ancient organisms are used to make new materials.

Carbon moves through the environment in the carbon cycle. Carbon dioxide in the atmosphere is absorbed by plants during photosynthesis. This carbon becomes a part of sugars, starches, and cellulose. Animals absorb carbon from the foods they eat. Some of this carbon is used to build the cells of the animal and some of it is returned to the atmosphere as carbon dioxide. When plants or animals die, they decompose and carbon returns to the environment. In some cases, carbon can fall out of the carbon cycle if the decaying organisms are trapped in mud or sediment. Carbon that fell out of the carbon cycle millions of years ago is the source of hydrocarbons used in today's world. Thus, carbon cycles through the environment, as shown in the diagram below.

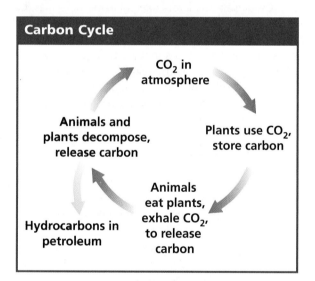

Carbon Cycle

CO_2 in atmosphere

Plants use CO_2, store carbon

Animals eat plants, exhale CO_2, to release carbon

Hydrocarbons in petroleum

Animals and plants decompose, release carbon

Hydrocarbons are molecules of carbon and hydrogen that are found in petroleum. The many hydrocarbons in petroleum can be separated according to differences in their boiling points through a process called distillation. Each fraction, such as lubricating oil or gasoline, is used for a different purpose.

2. Polymers contain repeating carbon-based units.

Polymers are very large carbon-based molecules made by linking together smaller repeating units called **monomers.** Polymers can be made by synthesis reactions that break bonds within two monomers and re-form to link the monomers together. The two types of synthesis reactions typically used to make polymers are called addition reactions and condensation reactions. The properties of a polymer are determined by its size and structure. The figure below shows how polymers are made.

Monomer

$$CH_2 = CH - CH_3$$

Monomers Are Linked Together

$$CH_2 = CH - CH_3 \quad CH_2 = CH - CH_3 \quad CH_2 = CH - CH_3$$

Polymer

$$CH_2 - CH - CH_2 - CH - CH_2 - CH ---$$
$$\qquad | \qquad\qquad | \qquad\qquad |$$
$$\quad CH_3 \qquad\quad CH_3 \qquad\quad CH_3$$

— = double bond changed to single bond
— = new bond formed between monomers

- Some useful polymers are **plastics.** Thousands of plastics with different structures and functions have been created.
- Chemists have developed polymers for specific functions. Nylon, Teflon, Kevlar, and Nomex are a few of these polymers.

Previewing Labs

EXPLORE the BIG idea

Structure and Function, p. 145
Students observe that structure influences function of a rubber ball.

TIME 10 minutes
MATERIALS hollow rubber ball, hacksaw, glue

Sweet Crackers, p. 145
Students observe that crackers become sweet when chewed for a long time.

TIME 10 minutes
MATERIALS unsalted cracker

Internet Activity: Polymers, p. 145
Students investigate how changes in the structures of virtual polymers affect their properties.

TIME 20 minutes
MATERIALS computer with Internet access

SECTION 5.1

INVESTIGATE Carbon Bonding, p. 149
Students use foam balls and toothpicks to model the different numbers of bonds that can be formed between carbon atoms.

TIME 10 minutes
MATERIALS marking pen, 2 large foam balls, 6 small foam balls, 7 toothpicks

---

SECTION 5.2

EXPLORE Carbon in Food, p. 154
Students burn food samples and infer that they contain carbon.

TIME 10 minutes
MATERIALS aluminum pie plate, candle, wooden matches, tongs, small marshmallow, piece of carrot

---

INVESTIGATE Organic Molecules, p. 158
Students test food samples to infer which ones contain starch.

TIME 20 minutes
MATERIALS 4 small jar lids, 3 eyedroppers, cornstarch solution, liquid gelatin, iodine solution, bread, tofu, stopwatch

---

SECTION 5.3

CHAPTER INVESTIGATION
Polymers, pp. 170–171
Students make a polymer and test its physical properties.

TIME 40 minutes
MATERIALS measuring spoons, 2 plastic containers, tap water, white glue, food coloring, borax, jar with lid, plastic spoon, zip-top plastic bags, scissors, 2 L plastic bottle with cap, ring stand with ring, stopwatch, straws

---

R **Additional INVESTIGATION,** Testing Simple Sugars, A, B, & C, pp. 318–326; Teacher Instructions, pp. 329–330

Previewing Chapter Resources

	INTEGRATED TECHNOLOGY	LABS AND ACTIVITIES

CHAPTER 5
Carbon in Life and Materials

 CLASSZONE.COM
- eEdition Plus
- EasyPlanner Plus
- Misconception Database
- Content Review
- Test Practice
- Simulation
- Visualization
- Resource Centers
- Internet Activity: Polymers
- Math Tutorial

 SCILINKS.ORG
SCI LINKS

 CD-ROMS
- eEdition
- EasyPlanner
- Power Presentations
- Content Review
- Lab Generator
- Test Generator

 AUDIO CDS
- Audio Readings
- Audio Readings in Spanish

 EXPLORE the Big Idea, p. 145
- Structure and Function
- Sweet Crackers
- Internet Activity: Polymers

 UNIT RESOURCE BOOK
Unit Projects, pp. 5–10

 Lab Generator CD-ROM
Generate customized labs.

SECTION
 5.1 Carbon-based molecules have many structures.
pp. 147–153

Time: 2 periods (1 block)
 Lesson Plan, pp. 270–271

- **SIMULATION,** 3–D Molecules
- **RESOURCE CENTER,** Nanotubes

 UNIT TRANSPARENCY BOOK
- Big Idea Flow Chart, p. T33
- Daily Vocabulary Scaffolding, p. T34
- Note-Taking Model, p. T35
- 3-Minute Warm-Up, p. T36
- "Carbon Chains and Carbon Rings" Visual, p. T38

- INVESTIGATE Carbon Bonding, p. 149
- Extreme Science, p. 153

 UNIT RESOURCE BOOK
- Datasheet, Carbon Bonding, p. 279

SECTION
 5.2 Carbon-based molecules are life's building blocks.
pp. 154–162

Time: 2 periods (1 block)
 Lesson Plan, pp. 281–282

- **RESOURCE CENTER,** Carbohydrates, Lipids, Proteins, and Nucleic Acids
- **MATH TUTORIAL**

 UNIT TRANSPARENCY BOOK
- Daily Vocabulary Scaffolding, p. T34
- 3-Minute Warm-Up, p. T36

- EXPLORE Carbon in Food, p. 154
- INVESTIGATE Organic Molecules, p. 158
- Math in Science, p. 162

 UNIT RESOURCE BOOK
- Datasheet, Organic Molecules, p. 290
- Math Support, p. 307
- Math Practice, p. 308
- Additional INVESTIGATION, Testing Simple Sugars, A, B, & C, pp. 318–326

SECTION
 5.3 Carbon-based molecules are in many materials.
pp. 163–171

Time: 4 periods (2 blocks)
 Lesson Plan, pp. 292–293

- **RESOURCE CENTER,** Petroleum and Hydrocarbons
- **VISUALIZATION,** Petroleum Distillation

 UNIT TRANSPARENCY BOOK
- Big Idea Flow Chart, p. T33
- Daily Vocabulary Scaffolding, p. T34
- 3-Minute Warm-Up, p. T37
- Chapter Outline, pp. T39–T40

 CHAPTER INVESTIGATION, Polymers, pp. 170–171

 UNIT RESOURCE BOOK
CHAPTER INVESTIGATION, Polymers, A, B, & C, pp. 309–317

READING AND REINFORCEMENT

 • Magnet Word Diagram, B24–25
• Supporting Main Ideas, C42
• Daily Vocabulary Scaffolding, H1–8

 UNIT RESOURCE BOOK
• Vocabulary Practice, pp. 304–305
• Decoding Support, p. 306
• Summarizing the Chapter, pp. 327–328

 Audio Readings CD
Listen to Pupil Edition.

Audio Readings in Spanish CD
Listen to Pupil Edition in Spanish.

 UNIT RESOURCE BOOK
• Reading Study Guide, A & B, pp. 272–275
• Spanish Reading Study Guide, pp. 276–277
• Challenge and Extension, p. 278
• Reinforcing Key Concepts, p. 280

 UNIT RESOURCE BOOK
• Reading Study Guide, A & B, pp. 283–286
• Spanish Reading Study Guide, pp. 287–288
• Challenge and Extension, p. 289
• Reinforcing Key Concepts, p. 291

UNIT RESOURCE BOOK
• Reading Study Guide, A & B, pp. 294–297
• Spanish Reading Study Guide, pp. 298–299
• Challenge and Extension, p. 300
• Reinforcing Key Concepts, p. 301
• Challenge Reading, pp. 302–303

ASSESSMENT

 • Chapter Review, pp. 173–174
• Standardized Test Practice, p. 175

 UNIT ASSESSMENT BOOK
• Diagnostic Test, pp. 79–80
• Chapter Test, A, B, & C, pp. 84–95
• Alternative Assessment, pp. 96–97
 • Unit Test, pp. 98–109
• Spanish Chapter Test, pp. 249–252
• Spanish Unit Test, pp. 253–256

 Test Generator CD-ROM
Generate customized tests.

Lab Generator CD-ROM
Rubrics for Labs

 Ongoing Assessment, pp. 147–148, 150, 152

 Section 5.1 Review, p. 152

UNIT ASSESSMENT BOOK
Section 5.1 Quiz, p. 81

 Ongoing Assessment, pp. 156–157, 159–161

 Section 5.2 Review, p. 161

 UNIT ASSESSMENT BOOK
Section 5.2 Quiz, p. 82

 Ongoing Assessment, pp. 164, 166–169

 Section 5.3 Review, p. 169

 UNIT ASSESSMENT BOOK
Section 5.3 Quiz, p. 83

STANDARDS

National Standards
A.1–8, A.9.a–c, A.9.e–f, B.1.c, C.1.a

See p. 144 for the standards.

National Standards
A.2–7, A.9.a–b, A.9.e–f, B.1.c

National Standards
A.2–8, A.9.a–c, A.9.e–f, C.1.a

National Standards
A.1–7, A.9.a–b, A.9.e–f, B.1.c

Previewing Resources for Differentiated Instruction

CHAPTER INVESTIGATION

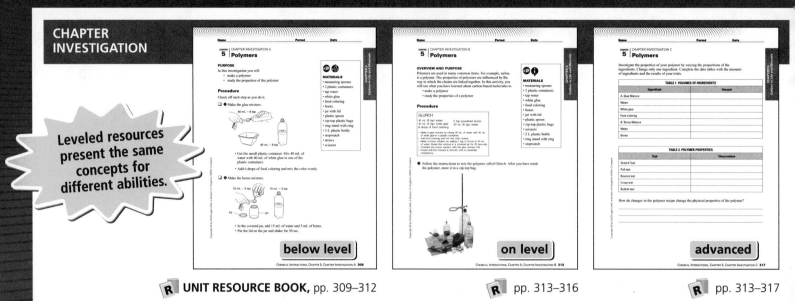

below level

UNIT RESOURCE BOOK, pp. 309–312

on level

pp. 313–316

advanced

pp. 313–317

Leveled resources present the same concepts for different abilities.

READING STUDY GUIDE

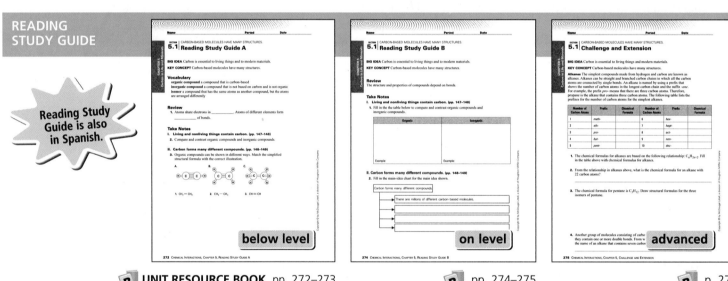

below level

UNIT RESOURCE BOOK, pp. 272–273

on level

pp. 274–275

advanced

p. 278

Reading Study Guide is also in Spanish.

CHAPTER TEST

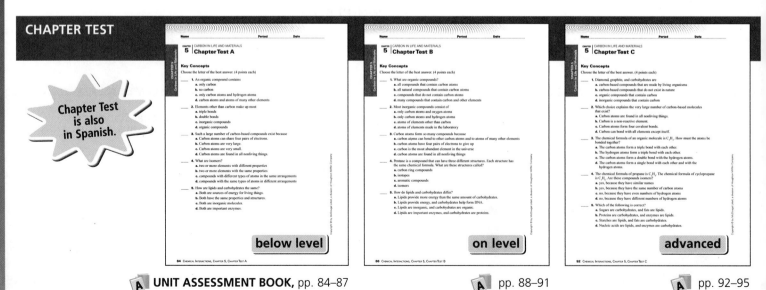

below level

UNIT ASSESSMENT BOOK, pp. 84–87

on level

pp. 88–91

advanced

pp. 92–95

Chapter Test is also in Spanish.

TECHNOLOGY

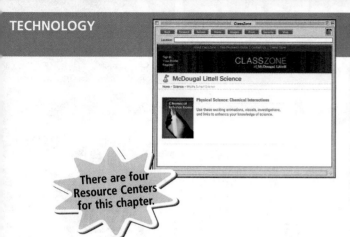

There are four Resource Centers for this chapter.

CLASSZONE.COM

CD/CD-ROMS

CLASSZONE.COM

VISUAL CONTENT

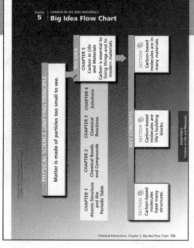

UNIT TRANSPARENCY BOOK, p. T33

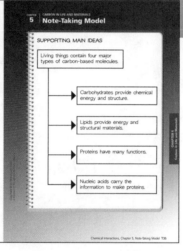

p. T35

p. T38

MORE SUPPORT

Reinforcing Key Concepts for each section

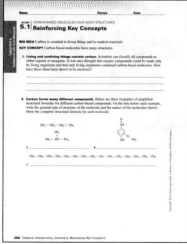

UNIT RESOURCE BOOK, p. 280

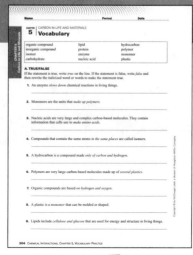

pp. 304–305

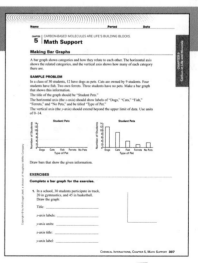

p. 307

Carbon in Life and Materials

CHAPTER

Carbon in Life and Materials

INTRODUCE

the **BIG** idea

Have students look at the photograph of the race car and pit crew and discuss how the question in the box links to the Big Idea:

- What do you think the clothing worn by the pit crew is made of?
- What are the car tires made of?
- What else shown in the photograph contains carbon?

National Science Education Standards

Content

B.1.c There are more than 100 known elements that combine in a multitude of ways to produce compounds, which account for the living and nonliving substances that we encounter.

C.1.a Living systems at all levels of organization demonstrate the complementary nature of structure and function. Important levels of organization for structure and function include cells, organs, tissues, organ systems, whole organisms, and ecosystems.

Process

A.1–8 Identify questions that can be answered through scientific investigation; design and conduct an investigation; use tools to gather and interpret data; use evidence to describe, predict, explain, model; think critically to make relationships between evidence and explanation; recognize different explanations and predictions; communicate scientific procedures and explanations; use mathematics.

A.9.a–c, A.9.e–f Understand scientific inquiry by using different investigations, methods, and explanations based on logic, evidence, and skepticism.

the **BIG** idea

Carbon is essential to living things and to modern materials.

Where in this photograph might you find carbon-based molecules?

Key Concepts

SECTION
5.1
Carbon-based molecules have many structures.
Learn why carbon forms many different compounds.

SECTION
5.2
Carbon-based molecules are life's building blocks.
Learn about the four main types of carbon-based molecules in living things.

SECTION
5.3
Carbon-based molecules are in many materials.
Learn how common materials are made from carbon-based molecules.

Internet Preview

CLASSZONE.COM
Chapter 5 online resources: Content Review, Simulation, Visualization, four Resource Centers, Math Tutorial, Test Practice

INTERNET PREVIEW

CLASSZONE.COM For student use with the following pages:

Review and Practice
- Content Review, pp. 146, 172
- Math Tutorial: Bar Graphs, p. 162
- Test Practice, p. 175

Activities and Resources
- Internet Activity: p. 145
- Simulation, p. 150
- Visualization, p. 164
- Resource Centers: Nanotubes, p. 153; Carbohydrates, Lipids, Proteins, and Nucleic Acids, p. 161; Petroleum and Hydrocarbons, p. 163

NSTA
scilinks.org
SCLINK

Organic Compounds
Code: MDL026

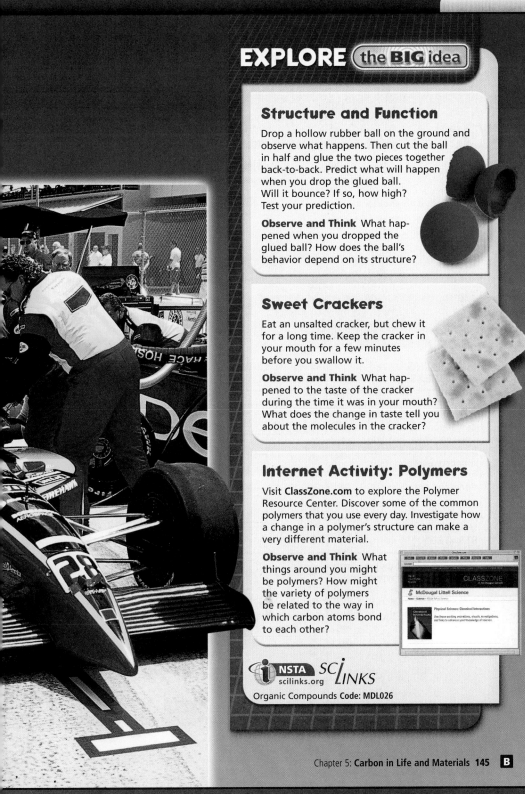

EXPLORE (the BIG idea)

Structure and Function

Drop a hollow rubber ball on the ground and observe what happens. Then cut the ball in half and glue the two pieces together back-to-back. Predict what will happen when you drop the glued ball. Will it bounce? If so, how high? Test your prediction.

Observe and Think What happened when you dropped the glued ball? How does the ball's behavior depend on its structure?

Sweet Crackers

Eat an unsalted cracker, but chew it for a long time. Keep the cracker in your mouth for a few minutes before you swallow it.

Observe and Think What happened to the taste of the cracker during the time it was in your mouth? What does the change in taste tell you about the molecules in the cracker?

Internet Activity: Polymers

Visit **ClassZone.com** to explore the Polymer Resource Center. Discover some of the common polymers that you use every day. Investigate how a change in a polymer's structure can make a very different material.

Observe and Think What things around you might be polymers? How might the variety of polymers be related to the way in which carbon atoms bond to each other?

NSTA scilinks.org **SCLINKS**

Organic Compounds Code: MDL026

Chapter 5: **Carbon in Life and Materials** 145 B

TEACHING WITH TECHNOLOGY

Digital Camera In "Investigate Organic Molecules" on p. 158, students could photograph a positive reaction when they test cornstarch, so they can compare the color when testing other substances.

Video Camera You may wish to videotape parts of the Chapter Investigation on pp. 170–171. Ask students to narrate the procedures as you tape them.

EXPLORE (the BIG idea)

These inquiry-based activities are appropriate for use at home or as a supplement to classroom instruction.

Structure and Function

PURPOSE To observe that structure influences function. Students observe a rubber ball's behavior before and after they cut it in half.

TIP *10 min.* Use a racquetball; an adult should cut the ball in half carefully with a hacksaw.

Answer: The glued ball should barely bounce, if at all. A change in the ball's structure changes its function completely.

REVISIT after p. 152.

Sweet Crackers

PURPOSE To observe that crackers contain sugar. Students observe that crackers begin to taste sweet when chewed for a long time.

TIP *10 min.* Unsalted crackers must be used. Oyster crackers work best.

Answer: It became sweet. Starch in the crackers is broken down into sugar.

REVISIT after p. 155.

Internet Activity: Polymers

PURPOSE To investigate polymers in common items.

TIP *20 min.* Students should realize that a polymer changes when its structure changes.

Answer: Living things, plastics, and fabrics are made of polymers. Because carbon atoms can form different numbers of bonds with each other and carbon-based molecules can include long chains of carbon atoms or carbon-based rings, a very large number of different polymers are possible.

REVISIT after p. 166.

PREPARE

◔ CONCEPT REVIEW

Activate Prior Knowledge

- Draw a diagram of a carbon atom. With an atomic number of 6, carbon has two electrons in its inner shell and four in its outer shell.
- Ask: How many electrons are available to form bonds in carbon atoms? *four*
- Ask: What type of bonds do carbon atoms form? *covalent*

▶ TAKING NOTES

Supporting Main Ideas

Students will find these charts analogous to an outline because of the way they organize main points and supporting information. The chart provides a meaningful way to organize information, especially for visual learners.

Vocabulary Strategy

Making side-by-side magnet word diagrams for two terms that can be compared, such as *organic compound* and *inorganic compound,* can be a valuable study tool.

Vocabulary and Note-Taking Resources

- Vocabulary Practice, pp. 304–305
- Decoding Support, p. 306

- Daily Vocabulary Scaffolding, p. T34
- Note-Taking Model, p. T35

- Magnet Word Diagram, B24–25
- Supporting Main Ideas, C42
- Daily Vocabulary Scaffolding, H1–8

◔ CONCEPT REVIEW

- Atoms share electrons when they form covalent bonds.
- Some atoms can form multiple bonds with another atom.
- Chemical reactions alter the arrangement of atoms.

◔ VOCABULARY REVIEW

electron p. 11
covalent bond p. 50
chemical reaction p. 69
catalyst p. 76

 **CONTENT REVIEW**
CLASSZONE.COM
Review concepts and vocabulary.

▶ TAKING NOTES

SUPPORTING MAIN IDEAS

Make a chart to show main ideas and the information that supports them. Copy each blue heading. Below each heading, add supporting information, such as reasons, explanations, and examples.

VOCABULARY STRATEGY

Think about a vocabulary term as a **magnet word** diagram. Write the other terms or ideas related to that term around it.

See the Note-Taking Handbook on pages R45–R51.

SCIENCE NOTEBOOK

Living and nonliving things contain carbon.

→ All life on Earth is based on carbon.

→ All organic compounds contain carbon.

→ Compounds that do not contain carbon are inorganic.

compounds that contain the same atoms in different places — ISOMER — carbon-based molecules

butane and isob[...]

B 146 Unit: Chemical Interactions

CHECK READINESS

Administer the Diagnostic Test to determine students' readiness for new science content and their mastery of requisite math skills.

 Diagnostic Test, pp. 79–80

Technology Resources

Students needing content and math skills should visit **ClassZone.com**.

- CONTENT REVIEW
- MATH TUTORIAL

 CONTENT REVIEW CD-ROM

KEY CONCEPT
5.1 Carbon-based molecules have many structures.

◀ **BEFORE, you learned**

- Atoms of one element differ from atoms of other elements
- The structure and properties of compounds depend on bonds
- Atoms of different elements can form different numbers of bonds

▶ **NOW, you will learn**

- About the importance of carbon in living things
- Why carbon can form many different compounds
- About different structures of carbon-based molecules

VOCABULARY

organic compound p. 147
inorganic compound
  p. 148
isomer p. 152

THINK ABOUT

Where can you find carbon?

The wood of a pencil consists of carbon-based molecules. These molecules are considered to be organic. The graphite in the center of the pencil is also made of carbon. In fact, graphite is pure carbon, but it is not considered to be organic. What makes the carbon in wood different from the carbon in graphite?

VOCABULARY
Remember to make a magnet word diagram for *organic compound* and for other vocabulary terms.

Living and nonliving things contain carbon.

Just about every substance that makes up living things contains carbon atoms. In fact, carbon is the most important element for life. Molecules containing carbon atoms were originally called organic because a large number of carbon-based molecules were found in living organisms. Sugars are organic. They are formed by plants, which are living organisms, and they contain carbon. Notice that the term *organic* is closely related to the term *organism*.

Organic compounds are based on carbon. Besides carbon, organic compounds often contain atoms of the elements hydrogen and oxygen, but they can also contain atoms of nitrogen, sulfur, and phosphorus. Scientists once thought that organic compounds could be made only in living organisms by an organism's life processes. Then an organic compound was made in a laboratory. This discovery showed that organic substances were not unique to living things. Instead, organic compounds could be made in a laboratory just like all other chemical compounds.

 CHECK YOUR READING Why were carbon compounds called organic compounds?

RESOURCES FOR DIFFERENTIATED INSTRUCTION

Below Level

UNIT RESOURCE BOOK
- Reading Study Guide A, pp. 272–273
- Decoding Support, p. 306

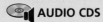 **AUDIO CDS**

Advanced

UNIT RESOURCE BOOK
Challenge and Extension, p. 278

English Learners

UNIT RESOURCE BOOK
Spanish Reading Study Guide, pp. 276–277

AUDIO CDS

- Audio Readings in Spanish
- Audio Readings (English)

5.1 FOCUS

◐ Set Learning Goals
Students will

- Recognize the importance of carbon in living things.
- Describe how carbon can form many different compounds.
- Identify different structures of carbon-based molecules.
- Make experimental models to demonstrate how two carbon atoms can form different numbers of bonds.

◐ 3-Minute Warm-Up

Display Transparency 36 or copy this exercise on the board:

Draw and label two diagrams, one of an ionic compound such as sodium chloride (NaCl) and one of a covalent compound such as methane (CH_4). Identify the difference in what happens to electrons in the two types of bonds. *Electrons are transferred from one atom to another in ionic bonds; pairs of electrons are shared by atoms in covalent bonds.*

T 3-Minute Warm-Up, p. T36

5.1 MOTIVATE

THINK ABOUT

PURPOSE To identify common objects that contain carbon-based molecules and that may be either organic or inorganic

DISCUSS Have students identify classroom items that contain carbon-based molecules. Compile a list on the board. Have students identify items that are organic and items that are inorganic.

Answer: The structure of carbon in wood is different from that of graphite.

Ongoing Assessment

CHECK YOUR READING *Answer: They were thought to be found only in living organisms.*

5.1 INSTRUCT

Teach Difficult Concepts

Explain that the definitions of *organic* and *inorganic* are difficult because they have arbitrary boundaries. The concepts of organic and inorganic arose when carbon-based molecules were believed to be formed by the life force of a living thing. Today, organic chemistry typically refers to the study of hydrocarbon compounds and their derivatives.

History of Science

The first organic compound to be made outside a living organism was urea, a waste product formed from the breakdown of proteins in the body. Kidneys filter urea out of the blood. In 1828, Friedrich Wöhler was trying to synthesize ammonium cyanate. The compound he made turned out to be identical to urea. Amazed, he wrote to a colleague, "I must tell you that I can make urea without the use of kidneys, either man or dog. Ammonium cyanate is urea."

Teach from Visuals

To help students interpret the bonding diagram, ask: How does the number of hydrogen atoms in a molecule change when two carbon atoms form a single, a double, or a triple bond? *Each carbon atom shares four pairs of electrons. The more pairs that are shared by the two carbon atoms, the fewer that can be shared with other atoms.*

Ongoing Assessment

Recognize that carbon is the basis of life.

Ask: Where do you find carbon in nature? *in diamond, graphite, cyanides, carbonates, and carbon dioxide, and mostly in every organism*

 Answer: one, two, or three

Organic Compound

Sugar, shown here as cubes, is organic and contains carbon atoms. It is made by plants from inorganic substances.

Inorganic Compound

Carbon dioxide, shown here as dry ice, is inorganic even though it contains carbon atoms. It is used by plants to make sugars.

READING TiP
The prefix *in-* means "not," so *inorganic* means "not organic."

There are several exceptions to the rule that carbon-based molecules are organic. These include diamond and graphite, which are made entirely of carbon but are not considered to be organic. The same is true of other compounds, such as cyanides (which contain a CN^- group), carbonates (which contain a CO_3^{2-} group), and carbon dioxide (CO_2). These carbon-containing compounds, among others, and all compounds without carbon are called **inorganic compounds.**

Carbon forms many different compounds.

Millions of different carbon-based molecules exist. Consider the number of molecules that make up living things and all of the processes that occur in living things. Carbon-based molecules are vital for all of them.

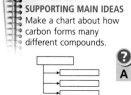

SUPPORTING MAIN IDEAS
Make a chart about how carbon forms many different compounds.

The large variety of carbon-based molecules results from the number of bonds that each carbon atom forms in a molecule and from a carbon atom's ability to form bonds with atoms of many different elements. In compounds, carbon atoms always share four pairs of electrons in four covalent bonds. This means that one carbon atom can form single bonds with up to four other atoms. Carbon atoms can also form multiple bonds—the atoms can share more than one pair of electrons—with other atoms including, most importantly, other carbon atoms. Different ways of showing the same carbon-based molecules are illustrated below and on page 149.

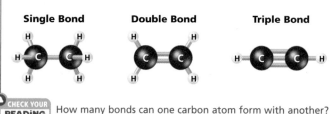

Single Bond   **Double Bond**   **Triple Bond**

CHECK YOUR READING How many bonds can one carbon atom form with another?

DIFFERENTIATE INSTRUCTION

? More Reading Support

A How many pairs of electrons does a carbon atom share in its bonds? *four*

English Learners Encourage students to be aware of words or phrases that make generalizations such as "Just about every substance that makes up living things contains carbon atoms" and "Besides carbon, organic compounds often contain atoms of the elements hydrogen and oxygen." "Just about" implies that most, but not all, of the substances that make up living things contain carbon. "Often" implies that organic compounds sometimes contain hydrogen and oxygen, but not always.

As you can see in the compounds shown, two carbon atoms can form single, double, or even triple bonds with one another. The compounds have different numbers of hydrogen atoms and different numbers of bonds between their carbon atoms. Count the bonds for each carbon atom. Each carbon atom makes a total of four bonds and always makes just one bond with each hydrogen atom.

Organic molecules are often shown in a simplified way. Instead of models that include all of a molecule's atoms and bonds, structural formulas—such as those shown below—can be used.

▼ REMINDER
One pair of electrons is shared in a single covalent bond.

Full Structural Formulas

$$H-\underset{\underset{H}{|}}{\overset{\overset{H}{|}}{C}}-\underset{\underset{H}{|}}{\overset{\overset{H}{|}}{C}}-H \qquad \underset{\underset{H}{|}}{\overset{\overset{H}{|}}{C}}=\underset{\underset{H}{|}}{\overset{\overset{H}{|}}{C}} \qquad H-C\equiv C-H$$

Simplified Structural Formulas

$$CH_3-CH_3 \qquad CH_2=CH_2 \qquad CH\equiv CH$$

Carbon-based molecules can have many different structures. Some of the most important structures are molecules shaped like chains and molecules shapes like rings.

INVESTIGATE Carbon Bonding

How do carbon-based molecules depend on the number of bonds between carbon atoms?

PROCEDURE

1. Label the large foam balls "C" for carbon, and label the small foam balls "H" for hydrogen.

2. Using a toothpick to represent a bond, construct a model of a molecule with two carbons, six hydrogens, and seven toothpicks. Carbon has four bonds and hydrogen has one.

3. Make a new model, using two carbons, two hydrogens, and five toothpicks.

WHAT DO YOU THINK?

- How many bonds are there between carbon atoms in the first model? in the second model?
- Which molecule might be more tightly held together? Why?

CHALLENGE In the model on the right, would it be possible for an additional hydrogen atom to bond to each carbon atom? Why or why not?

SKILL FOCUS
Modeling

MATERIALS
- marking pen
- 2 large foam balls
- 6 small foam balls
- toothpicks

TIME

10 minutes

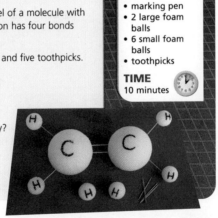

Chapter 5: **Carbon in Life and Materials** 149 **B**

DIFFERENTIATE INSTRUCTION

More Reading Support

B What is a simplified way of showing organic molecules? *structural formulas*

Inclusion "Investigate Carbon Bonding" is a good activity for students with visual impairments and tactile learners. Ask students to rephrase each direction in their own words to demonstrate their comprehension.

Teach from Visuals

To help students understand that the first molecule in each set of illustrations on pp. 148–149 is the same but is shown in different ways, ask:

- Does each set of illustrations show the same molecules? *yes*
- What is the difference between the full structural formulas and the simplified structural formulas? *The bonds between carbon and hydrogen atoms are not shown in the simplified structural formulas.*

INVESTIGATE Carbon Bonding

PURPOSE To model how two carbon atoms can form different numbers of bonds

TIP *10 min.* Marshmallows or colored gumdrops can be used instead of foam balls.

WHAT DO YOU THINK? *one; three; the molecule with the triple bond because the structure is more rigid*

CHALLENGE *No, because each carbon atom has already formed four bonds. However, if one of the bonds between the carbon atoms were broken, each carbon atom could form a bond with another hydrogen atom.*

R Datasheet, Carbon Bonding, p. 279

Technology Resources

Customize this student lab as needed or look for an alternative. Print rubrics to assess student lab reports.

🔬 **Lab Generator CD-ROM**

Metacognitive Strategy

Have students write a short paragraph answering the question, "Did making models help you understand bonding in carbon-based molecules? Why or why not?"

Teach from Visuals

To help students understand the structural formulas of the straight and branched chains, ask:

- In the part of the molecule shown as $CH_3—CH_2—CH_2$, which atoms are bonded to the first carbon atom? *3 hydrogen atoms and one carbon atom* the second carbon atom? *two carbon atoms and 2 hydrogen atoms*

- Are the hydrogen atoms located within the chain (between each successive carbon atom) or outside the chain (surrounding, but not between, carbon atoms)? *outside the chain (not between carbon atoms)*

Teach Difficult Concepts

The model showing the benzene molecule as a circle inside a hexagon is a better representation of the actual bonding that occurs within the molecule. The pairs of electrons in the carbon-carbon bonds are all shared equally by the six carbon atoms, and the bonds are in an intermediate state between single and double bonds.

Ongoing Assessment

Describe how carbon can form many different compounds.

Ask: Why are there so many structures of carbon-based molecules? *Carbon atoms can bond to each other and to atoms of other elements in many ways.*

CHECK YOUR READING *Answer: Because a carbon atom can form four bonds, a carbon atom can bond to up to four other carbon atoms.*

SIMULATION
CLASSZONE.COM

Observe and rotate three-dimensional models of carbon-based molecules.

C

Carbon Chains

Unlike atoms of other elements, carbon atoms have the unusual property of being able to bond to each other to form very long chains. One carbon chain might have hundreds of carbon atoms bonded together. A carbon chain can be straight or branched.

Straight Chain

$$CH_3 — CH_2 — CH_2 — CH_2 — CH_2 — CH_3$$

Branched Chain

$$\begin{array}{c} CH_3 \\ | \\ CH_2 \\ | \\ CH_3 — CH — CH_2 — CH_3 \end{array}$$

In a branched carbon chain, additional carbon atoms, or even other carbon chains, can bond to carbon atoms in the main carbon chain. Straight chains and branched chains are both results of carbon's ability to form four bonds.

CHECK YOUR READING How is it possible for carbon atoms to form both straight and branched chains?

Carbon Rings

Carbon-based molecules also can be shaped like rings. Carbon rings containing either 5 or 6 carbon atoms are the most common ones, and carbon rings containing more than 20 carbon atoms do not occur naturally.

Just as there are different types of carbon chains, there are different types of carbon ring molecules. One of the most important carbon-based ring molecules is a molecule called benzene (BEHN–ZEEN). Benzene contains six carbon atoms and six hydrogen atoms. Benzene differs from other carbon-based rings because it contains alternating single and double bonds between carbon atoms, as shown below. The benzene molecule is often shown as a circle inside a hexagon.

D

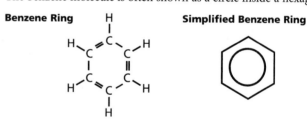

Many compounds are based on benzene's ring structure. These carbon-based molecules often have very strong smells, or aromas, and so are called aromatic compounds. One aromatic compound that contains a benzene ring is a molecule called vanillin. Vanillin is the molecule that gives vanilla its distinctive smell.

DIFFERENTIATE INSTRUCTION

 More Reading Support

C What are two types of carbon chains? *straight and branched*

D How is benzene different from other carbon rings? *It has alternating single and double bonds.*

Below Level Making models of various straight-chain and branched hydrocarbons will help students visualize how the atoms in hydrocarbon molecules are arranged and bonded. Model the benzene ring structure as well.

Carbon Chains and Carbon Rings

Carbon-based molecules shaped like chains or rings are found in the world around you.

Carbon Chains

One of the carbon chains in the diesel fuel for this locomotive has the formula $C_{15}H_{32}$. It contains 13 CH_2 groups between the CH_3 groups that are on both ends of the molecule. This molecule can be written as $CH_3(CH_2)_{13}CH_3$.

$$CH_3 - CH_2 - CH_2 - CH_2 - CH_2 - CH_2 - CH_2 - CH_2 - CH_2 - CH_2 - CH_2 - CH_2 - CH_2 - CH_2 - CH_3$$

Carbon Rings

Vanilla ice cream gets its flavor from vanilla, which is also used to enhance other flavors. The molecule that gives vanilla its strong smell is based on the benzene carbon ring.

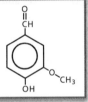

Carbon Chains and Rings

The molecules in polystyrene, which make up this foam container, contain carbon rings attached to a long carbon chain. The dashed lines at both ends of the structural formula tell you that the molecule continues in both directions.

$$---CH_2 - CH - CH_2 - CH - CH_2 - CH---$$

DIFFERENTIATE INSTRUCTION

Advanced Have interested students investigate how hydrocarbons are named. *Hydrocarbons with only single bonds end in –ane (for example, ethane). Hydrocarbons with at least one double bond end in –ene (for example, ethene). Hydrocarbons with at least one triple bond end in –yne (for example, ethyne).*

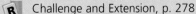 Challenge and Extension, p. 278

Teach from Visuals

To help students interpret the structural formulas on the page, ask:

• Is the structure of the carbon chain that is shown for diesel fuel a straight chain or a branched chain? *a straight chain*

• Do you think it is possible for that carbon chain to be branched? Why or why not? *Yes; carbon chains can be straight or branched by changing where carbon atoms bond to other carbon atoms in the molecule. A carbon atom can replace a hydrogen atom to form a branch off the main part of the chain.*

• Why do some of the carbon atoms in the vanillin structure (vanilla) seem to have only three bonds? *They actually have four bonds, but to save space the hydrogen atoms are not shown sticking out from the carbon. A CH in a structure represents a bond between carbon and hydrogen.*

• What do the dotted lines at both ends of the polystyrene structure (at bottom) indicate? *The molecule is much longer than shown. The chain continues in both directions.*

 This visual is also available as T38 in the Unit Transparency Book.

History of Science

In the mid-1800s, chemists were unable to identify a structure for benzene that had the correct number of carbon and hydrogen atoms and also accounted for benzene's unusual properties. A German chemist, Friedrich Kekulé, solved the mystery when he dreamed about a dancing snake that formed a ring by biting its own tail. Kekulé realized that benzene had a ring structure.

Identify different carbon-based molecules.

Ask: Can you know what the structure of an organic molecule is from its chemical formula? Explain. *No; the formula for both butane and isobutane is C_4H_{10}. You would have to see the structural formula to know how the molecule is arranged.*

 CHECK YOUR READING *Answer: They are each made up of the same atoms, but the atoms are arranged differently. The isomers have different properties.*

EXPLORE (the BIG idea)

Revisit "Structure and Function" on p. 145. Have students explain their results.

Reinforce (the BIG idea)

Have students relate the section to the Big Idea.

 Reinforcing Key Concepts, p. 280

5.1 ASSESS & RETEACH

Assess

 Section 5.1 Quiz, p. 81

Reteach

Have students write a short summary of the section. They should use heads, topic sentences, and vocabulary terms for guidance.

Technology Resources

Have students visit **ClassZone.com** for reteaching of Key Concepts.

 CONTENT REVIEW

 CONTENT REVIEW CD-ROM

Isomers

Another reason there is such a large number of carbon-based molecules is that carbon can form different molecules with the same atoms. The atoms in these molecules are in different places, and the molecules have different structures. Because the atoms are arranged differently, they are actually two different substances. Compounds that contain the same atoms, but in different places, are called **isomers.**

READING TIP
The prefix *iso-* means "equal," and the root *mer-* means "part."

The formulas below show a pair of compounds—butane and isobutane—that are isomers. Both molecules contain four carbon atoms and ten hydrogen atoms. However, butane molecules are straight chains of carbon atoms. Isobutane molecules are branched chains of carbon atoms. Even though both butane and isobutane contain the same atoms, the structures of the molecules are different, so they are isomers.

Butane

$$CH_3 - CH_2 - CH_2 - CH_3$$

Butane contains four carbon atoms and ten hydrogen atoms. It has a straight chain structure.

Isobutane

$$CH_3$$
$$|$$
$$CH_3 - CH - CH_3$$

Isobutane also contains four carbon atoms and ten hydrogen atoms. It has a branched chain structure.

Some carbon-based molecules can shift from one isomer to another, and then back to the original structure. For example, isomers of a molecule called retinal are necessary for your eyesight. When light strikes retinal, its structure changes from one isomer to another. The new isomer of retinal starts a process that sends a signal from the eye to the brain. After the retinal isomer starts the signaling process, the molecule shifts back to its original structure.

 CHECK YOUR READING If two substances are isomers of each other, how are they the same? different?

5.1 Review

KEY CONCEPTS

1. Why were carbon-based compounds first called organic? How has the understanding of organic compounds changed?
2. How is the way in which carbon atoms bond to each other important for the number of carbon-based compounds?
3. Describe three structures of carbon-based molecules.

CRITICAL THINKING

4. **Infer** Could the last carbon atom in a carbon chain make bonds with four hydrogen atoms? Why or why not?
5. **Synthesize** Do you think molecules based on carbon rings can have isomers? Why or why not?

CHALLENGE

6. **Communicate** A molecule called naphthalene consists of ten carbon atoms and eight hydrogen atoms in two linked benzene rings. Draw a diagram of a molecule that could be naphthalene. Be sure to include the atoms and the bonds between the atoms.

B 152 Unit: **Chemical Interactions**

ANSWERS

1. They were thought to be found only in organisms. Some carbon compounds are inorganic, and organic compounds can be made in a laboratory.

2. Two carbon atoms can form multiple bonds, making a great variety of molecules.

3. Possible answers: chains (carbon atoms bonded one after another), rings (carbon atoms on the ends of a chain are bonded), chains and rings (rings attached to a chain), isomers (same atoms, different structures)

4. No; it must make at least

one bond with a carbon atom.

5. Yes; the isomers would include atoms bonded to the ring in different places.

6. 2 carbon rings joined at 2 carbons that share a double bond; 8 hydrogens at the other 8 carbons

CARBON NANOTUBES

Stronger Than Steel

Can you imagine something that is much smaller than a human hair yet much stronger than steel? Welcome to the world of carbon nanotubes. Carbon nanotubes are made of pure carbon. They are 10 to 100 times as strong as the same weight of steel, but they can be 10,000 times smaller than a hair from your head. Carbon nanotubes were discovered in 1991 as a byproduct of a chemical reaction, but they may have many uses in the near future.

The Tiniest Test Tube

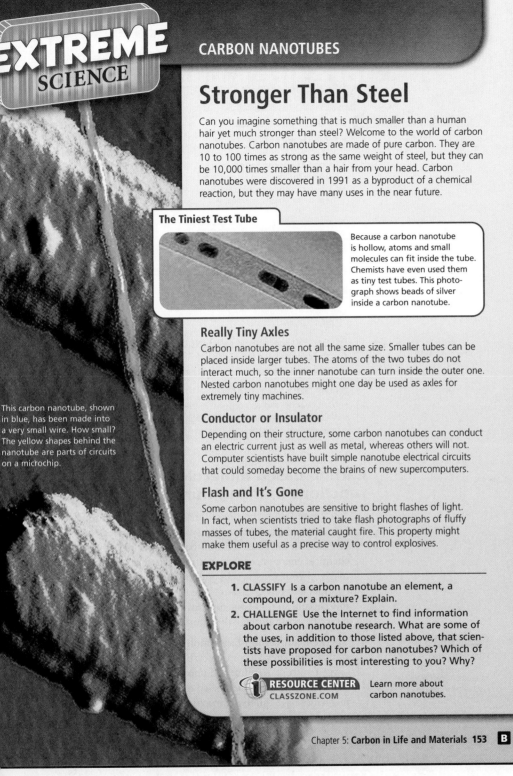

Because a carbon nanotube is hollow, atoms and small molecules can fit inside the tube. Chemists have even used them as tiny test tubes. This photograph shows beads of silver inside a carbon nanotube.

This carbon nanotube, shown in blue, has been made into a very small wire. How small? The yellow shapes behind the nanotube are parts of circuits on a microchip.

Really Tiny Axles

Carbon nanotubes are not all the same size. Smaller tubes can be placed inside larger tubes. The atoms of the two tubes do not interact much, so the inner nanotube can turn inside the outer one. Nested carbon nanotubes might one day be used as axles for extremely tiny machines.

Conductor or Insulator

Depending on their structure, some carbon nanotubes can conduct an electric current just as well as metal, whereas others will not. Computer scientists have built simple nanotube electrical circuits that could someday become the brains of new supercomputers.

Flash and It's Gone

Some carbon nanotubes are sensitive to bright flashes of light. In fact, when scientists tried to take flash photographs of fluffy masses of tubes, the material caught fire. This property might make them useful as a precise way to control explosives.

EXPLORE

1. **CLASSIFY** Is a carbon nanotube an element, a compound, or a mixture? Explain.
2. **CHALLENGE** Use the Internet to find information about carbon nanotube research. What are some of the uses, in addition to those listed above, that scientists have proposed for carbon nanotubes? Which of these possibilities is most interesting to you? Why?

RESOURCE CENTER Learn more about
CLASSZONE.COM carbon nanotubes.

Chapter 5: **Carbon in Life and Materials** 153 **B**

EXTREME SCIENCE
Fun and Motivating Science

Set Learning Goal
To learn about carbon nanotubes and their possible uses

Present the Science
Carbon nanotubes were discovered accidentally as a byproduct of making fullerenes (also called buckyballs), which are C_{60} molecules shaped like soccer balls. A carbon nanotube can be thought of as a single sheet of graphite that has been rolled into a tube shape. Therefore, the walls of the tube are one atom thick. Today, carbon nanotubes are made purposely, although it is very difficult for chemists to make them the size they want. NASA expects to use carbon nanotubes in space exploration in applications such as energy storage, life-support systems, electronics, sensors, and biomedicine.

Discussion Questions
- Why do you think these carbon tubes are called nanotubes? *They are only a few nanometers (one-billionth of a meter) in diameter. They can be very long compared with their diameter, however.*
- Why might carbon nanotubes be combined with fibers? What would the advantage of these fibers be? *Nanotubes could be woven together with fibers to strengthen them. The fibers would probably look like ordinary fibers but be extremely strong.*

Close
Ask: Why are carbon nanotubes likely to be an important part of future technology? *Electronic devices, aerospace equipment, and computers are becoming smaller and smaller. Nanotechnology can contribute to miniaturization.*

EXPLORE

1. *CLASSIFY It is an element. It does not contain atoms of other elements, so it cannot be a compound, and it is not a mixture for similar reasons.*

2. *CHALLENGE Answers will vary depending on each student's interests. Check students' research on nanotubes.*

● Set Learning Goals

Students will

- Describe the functions of carbohydrates and lipids in living things.
- Describe how protein structure determines protein function.
- Recognize that nucleic acids carry instructions for building proteins.
- Infer from an experiment how starches can be detected in samples of food.

◐ 3-Minute Warm-Up

Display Transparency 36 or copy this exercise on the board:

Decide if these statements are true. If not true, correct them.

1. Carbon atoms can form only one bond with each other. *one, two, or three bonds*

2. Carbon chains are always branched. *They can also be straight.*

3. Two isomers contain the same kind and number of atoms arranged in different ways. *true*

 3-Minute Warm-Up, p. T36

5.2 MOTIVATE

EXPLORE Carbon in Food

PURPOSE To test foods for the presence of carbon

TIP *10 min.* Samples should be held in the blue part closest to the wick because it is the hottest part of the flame.

WHAT DO YOU THINK? *The food samples turned black; combustion.*

5.2 Carbon-based molecules are life's building blocks.

5.2

◀ **BEFORE, you learned**

- Carbon is the basis of life on Earth
- Carbon atoms can form multiple bonds
- Carbon can form molecules shaped like chains or rings

▶ **NOW, you will learn**

- About the functions of carbohydrates and lipids in living things
- About structures and functions of proteins
- How nucleic acids carry instructions for building proteins

VOCABULARY

carbohydrate p. 155
lipid p. 156
protein p. 158
enzyme p. 159
nucleic acid p. 161

EXPLORE Carbon in Food

How can you see the carbon in food?

PROCEDURE

① Place the candle in the pie plate and light the candle.

② Use the tongs to hold each food sample in the candle flame for 20 seconds. Record your observations.

WHAT DO YOU THINK?

- What changes did you observe in the samples?
- What type of chemical reaction might have caused these changes?

MATERIALS

- aluminum pie plate
- candle
- wooden matches
- tongs
- small marshmallow
- piece of carrot

Carbon-based molecules have many functions in living things.

You depend on carbon-based molecules for all of the activities in your life. For example, when you play softball, you need energy to swing the bat and run the bases. Carbon-based molecules are the source of the chemical energy needed by your muscle cells. Carbon-based molecules make up your muscle cells and provide those cells with the ability to contract and relax. Carbon-based molecules carry oxygen to your muscle cells so that your muscles can function properly. Carbon-based molecules even provide the information for building new molecules.

The many carbon-based molecules in all living things have certain similarities. They all contain carbon and elements such as hydrogen, oxygen, nitrogen, sulfur, and phosphorus. Many of the molecules are also very large molecules called macromolecules. However, these molecules have different structures and different functions.

READING **TiP**

The prefix *macro-* means "large," so a macromolecule is a large molecule.

RESOURCES FOR DIFFERENTIATED INSTRUCTION

Below Level

UNIT RESOURCE BOOK
- Reading Study Guide A, pp. 283–284
- Decoding Support, p. 306

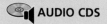 **AUDIO CDS**

 Additional INVESTIGATION,
Testing Simple Sugars, A, B, & C, pp. 318–326;
Teacher Instructions, pp. 329–330

Advanced

UNIT RESOURCE BOOK
Challenge and Extension, p. 289

English Learners

UNIT RESOURCE BOOK
Spanish Reading Study Guide, pp. 287–288

AUDIO CDS

- Audio Readings in Spanish
- Audio Readings (English)

Living things contain four major types of carbon-based molecules.

The organic molecules found in living things are classified into four major groups—carbohydrates, lipids, proteins, and nucleic acids. You may already be familiar with these types of molecules and their functions in living things.

Carbohydrates include sugars and starches found in foods such as bread and pasta. Many lipids are fats or oils. Proteins are necessary for many functions in the body, including the formation of muscle tissue. Nucleic acids are the molecules that carry the genetic code for all living things. As you read about each of these types of molecules, look for ways in which the molecule's function depends on its structure.

Carbohydrates

Carbohydrates (KAHR-boh-HY-DRAYTZ) include sugars, starches, and cellulose, and contain atoms of three elements—carbon, hydrogen, and oxygen. They serve two main functions. Carbohydrates are a source of chemical energy for cells in many living things. They are also part of the structural materials of plants.

One important carbohydrate is the sugar glucose, which has the chemical formula $C_6H_{12}O_6$. Cells in both plants and animals break down glucose for energy. In plants glucose molecules also can be joined together to form more complex carbohydrates, such as starch and cellulose. Starch is a macromolecule that consists of many glucose molecules, or units, bonded together. Many foods, such as pasta, contain starch. When starch is broken back down into individual glucose molecules, those glucose molecules can be used as an energy source by cells.

Modeling Glucose

The glucose molecule can be represented by a hexagon. The red O shows that an oxygen atom is in the ring.

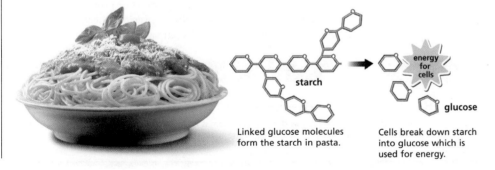

Linked glucose molecules form the starch in pasta.

Cells break down starch into glucose which is used for energy.

5.2 INSTRUCT

Develop Critical Thinking

INFER Ask: Why do athletes eat large amounts of pasta or other starches when they are in training, a process called carbo-loading? *They need large amounts of energy to train. Pasta, which is rich in carbohydrates, provides that energy.*

EXPLORE (the BIG idea)

Revisit "Sweet Crackers" on p. 145. Have students explain their results.

Integrate the Sciences

Animals, as well as plants, store glucose in a macromolecule. The macromolecule in animals is glycogen, which is also called animal starch. In glycogen, glucose molecules are bonded together in a highly branched arrangement. The liver and muscles store glycogen. Glycogen can be broken down into glucose molecules when cells need energy. It is also important to note that the glucose molecule is not the direct source of energy for cells. Through the process of cellular respiration, glucose is broken down to produce ATP (adenosine triphosphate), which is the molecule that provides energy.

Teach from Visuals

To help students interpret the diagrams of glucose and starch, ask:

- How is the hexagon that represents glucose different from the hexagon that represents benzene? *There is an oxygen atom in the glucose ring; there is no circle inside the ring to represent alternating double and single bonds.*

- Where is the energy in pasta stored? *in the glucose molecules that make up the starch in pasta*

- Why does starch have to be broken down into glucose before energy can be released? *Glucose, not starch, is broken down in the reactions that provide energy for cells.*

DIFFERENTIATE INSTRUCTION

More Reading Support

A Which major groups of organic molecules are found in living things? *carbohydrates, lipids, proteins, and nucleic acids*

English Learners Point out the word *macromolecule* on p. 155. Ask students what they thing the prefix *macro* means. Guide them to use the context of the sentence to find the answer. Then have students look up *macro* in the dictionary to find that it means "large."

Additional Investigation To reinforce Section 5.2 learning goals, use the following full-period investigation:

R Additional **INVESTIGATION**, Testing Simple Sugars, A, B, & C, pp. 318–326, 329–330

Teach from Visuals

To help students interpret the picture of the structure of cellulose, ask:

- How is the structure of cellulose different from the structure of starch as shown on p. 155? *The glucose molecules in cellulose are linked in a long chain, whereas the glucose molecules in starch are linked in branches.*

- How is the structure of cellulose similar to the structure of starch? *They both contain glucose molecules.*

Real World Example

Lipids are found in soaps and detergents. Many substances that need to be cleaned from clothes and skin, such as soil, grass, and grease, are nonpolar and organic. These substances are not easily removed by water alone because they do not dissolve in water, which is polar.

Teach from Visuals

To help students interpret the model of a fatty acid molecule, ask:

- What is the basic structure of a fatty acid? *It is a carbon chain.*

- What does the break in the middle of the model of the structure represent? Why? *Not all carbon atoms are shown in the chain because a fatty acid can be a very large molecule.*

Ongoing Assessment

Recognize that carbohydrates and lipids have similar functions.

Ask: Why are lipids better energy storage molecules than carbohydrates? *One gram of fat contains approximately twice as much energy as one gram of carbohydrate or protein.*

CHECK YOUR READING *Answer: to supply chemical energy; to supply chemical energy and structure*

Moss Leaf Cells

Cellulose is a long chain-like molecule that forms part of a plant's structure. **B**

Plants make their own glucose through a process called photosynthesis, which you read about in Chapter 3. Some of the glucose made during photosynthesis is used to make the complex carbohydrate molecules that form a plant's structure.

Cellulose

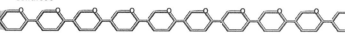

Unlike animal cells, plant cells have a tough, protective layer outside the cell membrane called the cell wall. Cellulose (SEHL-yuh-LOHS) is a macromolecule found in plant cell walls, and it is a large part of vegetables such as lettuce and celery. The illustration shows moss leaf cells with their cell walls, and a diagram of part of a cellulose molecule.

Cellulose and starch are both carbohydrates composed of glucose molecules, but the glucose molecules that make up these larger macromolecules are linked in different ways. Because of their different structures, starch and cellulose have different functions. In fact, this structural difference also prevents your body from breaking down and using cellulose as it would starch.

CHECK YOUR READING What are some functions of carbohydrates in animals? in plants?

VOCABULARY
Make a magnet word diagram for *lipid* and for other vocabulary terms.

Lipids

Lipids include fats and oils and are used mainly for energy and as structural materials in living things. Like carbohydrates, most lipids are made of carbon, hydrogen, and oxygen. Even though lipids and carbohydrates have many similarities, they have different structures and properties.

Animals store chemical energy in fat. Plants store chemical energy in oils, such as olive oil and peanut oil. Fats and oils store energy very efficiently—one gram of fat contains about twice as much energy as one gram of carbohydrate or protein. Fats and oils contain three carbon chains called fatty acids. The illustration below shows the general structure of a fatty acid.

Modeling Fatty Acids

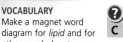

The carbon chains in lipids are called fatty acids. A carbon atom is at each bend of the zig-zag model above. The break in the middle of the chain shows that some carbon atoms have been left out.

B 156 Unit: Chemical Interactions

DIFFERENTIATE INSTRUCTION

More Reading Support

B Where is cellulose found? *plant cell walls*

C What are the main functions of lipids? *energy storage and as structural materials*

English Learners Help English learners understand how prepositions are used in phrasal verbs. The sentence below is an example—note the verb phrase *breaking down.*

"In fact, this structural difference also prevents your body from breaking down and using cellulose as it would starch."

Have students read this sentence and point out the verb phrase—be sure they do not interpret "down" as a literal direction.

You may have heard the terms *saturated* or *unsaturated* in relation to fats. If all of the bonds between carbon atoms in the fatty acids are single bonds, the lipid is a saturated fat. If one or more of these bonds is a double bond, the lipid is an unsaturated fat. Most animal fats are saturated, and most oils from plants are unsaturated. Diets high in saturated fats have been linked to heart disease. Lipids in the butter in the photograph on the right are saturated fats.

Fat Structure

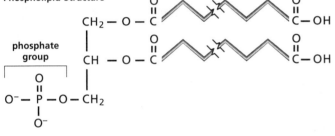

CHECK YOUR READING What is the difference between a saturated fat and an unsaturated fat?

Fats in butter contain three fatty acids and are used for energy. Butter contains saturated fats.

Some lipids are important parts of cell structure. Structural lipids often contain the element phosphorus and are called phospholipids. Phospholipids are a significant part of cell membranes such as the one shown in the photograph of the nerve cell on the right.

Phospholipid Structure

phosphate group

Some lipids in this nerve cell's membrane have two fatty acids and one phosphate group. These lipids are called phospholipids.

Another lipid involved in cell structure is cholesterol, which is a part of cell membranes. Cholesterol has other functions as well. It is necessary to make substances called hormones. Hormones, such as adrenaline, are chemical messengers in your body.

Your body makes some of the cholesterol that it needs, but it also uses cholesterol from foods you eat. Cholesterol is found in many foods that come from animals, such as meat and eggs. Even some plant products, such as coconut oil, can increase the amount of cholesterol in your body. Although you need cholesterol, eating too much of it—just like eating too much saturated fat—can lead to heart disease.

Chapter 5: **Carbon in Life and Materials 157** B

Integrating the Sciences

A very important fat in the human nervous system is myelin. Myelin forms a sheath that covers many nerve cells, which increases the speed of transmission of nerve signals. Without myelin, nerves cannot transmit signals. In the disease multiple sclerosis, the body's own immune system destroys the myelin sheath that covers nerve cells.

Language Arts Connection

Many lipids are based on fatty acids that are linked to a molecule called glycerol. The fat illustrated in butter is called a triglyceride; the phospholipid is called a diglyceride. Ask: What does this difference tell you about the structure of a lipid molecule? (If necessary, instruct students to look up the prefixes *tri-* and *di-* in a dictionary.) *A triglyceride has three fatty acids, whereas a diglyceride has two fatty acids.*

Teach from Visuals

To help students interpret the diagrams of different kinds of fats, ask: Where is the phosphate group in a phospholipid? *It takes the place of one fatty acid.*

Ongoing Assessment

CHECK YOUR READING *Answer: saturated fat—all bonds between carbon atoms in fatty acids are single bonds; unsaturated fat—one or more bonds between carbon atoms are double bonds.*

DIFFERENTIATE INSTRUCTION

More Reading Support

D Where are phospholipids in the body? *in cell membranes*

E Why is cholesterol important? *for cell membrane structure and in making hormones*

Advanced Students could extend their knowledge of phospholipids by researching how they participate in the structure of cell membranes. The fluid mosaic model of membrane structure is not difficult to understand and will enhance students' understanding of phospholipids.

R Challenge and Extension, p. 289

INVESTIGATE Organic Molecules

PURPOSE To discover what foods contain starch by testing them with iodine

TIP *20 min.* Use plain gelatin. The liquid gelatin and tofu are included as substances that do not contain starch. They are both proteins.

WHAT DO YOU THINK? *Iodine turned blue-black with cornstarch but not with gelatin. The bread contains starch but the tofu does not, as shown by a color change with bread but not with tofu.*

CHALLENGE *crust and sauce*

 Datasheet, Organic Molecules, p. 290

Technology Resources

Customize this student lab as needed or look for an alternative. Print rubrics to assess student lab reports.

 Lab Generator CD-ROM

Teaching with Technology

Students may want to photograph a positive reaction when testing cornstarch, so they can compare the color when testing other substances.

Real World Example

A change of amino acids in hemoglobin, the protein responsible for carrying oxygen in the blood, changes the shape of the protein. Sickle-cell anemia is a disease that results from this kind of change in protein structure. The change in the protein from hemoglobin A to hemoglobin S changes the shape of red blood cells and leads to many health problems.

INVESTIGATE Organic Molecules

Where can you find organic molecules?

SKILL FOCUS
Inferring

MATERIALS
- 4 small jar lids
- 3 eyedroppers
- cornstarch solution
- liquid gelatin
- iodine solution
- bread
- tofu
- stopwatch

TIME
20 minutes

PROCEDURE

1. Place a dropper of cornstarch solution into one jar lid and a dropper of liquid gelatin into a second jar lid.

2. Add a drop of iodine solution to the cornstarch sample and to the gelatin sample.

3. Examine the jar lids after 1 minute. Record your observations.

4. Using the remaining two jar lids, repeat steps 2 and 3 with the bread and the tofu instead of the cornstarch and gelatin.

WHAT DO YOU THINK?

- What changes occurred after the addition of iodine to the cornstarch and to the gelatin?

- Iodine can be used to detect the presence of starches. What carbon-based molecules might be in the bread and tofu? How do you know?

CHALLENGE Suppose you tested a piece of pepperoni pizza with iodine. Which ingredients (crust, sauce, cheese, pepperoni) would likely contain starch?

Proteins

Proteins are macromolecules that are made of smaller molecules called amino acids. Proteins, like carbohydrates and lipids, contain carbon, hydrogen, and oxygen. However, proteins differ from carbohydrates and lipids in that they also contain nitrogen, sulfur, and other elements. Unlike carbohydrates and lipids, which are used primarily for energy and structure, proteins have many different functions.

Think of a protein as being like a word, with amino acids as the letters in that word. The meaning of a word depends on the order of letters in the word. For example, rearranging the letters in the word "eat" makes different words with different meanings. Similarly, proteins depend on the order of their amino acids.

Linked Amino Acids

tyrosine   lysine   cysteine   serine   leucine

DIFFERENTIATE INSTRUCTION

 More Reading Support

F Which small molecules make up a protein? *amino acids*

Alternative Assessment Have students make a poster with colored pictures showing their results in the investigation. They could extend the activity by including a section with pictures of other foods and their predictions of the results when these foods are tested with iodine. Students may want to test these foods after they make their predictions.

Just as 26 letters of the alphabet make up all words in the English language, 20 amino acids make up all of the proteins in your body. The structure of a protein is determined by the order of its amino acids. If two amino acids change places, the entire protein changes.

The function of a protein depends on its structure. There are at least 100,000 proteins in your body, each with a different structure that gives them a specific function. Some proteins are structural materials, some control chemical reactions, and others transport substances within cells and through the body. Still others are a part of the immune system, which protects you from infections.

 CHECK YOUR READING How does the function of a protein depend on its structure?

Proteins that are part of the structure of living things are often shaped like coils. One coil-shaped protein, keratin, is part of human hair as shown on the left below. Proteins called actin and myosin are coil-shaped proteins that help your muscles contract.

Other types of proteins have coiled regions but curl up into shapes like balls. One example is hemoglobin, shown on the right below. Hemoglobin is a transport protein that carries oxygen in the blood.

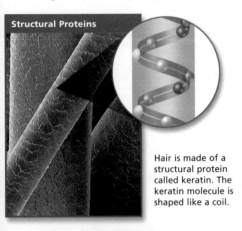

Structural Proteins

Hair is made of a structural protein called keratin. The keratin molecule is shaped like a coil.

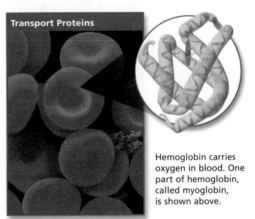

Transport Proteins

Hemoglobin carries oxygen in blood. One part of hemoglobin, called myoglobin, is shown above.

Some proteins that curl up into a shape like a ball are enzymes. An **enzyme** (EHN-zym) is a catalyst for a chemical reaction in living things. Catalysts increase the rate of chemical reactions. Enzymes are necessary for many chemical reactions in your body. Without enzymes, these reactions would occur too slowly to keep you alive.

It is important to have proteins in your diet so that your body can make its own proteins. Proteins in foods such as meats, soybeans, and nuts are broken down into amino acids by your body. These amino acids are then used by your cells to make new proteins.

Chapter 5: Carbon in Life and Materials **159** **B**

DIFFERENTIATE INSTRUCTION

More Reading Support

G What are enzymes? *catalysts for chemical reactions in living things*

Below Level To help students understand the three-dimensional structure of proteins, provide a string of large beads (pop beads are good for this purpose) and some two-sided adhesive tape. Students can model a three-dimensional molecule by creating "attractive forces" between the beads with the tape.

Teach Difficult Concepts

Students often have trouble visualizing how a chain of amino acids can form a three-dimensional structure. Explain that the polarity of amino acids can either attract or repel other amino acids. These attractive and repulsive forces cause the chain to fold into a specific three-dimensional structure.

Address Misconceptions

IDENTIFY Ask: What are energy sources for your cells? If students answer carbohydrates and lipids, they may think that proteins cannot be used for energy.

CORRECT Have students examine a nutrition label that lists Calories per gram. (Not all labels list this information.) Proteins contain 4 Calories per gram, the same number of calories as carbohydrates.

REASSESS Ask: How can the body get energy if carbohydrates and lipids are not available? *It can break down proteins for energy, but usually only as a last resort.*

Technology Resources

Visit **ClassZone.com** for background on common student misconceptions.

MISCONCEPTION DATABASE

Ongoing Assessment

Describe how a protein's structure determines its function.

Ask: What determines the structure and function of a protein? *the order of the amino acids in the molecule*

CHECK YOUR READING *Answer: The order of amino acids gives a protein a specific structure, which also gives it a specific function.*

Teach from Visuals

To help students interpret the visual of the structure of DNA, ask:

- Which molecules make up the sides of the "ladder"? *alternating 5-carbon sugar molecules and phosphate groups*

- Which molecules make up the ladder's rungs? *the four bases (A, T, G,and C)*

- Which bases are always paired with each other? *A (adenine) always pairs with T (thymine); C (cytosine) always pairs with G (guanine).*

- Which part of DNA contains the code for an amino acid? *a sequence of three DNA bases*

History of Science

James Watson and Francis Crick discovered the double-helix structure and bonding of base pairs in the DNA molecule in 1953. The story behind the discovery and a personal look at the scientific process is described in the book *The Double Helix,* written by James Watson.

Ongoing Assessment

Explain how proteins are put together from the information contained in nucleic acid molecules.

Ask: How does the fact that DNA is a long chain of subunits help to create a protein molecule? *A protein molecule also is a long chain of subunits. The process is linear, with one sequence of three bases coding for an amino acid.*

READING ViSUALS *Answer: It provides the code for correctly assembling proteins.*

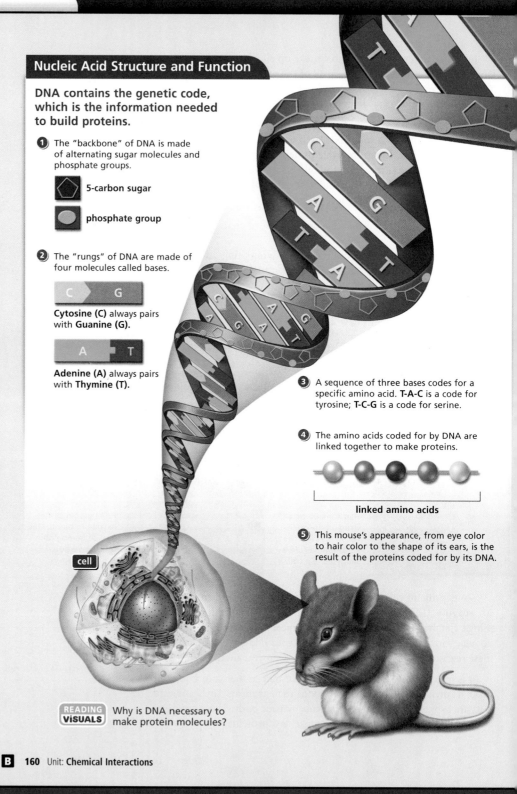

Nucleic Acid Structure and Function

DNA contains the genetic code, which is the information needed to build proteins.

1 The "backbone" of DNA is made of alternating sugar molecules and phosphate groups.

- ⬠ 5-carbon sugar
- ⬭ phosphate group

2 The "rungs" of DNA are made of four molecules called bases.

C **G**

Cytosine (C) always pairs with **Guanine (G).**

A **T**

Adenine (A) always pairs with **Thymine (T).**

3 A sequence of three bases codes for a specific amino acid. **T-A-C** is a code for tyrosine; **T-C-G** is a code for serine.

4 The amino acids coded for by DNA are linked together to make proteins.

linked amino acids

5 This mouse's appearance, from eye color to hair color to the shape of its ears, is the result of the proteins coded for by its DNA.

cell

READING ViSUALS Why is DNA necessary to make protein molecules?

DIFFERENTIATE INSTRUCTION

Inclusion Students with visual impairments may better understand how the DNA molecule is constructed if you let them handle a zipper. The sides of the zipper represent the sugar-phosphate backbones of the DNA molecule. The zipper's teeth represent the bases. When the zipper is closed, hydrogen bonds have formed between the bases to hold the two halves of the DNA molecule together. Twisting the zipper into a helix also helps students visualize the molecule's structure.

Nucleic Acids

Nucleic acids (noo-KLEE-ihk AS-ihdz) are huge, complex carbon-based molecules that contain the information that cells use to make proteins. These macromolecules are made of carbon, hydrogen, and oxygen, as well as nitrogen and phosphorus. Each of the cells in your body contains a complete set of nucleic acids. This means that each cell has all of the instructions necessary for making any protein in your body.

The illustration on page 160 shows part of a nucleic acid molecule called DNA, which looks like a twisted ladder. The sides of the ladder are made of sugar molecules and phosphate groups. Each rung of the ladder is composed of two nitrogen-containing molecules called bases. DNA has four types of bases, represented by the letters A, C, T, and G. The order of the bases in a DNA molecule is the way in which DNA stores the instructions for making proteins. How do just four molecules—A, C, T, and G—carry all of this important information?

Recall that a protein is composed of amino acids that have to be linked in a certain order. Each of the 20 amino acids is represented by a particular series of three DNA bases. For example, the sequence T–A–C corresponds to—or is a code for—the amino acid tyrosine. There are 64 different three-base sequences in DNA, all of which have a specific meaning. This genetic code works in the same way in every living thing on Earth. It provides a complete set of instructions for linking amino acids in the right order to make each specific protein molecule. The DNA code is only one part of making proteins, though. Other types of nucleic acids, called RNA, are responsible for reading the code and assembling a protein with the correct amino acids.

READING TiP

The *NA* in DNA stands for nucleic acid. The *D* stands for deoxyribose, which is the type of sugar in the molecule.

 RESOURCE CENTER
CLASSZONE.COM

Find out more about carbohydrates, lipids, proteins, and nucleic acids.

CHECK YOUR READING How many different types of bases make up the genetic code in DNA?

5.2 Review

KEY CONCEPTS

1. How does the function of a lipid depend on its structure?
2. What determines the structure of a protein?
3. What role does DNA perform in the making of proteins?

CRITICAL THINKING

4. **Synthesize** Give two examples of carbon-based molecules in living things that are based on a chain structure. Explain.
5. **Compare and Contrast** How are carbohydrates and lipids similar? How are they different?

CHALLENGE

6. **Infer** Suppose the order of bases in a DNA molecule is changed. What do you think will happen to the structure of the protein that is coded for by that region of DNA? Why?

Chapter 5: Carbon in Life and Materials **161** **B**

ANSWERS

1. Sample answer: A lipid with a phosphate group is used for cell structure.

2. The order of amino acids in a protein gives the protein its structure and function.

3. DNA carries instructions for lining up amino acids in a protein.

4. Starch and cellulose are chains of glucose; fatty acids in lipids contain chains; proteins are long chains of amino acids.

5. They are both organic; contain carbon, hydrogen, and oxygen; are based on carbon chains; are used for energy

and structure. However, lipids include fats and oils, and carbohydrates include sugars and starches; lipids can also contain other elements.

6. The protein would change because the code for amino acids is changed.

CHECK YOUR READING   *Answer: four*

Integrate the Sciences

DNA is one of two nucleic acids. DNA contains the genetic code, but a different type of nucleic acid, called RNA (ribonucleic acid), is necessary for making a protein. There are several kinds of RNA. Unlike DNA, which is double-stranded, RNA molecules are single-stranded. Also, RNA molecules contain the base uracil (U) instead of thymine (T).

Reinforce (the **BIG** idea)

Have students relate the section to the Big Idea.

R   Reinforcing Key Concepts, p. 291

5.2 ASSESS & RETEACH

Assess

A   Section 5.2 Quiz, p. 82

Reteach

Have students examine nutrition labels on food products at home for information about carbohydrates, lipids (fats), and proteins. Have students describe the general structures of these organic molecules in the food. Students should identify the difference between sugar and fiber (cellulose) and between saturated and unsaturated fats.

Technology Resources

Have students visit **ClassZone.com** for reteaching of Key Concepts.

 CONTENT REVIEW

 CONTENT REVIEW CD-ROM

Chapter 5 **161** **B**

Set Learning Goal

To make a bar graph that shows the daily dietary recommendations for healthful nutrition

Present the Science

The data in the table are based on the U.S. Department of Agriculture's (USDA) food guide pyramid. The USDA compiled the pyramid from research on what foods Americans eat, which nutrients are in these foods, and how Americans should make the best food choices to promote good health. Although it outlines what Americans should eat every day, the pyramid is a guideline rather than a rigid prescription.

Develop Graphing Skills

- Explain that a bar graph is the best choice of graphs to represent data that are not continuously changing.
- The independent variable is usually plotted on the x-axis. The dependent variable is usually plotted on the y-axis.

DIFFERENTIATION TIP Review independent and dependent variables with students. Be sure they understand why the axes are labeled as they are in the bar graph.

Close

Which type of organic molecule (carbohydrates, lipids, or proteins) accounts for the majority of recommended servings in a healthy diet? How do you know? *Carbohydrates; grains, vegetables, and fruits make up more of a healthy diet than dairy products, meats, and beans.*

- Math Support, p. 307
- Math Practice, p. 308

Technology Resources

Students can visit **ClassZone.com** for practice making bar graphs.

 MATH TUTORIAL

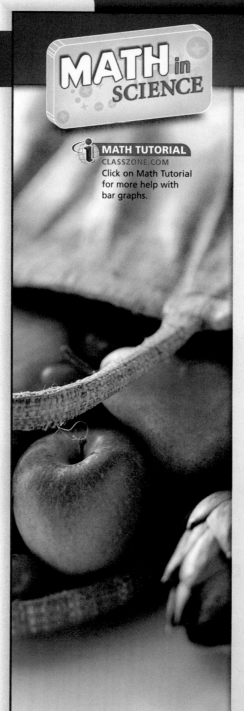

MATH in SCIENCE

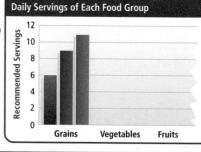

MATH TUTORIAL
CLASSZONE.COM
Click on Math Tutorial for more help with bar graphs.

SKILL: MAKING BAR GRAPHS

Graphing Good Food

People need to eat carbohydrates, proteins, and lipids to have a healthy diet. Different amounts of each type of organic molecule are recommended for different groups of people. In general, grains, vegetables, and fruits contain carbohydrates. Dairy products, meats, and beans contain proteins and lipids. The table on the right shows dietary recommendations. The information could also be shown in a bar graph.

Recommended Servings

Food group	Young children	Teen girls	Teen boys
Grains	6	9	11
Vegetables	3	4	5
Fruits	2	3	4
Dairy	3	3	3
Meats, beans	2	2	3

SOURCE: *U.S. Department of Agriculture, Home and Garden Bulletin Number 252, 1996*

Example

Create a bar graph that shows the dietary recommendation of grains for each group.

(1) Use the height of the bar to indicate the numerical value of a piece of data.

(2) Show the number of servings on the vertical axis. Label each group of bars on the horizontal axis.

(3) Use a different color for each group.

ANSWER

- ■ Young Children
- ■ Teen Girls
- ■ Teen Boys

Daily Servings of Each Food Group

Use the Recommended Servings table above to answer the following questions.

1. Copy and complete the bar graph to show the dietary recommendations for the other four food groups.

2. Which group has the tallest bars on the graph? the shortest?

CHALLENGE Choose one group of people and make a pie graph showing recommendations for them. (**Hint:** First convert the numbers of servings into fractions of the whole diet.)

B 162 Unit: **Chemical Interactions**

ANSWERS

1. See students' graphs.

2. tallest: teen boys; shortest: young children

CHALLENGE Students' pie charts will vary with the age group chosen. The fractions should add up to one.

KEY CONCEPT

5.3 Carbon-based molecules are in many materials.

BEFORE, you learned

- Organic compounds are based on carbon
- Carbon can form molecules shaped like chains or rings
- Four types of carbon-based compounds are common in living things

NOW, you will learn

- How carbon-based molecules are obtained from petroleum
- How carbon-based molecules are designed for specific uses
- How a material's properties depend on its molecular structure

VOCABULARY

hydrocarbon p. 163
polymer p. 166
monomer p. 166
plastic p. 167

THINK ABOUT

What do windbreakers and motor oil have in common?

Motor oil and windbreakers are both made of carbon-based molecules. The motor oil is composed mostly of carbon and hydrogen. The nylon windbreaker is also composed mostly of carbon and hydrogen. How can two materials that are so different be made of similar molecules?

Carbon-based compounds from ancient organisms are used to make new materials.

Many of the things you see around you every day contain carbon-based molecules. Some, such as people, plants, and animals, are easy to spot. Others are not so easy to identify. These objects include clothing, furniture, packing materials, sports equipment, and more. You have read that a large number of substances that make up living things are based on carbon. Where do we get carbon-based molecules that we use to make modern materials?

The carbon-based compounds that are the basis of many materials are called hydrocarbons. A **hydrocarbon** is simply a compound made of only carbon and hydrogen. Many different hydrocarbons are found in large deposits underground and under the sea. The story of how they got there began a long time ago and is related to the way carbon moves through the environment in a cycle.

RESOURCE CENTER
CLASSZONE.COM
Find out more about petroleum and hydrocarbons.

Chapter 5: **Carbon in Life and Materials 163** B

RESOURCES FOR DIFFERENTIATED INSTRUCTION

Below Level
UNIT RESOURCE BOOK
- Reading Study Guide A, pp. 294–295
- Decoding Support, p. 306

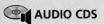 **AUDIO CDS**

Advanced
UNIT RESOURCE BOOK
- Challenge and Extension, p. 300
- Challenge Reading, pp. 302–303

English Learners
UNIT RESOURCE BOOK
Spanish Reading Study Guide, pp. 298–299

AUDIO CDS
- Audio Readings in Spanish
- Audio Readings (English)

5.3 FOCUS

▶ Set Learning Goals
Students will
- Recognize how carbon-based molecules are obtained from petroleum.
- Explain how carbon-based molecules are designed for specific uses.
- Describe how a material's properties depend on its molecular structure.

◀ 3-Minute Warm-Up
Display Transparency 37 or copy this exercise on the board:

Match the definition to the correct term.

Definitions
1. type of molecule made of linked units called amino acids *d*
2. type of molecule that contains carbon chains called fatty acids *c*
3. a class of compounds based on carbon *a*

Terms
a. organic compound
b. inorganic compound
c. lipid
d. protein
e. carbohydrate

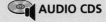 3-Minute Warm-Up, p. T37

5.3 MOTIVATE

THINK ABOUT

PURPOSE To introduce the concept of hydrocarbons as the basis for many materials

DEMONSTRATE To demonstrate the importance of petroleum-based materials in modern life, have students take five minutes to list as many items that are made of petroleum-based materials as they can.

Sample answer: Different materials of similar molecules probably have different molecular structures.

Chapter 5 **163** B

Teach from Visuals

To help students interpret the diagram of the carbon cycle, ask:

- What type of organic molecule is made during the part of the cycle labeled "Plants use CO_2"? *carbohydrate*

- In which two ways does carbon return to the atmosphere? *exhalation of carbon dioxide by animals, decomposition of plants and animals*

- How is carbon removed from the carbon cycle? *Decaying organisms are trapped in the ground, where their carbon-based molecules become the hydrocarbons in petroleum.*

Teach Difficult Concepts

Students may have trouble understanding how distillation works. Remind them that steam forms when a kettle of water boils. Tell them that if the steam can be captured in a container and cooled, it becomes liquid water again. This process is called distillation. Ask whether distillation involves a physical or chemical change. *physical* To help students understand, you might try the following demonstration.

Teacher Demo

Boil some water in a beaker. After it begins to boil, place a cool glass plate over the top of the beaker. When the glass is covered with condensed water, lift it off the beaker while using thermal gloves, and drain the condensed water into another beaker.

Ongoing Assessment

Recognize why many items used in modern life contain carbon.

Ask: Which natural resource is the source of modern carbon-based materials? *petroleum or crude oil*

CHECK YOUR READING *Answer: by using differences in their boiling points*

SUPPORTING MAIN IDEAS
Make a chart about the carbon-based molecules in modern materials.

READING **TiP**
As you read the numbered steps, follow the process shown on page 165.

VISUALIZATION CLASSZONE.COM
Observe the process involved in the separation of the hydrocarbons in petroleum.

In the carbon cycle, plants use carbon dioxide from the air to make carbohydrates. Animals eat plants and absorb carbon, and then release carbon dioxide into the air when they exhale. When animals and plants die, they decompose and carbon returns to the environment. Most carbon returns to the atmosphere as part of the carbon cycle, but some does not. This carbon is the source of the carbon-based compounds that are so important for modern life.

Carbon Cycle

CO_2 in atmosphere → Plants use CO_2 and store carbon → Animals eat plants, exhale CO_2 to release carbon → Hydrocarbons in petroleum → Animals and plants decompose, release carbon

1. **Obtaining Petroleum** Some living things that died hundreds of millions of years ago fell into mud or sediment. Instead of returning to the atmosphere, the carbon-based molecules from these organisms became trapped in the ground. Over time, and through a process of chemical changes, some of this organic material became petroleum, which is a mixture of hundreds of different hydrocarbons. People pump petroleum from underground and undersea deposits. Liquid petroleum is called crude oil.

2. **Refining Petroleum** In its raw form, petroleum is not a useful substance. However, when it is processed at a refinery, it is separated into many useful parts.

3. **Using Products Made from Petroleum** Many products, including gasoline, plastics, and fibers such as nylon are made from the separated parts of petroleum.

The refining of petroleum is an example of the separation of a mixture based on physical properties. Each type of hydrocarbon in petroleum has a different boiling point. In general, each boiling point depends on the number and arrangement of carbon atoms in the molecule. For example, lubricating oil contains long carbon chains and boils at temperatures above 350°C. The hydrocarbons that make up gasoline are smaller molecules, and they boil between 35°C and 220°C.

At a refinery, petroleum is heated until all but the largest of the hydrocarbons are in gaseous form. This gaseous petroleum is released into a distillation tower. As the gases rise in the tower, they gradually cool. When each specific hydrocarbon cools to its boiling point, it condenses back into a liquid. Thus, lubricating oil cools to its boiling point quickly and is collected from a low level in the tower. Hydrocarbons in gasoline take longer to become liquids and are collected higher in the tower.

CHECK YOUR READING How are hydrocarbons in petroleum separated from each other?

DIFFERENTIATE INSTRUCTION

? More Reading Support

A Is petroleum formed by physical changes or chemical changes? *chemical changes*

English Learners Place the terms *hydrocarbon, polymer,* and *plastic* on the classroom's Science Word Wall with a brief definition for each. English learners may not have prior knowledge of windbreakers on p. 163, Teflon on p. 168, and the practice of using police dogs shown on p. 169.

Using Petroleum

Carbon-based compounds in petroleum are used to make a wide range of products.

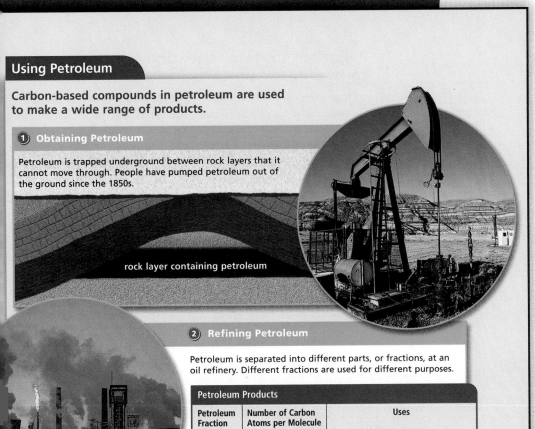

① Obtaining Petroleum

Petroleum is trapped underground between rock layers that it cannot move through. People have pumped petroleum out of the ground since the 1850s.

rock layer containing petroleum

② Refining Petroleum

Petroleum is separated into different parts, or fractions, at an oil refinery. Different fractions are used for different purposes.

Petroleum Products

Petroleum Fraction	Number of Carbon Atoms per Molecule	Uses
Gases	1 to 4	Cooking, heating, manufacturing
Gasoline	5 to 12	Automobile fuel
Kerosene	12 to 16	Airplane fuel
Fuel oil	15 to 18	Diesel fuel, heating oil
Greases	16 to 20	Lubrication

SOURCE: *Mortimer, Chemistry, 6th edition*

③ Using Products Made from Petroleum

The gas fraction of petroleum is often used to make such products as fibers and plastics. The gasoline fraction is often used as a fuel for cars.

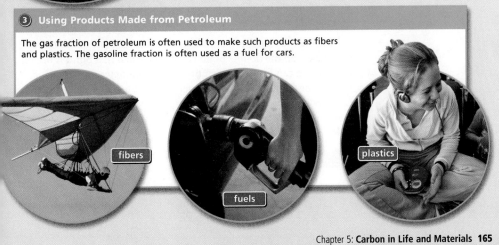

fibers

fuels

plastics

Chapter 5: **Carbon in Life and Materials 165** **B**

DIFFERENTIATE INSTRUCTION

Advanced Students might want to learn more about where crude oil deposits are found on Earth. If you can provide students with an outline of a world map, they can conduct research and color a copy of the map in to show the oil-rich regions.

R Challenge and Extension, p. 300

Teach from Visuals

To help students interpret the visuals showing how petroleum is processed and used, ask:

- Where are oil deposits found? *trapped between rock layers through which it cannot move*

- Why must the different compounds in petroleum be separated? *They each have different uses, based on their properties. The original mixture of compounds is not a useful product.*

Teach Difficult Concepts

The many hydrocarbons in petroleum can be separated by simple distillation only if their boiling points differ by 50°C or more. If two compounds have similar boiling points, the distillate will contain a mixture of the two. To separate these compounds, the mixture is redistilled over and over in a fractionating tower. The condensed liquid is boiled and recondensed several times. These are all physical changes; no chemical changes are involved.

Social Studies Connection

Recent estimates indicate the world's petroleum reserves amount to approximately 1 trillion barrels (1 barrel = 42 gallons). An even larger quantity may be available from a different source—oil shale. Oil shale is a type of rock that contains a relatively large amount of organic material, from which oil and gas can be extracted. It has been estimated that there could be 2.6 trillion barrels of oil that could be extracted from oil shale. However, due to the higher cost of mining and refining oil shale, only a few countries currently obtain oil from it.

Revisit "Internet Activity: Polymers" on p. 145. Have students explain how changes in a polymer's structure alter the material.

Develop Critical Thinking

ANALYZE Simple sugars, such as glucose, are called monosaccharides, and complex carbohydrates, such as starch, are called polysaccharides. Ask: In terms of monomers and polymers, what is the relationship between simple sugars and complex carbohydrates? *Simple sugars are monomers and complex carbohydrates are polymers made by linking simple sugars.*

Ongoing Assessment

CHECK YOUR READING *Answer: They are both polymers that contain repeating, smaller units.*

CHECK YOUR READING *Answer: Polypropylene is a polymer that is a strong, solid plastic, but the monomer propylene is a gas.*

Polymers contain repeating carbon-based units.

READING TIP

The prefix *mono-* means "one," and the prefix *poly-* means "many."

Many everyday materials made of carbon-based molecules contain macromolecules called polymers. **Polymers** are very large carbon-based molecules made of smaller, repeating units. These small, repeating units, called **monomers,** are linked together one after another. By themselves, monomers are also carbon-based molecules.

Some of the carbon-based molecules that you read about in the previous section are polymers. For example, both cellulose and starch are polymers. They are chains of linked glucose units. The glucose units are the monomers. Many common materials that are manufactured for specific purposes are polymers. Plastics and fibers are two examples of these kinds of polymers.

CHECK YOUR READING   What do starch and plastic have in common?

Formation of Polymers

The properties of a polymer depend on the size and structure of the polymer molecule. The size and structure of a polymer depend both on its particular monomers and how many monomers are added together to make the final product.

The process of making a polymer involves chemical reactions that bond monomers together. Think back to the different types of reactions described in Chapter 3. The process of making a polymer is a synthesis reaction that yields a more complex product from simpler reactants.

The diagram on the top of page 167 shows one way in which a polymer can be made. The monomer that is the building block of the polymer in the illustration is called propylene (PROH-puh-LEEN). The propylene molecule consists of three carbon atoms and six hydrogen atoms. By itself, propylene is a gas. Notice that propylene has a double bond between two of its carbon atoms.

During the reaction that links the monomers, one of the bonds that makes up the double bond is broken. When that bond is broken, a new bond can form between two of the monomer units. A large number of the propylene monomers bonded together form a polymer called polypropylene. Polypropylene is a strong, solid plastic used to make such items as plastic crates, toys, bicycle helmets, and even indoor-outdoor carpeting.

CHECK YOUR READING   How do the properties of polypropylene differ from those of propylene?

DIFFERENTIATE INSTRUCTION

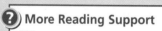

? **More Reading Support**

B What are polymers made of? *repeating monomers*

C What type of chemical reaction creates polymers? *a synthesis reaction*

Below Level Students can put together pop beads to model how a polymer is made out of monomers. As an alternative, they can use marshmallows or gumdrops and toothpicks to model the production of polymers.

Building Polymers

Polymers such as polypropylene are made by linking many monomers together.

This bicycle helmet is made of polypropylene.

1 Propylene (C_3H_6) can be used as a monomer.

Monomer

$$CH_2 = CH - CH_3$$

2 Propylene monomers are linked together.

Monomers Are Linked Together

$$CH_2 = CH - CH_3 \quad CH_2 = CH - CH_3 \quad CH_2 = CH - CH_3$$

3 Linked propylene monomers make polypropylene.

Polymer

$$CH_2 - CH - CH_2 - CH - CH_2 - CH ---$$
$$\qquad\; | \qquad\qquad | \qquad\qquad |$$
$$\qquad CH_3 \qquad\quad CH_3 \qquad\quad CH_3$$

— = double bond changed to single bond
— = new bond formed between monomers

Polymers may be composed of more than one type of monomer. Polyester fabric is an example of a polymer that contains two different monomers. Protein is another example. In fact, as you read earlier, proteins in living things contain several different monomers. The monomers in proteins are amino acids.

Plastics

Polypropylene is one of many polymers that are called plastics. As an adjective, *plastic* means "capable of being molded or shaped." A **plastic** is a polymer that can be molded or shaped. If you look around, you can see how common plastics are in everyday life.

The first plastic made by chemists was celluloid, which was patented in 1870. It was based on cellulose molecules from cotton plants and was used to make such things as billiard balls and movie film. Celluloid is different from many of the plastics that are made today. It was made by chemically changing an existing, naturally occurring polymer—cellulose.

Many of today's plastics are made artificially, by building polymers from monomers. The first completely artificial polymer made by scientists was a plastic called Bakelite, which was invented in 1907. Chemists made Bakelite by linking individual monomers together. Because Bakelite is moldable and nonflammable, it was used for many household items such as pot handles, jewelry, lamps, buttons, and radio cases.

CHECK YOUR READING How does the term plastic describe a polymer's properties?

Chapter 5: **Carbon in Life and Materials** 167 **B**

Teach from Visuals

To help students interpret the steps in building polymers, ask:

- What does the name *polypropylene* mean? *many propylene molecules*
- Which carbon atoms in propylene are linked together to make polypropylene? *The middle one is linked to the first carbon of the next monomer.*
- How many propylene monomers were joined to make the section of polymer shown in step 3? *three*
- What happens to the double bond in each monomer? *One of the bonds is broken so one monomer can bond to another.*

Integrating the Sciences

Celluloid was produced by treating nitrocellulose molecules from cotton plants with camphor. That discovery began the era of plastics. Even though celluloid is closely linked with the cinema and the film industry, it is at least partly responsible for helping to save African elephants. The ivory of elephant tusks was used to make products such as billiard balls and piano keys, and the killing of elephants for their tusks dramatically decreased their numbers. The use of celluloid as a substitute for ivory in billiard balls and piano keys decreased the hunting of elephants for those purposes.

Ongoing Assessment

Explain how carbon-based molecules are designed for specific uses.

Ask: What properties would you want a plastic to have if it were to be used for cookware handles? *nonflammable, heat-resistant, lightweight*

 CHECK YOUR READING *Answer: The polymer can be molded and shaped.*

To help students interpret the table of plastics, make a recycling display in your classroom. Have students bring in samples of each kind of plastic so they can examine them. Have students identify the monomers in each type of plastic.

Integrating the Sciences

In 1998, Americans recycled 1.45 billion pounds of plastic bottles. At least 94 percent of this recycled plastic was PET and HDPE, types 1 and 2, respectively. Recycled plastics can be used to make thousands of products. However, each time a plastic is recycled, it loses some of its strength. Often, after a plastic has been recycled about five times, it is of such poor quality that it must be discarded. Disposal of this old plastic is difficult.

Teacher Demo

Create a polymer gel that climbs the side of a beaker. Mix 20–25 mL of methyl or ethyl alcohol with 3–4 g of polyethylene oxide in a clean, dry 600 mL beaker. Swirl the mixture to completely wet it with the alcohol. Add 350–400 mL of tap water to the mixture all at once. Stir until the polymer has disappeared and the mixture is blended and thick. Pour the gel into a second 600 mL beaker and then pour it back and forth between the two beakers. The gel can be made to siphon "uphill" out of a beaker and against gravity. To start the process, raise the beaker with the gel above an empty beaker. Once the gel starts to pour, turn the raised beaker upright again. The gel will move up the sides of the raised beaker as a thin film and then form thick strands as it falls. This process can be repeated indefinitely.

Ongoing Assessment

Describe how a material's properties depend on its molecular structure.

Ask: Why do isomers have different properties? *Their structure is different.*

Recycling Plastics

Code	Chemical Name	Monomers	Properties	Uses
1	Polyethylene terephthalate (PET or PETE)	$C_2H_6O_2$ and $C_8H_6O_4$	Transparent, high strength, does not stretch	Clothing, soft-drink audiotapes, videota...
2	High Density Polyethylene (HDPE)	C_2H_4	Similar to LDPE (code 4) but denser, tougher, more rigid	Milk and water jugs gasoline tanks, cups
3	Polyvinyl chloride (PVC)	C_2H_3Cl	Rigid, transparent, high strength	Shampoo bottles, g... hoses, plumbing pip...
4	Low Density Polyethylene (LDPE)	C_2H_4	White, soft, subject to cracking	Plastic bags, toys, electrical insulation
5	Polypropylene (PP)	C_3H_6	High strength and rigidity, impermeable to liquids and gases	Battery cases, indo... carpeting, bottle ca...
6	Polystyrene (PS)	C_8H_8	Glassy, rigid, brittle	Insulation, drinking packing materials

SOURCE: *American*

The chart above lists some common plastics and their uses. You may have noticed the symbols shown in the first column of the chart on plastic bottles and containers that you have around your home. The numbers stand for different types of plastic, with different uses. After a plastic has been used in a certain way, such as in soft drink bottles, it can recycled and used again. When a plastic is recycled it can be made into a new product that has a different use. For example, recycled soft drink bottles can be made into fibers for carpeting.

Designing Materials

Chemists have been designing and making polymers for many years. However, new polymers are always being developed. One way in which scientists make new polymers is by chemically changing an original monomer. When scientists change a monomer and then link the monomers together, a new material is produced.

Teflon is a common polymer that is made by chemically changing a hydrocarbon monomer. Chemists replace the hydrogen atoms in the monomer with fluorine atoms and link the monomers together. You have probably seen Teflon as a nonstick coating on pots and pans. Teflon is also very strong, and it was used as a part of the structure of the stadium's dome in the photograph on the left.

Teflon is very strong and light in weight. It was used to make the roof of this stadium in Minneapolis.

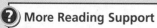

$$--- CF_2 - CF_2 - CF_2 - CF_2 ---$$

DIFFERENTIATE INSTRUCTION

? More Reading Support

E How can a new polymer be developed? *by chemically changing an original monomer*

Alternative Assessment Students who have difficulty reading can make a poster that summarizes the main ideas of the section. Students could draw or cut out pictures from magazines to illustrate the parts of the carbon cycle and how carbon that was removed from the cycle is used in the world.

Advanced Have students who are interested in molecules read the following article:

 Challenge Reading, pp. 302–303

Nomex

Nomex is used to make fireproof clothing worn by firefighters.

Kevlar

Kevlar is used in bulletproof vests worn by police officers and police dogs.

Chemists have developed several other materials by changing the monomers in nylon, a polymer that is often used in clothing. By adding a carbon-based ring to the nylon monomer, chemists made Nomex, which is used in fireproof clothing. Then scientists changed the placement of this ring on the monomer. This change resulted in the polymer called Kevlar, which is used in bulletproof vests.

Nomex and Kevlar are isomers. The monomers that are the basis of each contain the same atoms but have different structures. As a result, the polymers have different properties and uses. The structure of the polymer in Nomex gives the fibers flexibility along with fire resistance. As a result, Nomex can be made into relatively comfortable fireproof clothing for firefighters and race car drivers. Kevlar, however, is very rigid because of its structure and is the strongest known fiber. In fact, Kevlar is five times stronger than an equal amount of steel, which helps explain why Kevlar is used in bulletproof vests.

 CHECK YOUR READING Why are Nomex and Kevlar isomers?

5.3 Review

KEY CONCEPTS

1. What physical property allows the hydrocarbons in petroleum to be separated? Explain.

2. What are monomers? How are monomers related to polymers?

3. How are polymers such as Kevlar and Nomex similar to each other? How are they different?

CRITICAL THINKING

4. **Synthesize** What general type of chemical reaction makes polymers from monomers? Explain.

5. **Compare and Contrast** How are plastics such as polypropylene similar to celluloid, the first plastic? How are they different?

⚠ CHALLENGE

6. **Synthesize** A petroleum deposit is full of carbon compounds. How did that carbon get out of the atmosphere and into the petroleum?

Chapter 5: **Carbon in Life and Materials** 169 **B**

ANSWERS

. boiling point; larger molecules have higher boiling oints

. small carbon-based olecules; linked to ake polymers.

. They are both based on ylon; they are isomers ecause their carbon rings re in different positions.

4. synthesis reactions; a more complex molecule is made by combining simpler molecules

5. They are organic, and they are polymers. Celluloid is based on cellulose, a natural material, but modern plastics are artificial.

6. Plants removed carbon dioxide from the air during photosynthesis, and animals ate the plants. Petroleum was formed from the remains of the plants and animals.

Ongoing Assessment

CHECK YOUR READING *Answer: The monomers have the same atoms but different structures.*

History of Science

In the early 1930s, E. I. Du Pont de Nemours and Company hired Wallace Carothers to find a way of making artificial fibers. Carothers and his team soon created fibers that were long, strong, and very elastic. In 1935, this product was patented as nylon. The chemists called it Nylon 66 because it was made from two monomers that each had six carbon atoms. Each molecule of nylon consisted of 100 or more repeating units of two monomers strung in a chain. A nylon filament may have more than a million linked molecules of monomers.

Reinforce (the **BIG** idea)

Have students relate the section to the Big Idea.

R Reinforcing Key Concepts, p. 301

5.3 ASSESS & RETEACH

Assess

A Section 5.3 Quiz, p. 83

Reteach

Divide the class into small groups. Give each group 20 paper clips, and ask them to link the clips differently to make two different polymers of 10 paper clips each. Ask: If each of the 20 paper clips represents a molecule of the same monomer, are the two polymers isomers of each other? Explain. *Yes; they differ only in the way the monomers are linked.*

Technology Resources

Have students visit **ClassZone.com** for reteaching of Key Concepts.

 CONTENT REVIEW

 CONTENT REVIEW CD-ROM

Chapter 5 **169** **B**

Focus

PURPOSE To make a polymer and observe its properties

OVERVIEW Students will make a polymer by using white glue and borax and will test the polymer's properties. Students should find that their polymer

- is very stretchable and feels like both a solid and a liquid
- stretches when pulled slowly but breaks when pulled quickly
- does not bounce
- flows slowly and
- can be filled with air to make a bubble.

Lab Preparation

- Students can bring in many of the materials from home.
- Prior to the investigation, have students read through the investigation and prepare their data tables. You may wish to copy and distribute datasheets and rubrics.

    R UNIT RESOURCE BOOK, pp. 309–317

    SCIENCE TOOLKIT, F15

Lab Management

- The creep test can be done as a demonstration that students observe periodically during the class period.
- Do not allow students to take their polymers out of the classroom.
- Warn students not to eat their polymer or put it into their mouths.

Teaching with Technology

If a video camera is available, students may want to tape this lab and narrate their discoveries.

CHAPTER INVESTIGATION

Polymers

OVERVIEW AND PURPOSE Polymers are used in many common items. For example, the substance that is being pulled out of the beaker in the photograph on the left is raw nylon, which is a polymer. The properties of polymers are influenced by the way in which the chains are linked together. In this activity, you will use what you have learned about carbon-based molecules to
- make a polymer
- study the properties of a polymer

▶ Procedure

1. Create a data table like the one shown on the sample notebook page.

2. Follow the instructions to mix the polymer called Glurch. After you have made the polymer, store it in a zip-top bag.

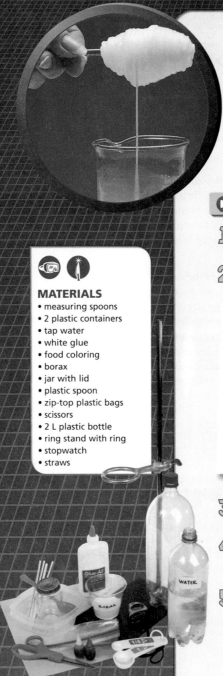

MATERIALS
- measuring spoons
- 2 plastic containers
- tap water
- white glue
- food coloring
- borax
- jar with lid
- plastic spoon
- zip-top plastic bags
- scissors
- 2 L plastic bottle
- ring stand with ring
- stopwatch
- straws

GLURCH

40 mL (8 tsp) water   2 tsp powdered borax
40 mL (8 tsp) white glue   30 mL (6 tsp) water
6 drops of food coloring

- Make a glue mixture by mixing 40 mL of water with 40 mL of white glue in a plastic container.
- Add food coloring and mix the color evenly.
- Make a borax solution by adding 2 tsp of borax to 30 mL of water. Shake the mixture in a covered jar for 30 seconds.
- Combine the borax solution with the glue mixture. Stir.
- Knead until the mixture is smooth, with a rubberlike consistency.

3. Remove some of the polymer from the bag. Try each test below in order. Observe each test and record your results.

4. **SQUEEZE TEST** Put some of the polymer in your hand. Squeeze the polymer to test its shape and its feel. Is it a solid, a liquid, or a little like both? Record your observations.

5. **PULL TEST** Hold the polymer between your hands and pull it apart slowly. Try again, and pull it apart very quickly. What happens to the polymer? Record the results of this test.

 B 170 Unit: **Chemical Interactions**

INVESTIGATION RESOURCES

 R CHAPTER INVESTIGATION, Polymers
- Level A, pp. 309–312
- Level B, pp. 313–316
- Level C, p. 317

Advanced students should complete Levels B & C.

 Writing a Lab Report, D12–13

Technology Resources

Customize this student lab as needed or look for an alternative. Print rubrics to assess student lab reports.

Lab Generator CD-ROM

6. **BOUNCE TEST** Roll some of the polymer into a ball between your palms. Test whether the polymer ball will bounce. Record your observations about the polymer's behavior.

7. **CREEP TEST**
Setup Cut the top off a 2 L bottle. Keep the bottle cap on. Set up a ring stand. To use the bottle top as a funnel, place it upside down in the ring. Put a plastic container under the funnel.

step 7

Trials Place approximately 100 mL of the polymer in the funnel. Remove the bottle cap and time how long it takes for the polymer to flow completely through the funnel. Record the time. If time allows, conduct one or two more trials.

8. **BUBBLE TEST** Take a small amount of the polymer and roll it into a ball around the end of a straw. Pinch the polymer closed around the straw. Hold the ball in one hand as you gently blow into the other end of the straw. Try to make a bubble by filling the polymer with air. Record your observations.

▶ Observe and Analyze Write It Up

1. **RECORD OBSERVATIONS** Be sure that your data table is complete. Describe how the polymer feels, as well as its state, shape, and behavior.

2. **COMMUNICATE** Include drawings of any observations for which a picture is helpful in understanding your results.

3. **INTERPRET DATA** Make a list of the polymer's physical properties. Which test was most helpful in identifying these properties? Which test was the least helpful?

▶ Conclude Write It Up

1. **INFER** The more complex a polymer is, the more rigid it is. Do you think that the polymer you made contains molecules with extremely long or complex carbon chains? Why or why not? What properties of your polymer provide evidence for your answer?

2. **EVALUATE** What limitations or difficulties did you experience in interpreting the results of your tests or other observations?

3. **APPLY** Based upon your results, what uses could you suggest for the polymer? What further tests would you need to do to make sure it would stand up to the demands of that use?

▶ INVESTIGATE Further

CHALLENGE Investigate the properties of your polymer by varying the proportions of the ingredients. Change only one ingredient. Be sure to record the change you made in the polymer. Make a new data table. Record the results of the experimental tests. How do changes in the polymer recipe change the physical properties of the polymer?

Polymers
Observe and Analyze

Table 1. Polymer Properties

Test	Observations
Squeeze test	
Pull test	
Bounce test	
Creep test	Trial 1 (time) Trial 2 (time) Trial 3 (time)
Bubble test	

Conclude

▶ Observe and Analyze   Write It Up

1. The polymer should be somewhat sticky and slimy and very malleable.

2. The diagram most likely to be helpful would show the flow of the polymer in the creep test.

3. The polymer should be sticky, slimy, fluid, and malleable. The stretch test, pull test, and creep test should be most helpful; the bounce test and bubble test the least helpful.

▶ Conclude   Write It Up

1. No; the polymer is not rigid but is instead, very malleable and somewhat fluid, as shown by the test results.

2. Answers will vary but could include difficulties using subjective rather than objective observations for most of the tests, changes in the polymer properties over the duration of the investigation, and problems associated with making an error while making the polymer.

3. Sample answer: Possible uses could be as a putty or a sealer. It could be used as a toy or as bubble gum if it is tested and found to be nontoxic and safe for use by children.

▶ INVESTIGATE Further

CHALLENGE Answers will depend on the proportions that are used and the ingredient that is changed. Changing the recipe will most likely change the polymer's properties and various test results.

Post Lab Discussion

Ask: Why was it important to follow the recipe exactly? *The recipe was developed to produce a certain polymer. Changes in the recipe would change the properties of the polymer.*

Describe how the testing of a polymer's properties might differ if the polymer was designed for a specific function or if it was discovered by accident. *If the polymer was designed for a particular function, tests would be carried out to see how well it performed that function. If the polymer was discovered by chance, tests would be carried out to see what it could be used for.*

BACK TO

the BIG idea

Have students discuss the following questions:

• Atoms of which element are the basis of all life on Earth? *carbon*

• How is carbon from living things used to make many modern materials? *Hydrocarbons from petroleum, formed from the carbon-based molecules of living things from hundreds of millions of years ago, are refined to make materials such as fibers and plastics.*

◑ KEY CONCEPTS SUMMARY

SECTION 5.1

Ask: Why can carbon atoms form a large number of molecules with different structures such as chains, rings, and isomers? *Carbon atoms form four bonds, including single, double, and triple bonds, with other atoms, including other carbon atoms.*

SECTION 5.2

Ask: Give an example of how each of the four carbon-based macromolecules is used in the body. *Sample answer: Carbohydrates are a source of chemical energy for cells; lipids are a part of cell membranes; proteins make up muscle tissue; nucleic acids carry genetic information.*

SECTION 5.3

Ask: How are polymers made? *A polymer is made from small repeating monomers. As in all chemical reactions, bonds within the reactants (the monomers) are broken and new bonds form in the products (the polymer).*

Review Concepts

• Big Idea Flow Chart, p. T33
• Chapter Outline, pp. T39–T40

---

5 Chapter Review

the BIG idea

Carbon is essential to living things and to modern materials.

 CONTENT REVIEW
CLASSZONE.COM

◑ KEY CONCEPTS SUMMARY

5.1 Carbon-based molecules have many structures.

Carbon forms a large number of different compounds because of the number of bonds it can make with other atoms.

Carbon can form chains and rings.

Hexane

$$CH_3 - CH_2 - CH_2 - CH_2 - CH_2 - CH_3$$

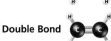
Single Bond

Double Bond

Triple Bond

Vanillin

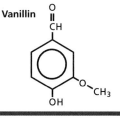

VOCABULARY
organic compound p. 147
inorganic compound p. 148
isomer p. 152

5.2 Carbon-based molecules are life's building blocks.

There are four main types of carbon-based molecules in living things.

Carbon-Based Molecules

Carbohydrates	Lipids	Proteins	Nucleic Acids
• include sugars and starches	• include fats and oils	• function depends on order of amino acids	• DNA
• energy for cells	• energy for cells	• structure, transport, immune system, enzymes	• carries genetic code
• plant cell walls	• cell membranes		• sequence of three DNA bases is the code for an amino acid

VOCABULARY
carbohydrate p. 155
lipid p. 156
protein p. 158
enzyme p. 159
nucleic acid p. 161

5.3 Carbon-based molecules are in many materials.

Carbon from ancient organisms is used to make many common items, such as clothing and plastics. These items are based on polymers.

Monomer $CH_2 = CH - CH_3$

Polymer
$$CH_2 - CH - CH_2 - CH - CH_2 - CH --- $$
$$\quad\quad\quad | \quad\quad\quad\quad | \quad\quad\quad\quad |$$
$$\quad\quad\quad CH_3 \quad\quad\quad CH_3 \quad\quad\quad CH_3$$

VOCABULARY
hydrocarbon p. 163
polymer p. 166
monomer p. 166
plastic p. 167

Technology Resources

Have students visit **ClassZone.com** or use the CD-ROM for a cumulative review of concepts.

 CONTENT REVIEW

 CONTENT REVIEW CD-ROM

Engage students in a whole-class interactive review of Key Concepts. Edit content as you wish.

 POWER PRESENTATIONS

Reviewing Vocabulary

Copy and complete the chart below. Fill in the blanks with the missing term, example, or function. See the example in the chart.

Term	Example	Function
inorganic compound	carbon dioxide	used by plants to make glucose
1. organic compound	glucose	
2. carbohydrate	sugar	
3.	fat	stores chemical energy
4.	keratin	found in hair and feathers
5. nucleic acid		instructions for proteins
6. plastic	polypropylene	

Greek Roots *Describe how each of the following terms is related to one or more of the following Greek roots.*

iso- means "equal"   *mono-* means "one"
-mer means "part"   *poly-* means "many"

7. isomer

8. polymer

9. monomer

10. polyunsaturated

Reviewing Key Concepts

Multiple Choice *Choose the letter of the best answer.*

11. All life on Earth is based on atoms of which element?
    a. oxygen
    b. nitrogen
    c. carbon
    d. hydrogen

12. Carbon atoms can form large numbers of compounds because a carbon atom forms
    a. two bonds with a hydrogen atom
    b. four bonds in its compounds
    c. ionic bonds in its compounds
    d. bonds with up to five hydrogen atoms

13. Which of the following is not found in living things?
    a. proteins   c. lipids
    b. petroleum   d. carbohydrates

14. What functions do carbohydrates and lipids perform in living things?
    a. They provide energy and instructions.
    b. They provide water and oxygen.
    c. They provide water and immunity.
    d. They provide energy and structure.

15. The molecules that carry instructions to make other molecules are called
    a. nucleic acids   c. carbohydrates
    b. proteins   d. lipids

16. Which kinds of molecules are best for storing chemical energy in living things?
    a. enzymes   c. lipids
    b. proteins   d. nucleic acids

17. The properties of artificial polymers are determined by
    a. the structure of the molecule
    b. the reaction used to make the polymer
    c. the time it took to make the polymer
    d. a series of DNA bases

Short Answer *Write a short answer to each question.*

18. Explain how carbon's ability to form isomers is related to the large number of carbon-based molecules that exist.

19. Describe the movement of carbon through the environment in a cycle. How does a break in the cycle provide carbon for modern materials?

Reviewing Vocabulary

1. provides energy for cells

2. provides energy in most living things

3. lipid

4. protein

5. DNA

6. used to make products such as containers, toys, and indoor-outdoor carpeting

7. Equal parts: isomers have the same kind and number of atoms.

8. Many parts: polymers contain many monomers.

9. One part: monomers are one part of a polymer.

10. many (more than one) double bonds in the fatty acids of a lipid

Reviewing Key Concepts

11. *c*

12. *b*

13. *b*

14. *d*

15. *a*

16. *c*

17. *a*

18. *Because carbon can form molecules with the same atoms but in different arrangements, the number of possible compounds increases greatly.*

19. *In the carbon cycle, plants use carbon dioxide from the atmosphere to create food; animals eat plants and exhale carbon dioxide; animals and plants decompose and release carbon into the environment. Petroleum comes from living things that died hundreds of millions of years ago. Petroleum can be separated into different parts that can be used to make many things, including plastics and fabrics.*

SSESSMENT RESOURCES

 UNIT ASSESSMENT BOOK
• Chapter Test A, pp. 84–87
• Chapter Test B, pp. 88–91
• Chapter Test C, pp. 92–95
• Alternative Assessment, pp. 96–97
• Unit Test, pp. 98–109

 SPANISH ASSESSMENT BOOK
• Spanish Chapter Test, pp. 249–252
• Spanish Unit Test, pp. 253–256

Technology Resources

Edit test items and answer choices.

 Test Generator CD-ROM

Visit **ClassZone.com** to extend test practice.

 Test Practice

Thinking Critically

20. Amino acids could be thought of as monomers, which are linked to form a protein.

21. The sequence of bases in a DNA molecule codes for a particular sequence of amino acids.

22. No, because the order of amino acids gives a protein its structure, which, in turn, gives the protein its function.

23. Sample answer: structural; an enzyme

24. Chains and rings: both contain carbon and are found in living things and in modern products; they differ in structure, bonding, and properties. Proteins and lipids: both contribute to the structure of living things; they differ in that proteins are not typically used for energy, as are lipids, and have many more functions than lipids. Glucose and amino acids: both are found in living things and are monomers; they contain different elements and form different polymers.

Using Math Skills in Science

25. approximately 53 from fats, 88 from carbohydrates, and 9 from proteins

26. Check students' charts; the portions of the pie chart should show 211° of the circle as carbohydrate, 127° as fat, and 22° as protein.

27. It is relatively close, but not exact—59% carbohydrates, 35% fats, and 6% protein. To be the recommended balance, it should have more protein and slightly less carbohydrates and fats.

the BIG idea

28. Answers will vary.

29. Answers will vary.

Thinking Critically

The illustration below models linked amino acids. Use the illustration to answer the next four questions.

20. **SYNTHESIZE** Why can the model shown by the illustration be considered to be a polymer?

21. **CONCLUDE** What would cause the amino acids in the illustration to be placed in that particular order? Explain.

22. **APPLY** If the order of amino acids shown in the illustration changes, would the protein formed likely still have the same function? Why or why not?

23. **PREDICT** Suppose the protein formed by the amino acids has a coiled shape. What might be the general function of that protein? What if the protein is coiled but also curls up into a ball?

24. **COMPARE AND CONTRAST** Copy and complete the chart below. Provide two similarities and two differences for each pair of items.

Items	Similarities	Differences
starch/cellulose	both carbohydrates; both polymers	starch used for energy, cellulose for structure; starch molecule branched, cellulose molecule straight
carbon chains/carbon rings		
proteins/lipids		
glucose/amino acids		

Using Math Skills in Science

The nutrition label below shows the Calories and the amount of fat, carbohydrates, and protein in a type of cracker. Use the information on the label to answer the following three questions.

Nutrition Facts

Servings Per Container   about 15

Amount Per Serving

Calories	150
Total Fat	6g
Total Carbohydrates	20g
Protein	2g

25. Fats contain about twice as many Calories per gram as carbohydrates and proteins. Assume that all of the Calories on the label come from the carbohydrates, fats, and proteins. About how many Calories come from each substance?

26. Make a pie chart that compares the number of Calories from carbohydrates, fats, and proteins contained in this food.

27. Adult athletes are recommended to eat a diet that provides 15% of its Calories from protein, 30% from fats, and 55% from carbohydrates. Does this food have the recommended balance of nutrients? Why or why not?

the BIG idea

28. **DRAW CONCLUSIONS** Look at the photographs on pages 144–145. Describe three ways in which carbon is important in the activities taking place.

29. **SYNTHESIZE** Write one or more paragraphs describing how plants, animals, and plastics are related to each other.

UNIT PROJECTS

Evaluate all the data, results, and information from your project folder. Prepare to present your project.

MONITOR AND RETEACH

If students have trouble applying the concept of polymer formation as a chemical reaction in items 20–23, review chemical reactions in general and synthesis reactions in particular. Refer students to the top of p. 167. Use ball-and-stick or gumdrop-and-toothpick models to demonstrate the process.

Students may benefit from summarizing one or more sections of the chapter.

R Summarizing the Chapter, pp. 327–328

Standardized Test Practice

For practice on your state test, go to . . .

TEST PRACTICE
CLASSZONE.COM

Interpreting Tables

The following table contains information about some of the different products that can be separated from petroleum. Use the information in the table to answer questions 1–5.

Characteristics of Petroleum Products		
Product	Number of Carbon Atoms per Molecule	Boiling Point (°C)
Natural gas	1 to 4	lower than 20
Gasoline	5 to 12	35 to 220
Kerosene	12 to 16	200 to 315
Jet fuel	12 to 16	200 to 315
Diesel fuel	15 to 18	250 to 375
Heating oil	15 to 18	250 to 375
Lubricating oil	16 to 20	350 and higher
Asphalt	More than 25	600 and higher

SOURCE: Mortimer, Chemistry, 6th edition

1. Which petroleum product has the lowest boiling point?
   a. diesel fuel
   b. gasoline
   c. kerosene
   d. natural gas

2. Which petroleum product has the highest boiling point?
   a. asphalt
   b. heating oil
   c. jet fuel
   d. kerosene

3. Petroleum is heated and turned into gas. The gas rises in a distillation tower. The lightest gases—those with the smallest molecules—rise highest. The heaviest gases—those with the largest molecules—stay lowest. Which of the following products would be found lowest in the tower?
   a. diesel fuel
   b. kerosene
   c. lubricating oil
   d. natural gas

4. Petroleum is split into fractions. Each fraction includes all the products that have the same boiling point. Which of the following pairs of products are in the same fraction?
   a. gasoline and natural gas
   b. jet fuel and kerosene
   c. lubricating oil and diesel fuel
   d. natural gas and asphalt

5. What might be the boiling point of a petroleum product that contains 22 carbon atoms?
   a. 100°C
   b. 300°C
   c. 500°C
   d. 700°C

Extended Response

Answer the following two questions in detail. Include some of the terms from the list in the box at right. Underline each term that you use in your answers.

atoms	carbon chains	carbon rings
molecules	properties	structure
function	monomer	

6. How are polymers made? Give examples of a natural polymer and an artificial polymer.

7. Carbohydrates, lipids, and proteins are all carbon-based molecules. How are they similar? How are they different?

Interpreting Tables

1. d   2. a   3. c   4. b   5. c

Extended Response

6. RUBRIC

4 points for a response that correctly answers the question, gives two correct examples, and uses the following terms accurately:

- carbon chains
- structure
- carbon rings
- monomer
- molecules

Sample: Polymers are large _molecules_ that are made of smaller _monomers_ linked together. A polymer can have a _structure_ made of _carbon chains_, _carbon rings_, or a combination of the two. Cellulose, proteins, and starches are natural polymers; plastics and fibers are artificial polymers.

3 points for a response that correctly answers the question, gives two correct examples, and uses at least three terms correctly

2 points for a response that correctly answers the question, gives two correct examples, and uses at least two terms correctly

1 point for a response that correctly answers the question and either gives one correct example or uses one term correctly

7. RUBRIC

4 points for a response that correctly answers both questions and uses the following terms accurately:

- atoms
- structure
- molecules
- function
- properties

Sample: Carbohydrates, lipids, and proteins are similar in that they all are large _molecules_ that contain carbon _atoms_. They are different in that they have different _structures_, _properties_, and _functions_.

3 points for a response that correctly answers both questions and uses at least three terms correctly

2 points for a response that correctly answers both questions and uses at least one term correctly

1 point for a response that correctly answers one question and uses at least one term correctly

METACOGNITIVE ACTIVITY

Have students answer the following questions in their **Science Notebook:**

1. Think of questions to address what you found most challenging to understand about carbon-based molecules. How can you answer these questions?

2. Which topics in this chapter would you like to learn more about?

3. What have you learned from your research on your Unit Project?

Student Resource Handbooks

Making Observations

An **observation** is an act of noting and recording an event, characteristic, behavior, or anything else detected with an instrument or with the senses.

Observations allow you to make informed hypotheses and to gather data for experiments. Careful observations often lead to ideas for new experiments. There are two categories of observations:

- **Quantitative observations** can be expressed in numbers and include records of time, temperature, mass, distance, and volume.

- **Qualitative observations** include descriptions of sights, sounds, smells, and textures.

EXAMPLE

A student dissolved 30 grams of Epsom salts in water, poured the solution into a dish, and let the dish sit out uncovered overnight. The next day, she made the following observations of the Epsom salt crystals that grew in the dish.

Table 1. Observations of Epsom Salt Crystals

To determine the mass, the student found the mass of the dish before and after growing the crystals and then used subtraction to find the difference.

The student measured several crystals and calculated the mean length. (To learn how to calculate the mean of a data set, see page R36.)

Quantitative Observations	Qualitative Observations
• mass = 30 g • mean crystal length = 0.5 cm • longest crystal length = 2 cm	• Crystals are clear. • Crystals are long, thin, and rectangular. • White crust has formed around edge of dish.

Photographs or sketches are useful for recording qualitative observations.

 Epsom salt crystals

MORE ABOUT OBSERVING

- Make quantitative observations whenever possible. That way, others will know exactly what you observed and be able to compare their results with yours.

- It is always a good idea to make qualitative observations too. You never know when you might observe something unexpected.

Predicting and Hypothesizing

A **prediction** is an expectation of what will be observed or what will happen. A **hypothesis** is a tentative explanation for an observation or scientific problem that can be tested by further investigation.

EXAMPLE

Suppose you have made two paper airplanes and you wonder why one of them tends to glide farther than the other one.

1. Start by asking a question.
2. Make an educated guess. After examination, you notice that the wings of the airplane that flies farther are slightly larger than the wings of the other airplane.
3. Write a prediction based upon your educated guess, in the form of an "If . . . , then . . ." statement. Write the independent variable after the word *if*, and the dependent variable after the word *then*.
4. To make a hypothesis, explain why you think what you predicted will occur. Write the explanation after the word *because*.

1. Why does one of the paper airplanes glide farther than the other?

2. The size of an airplane's wings may affect how far the airplane will glide.

3. Prediction: If I make a paper airplane with larger wings, then the airplane will glide farther.

> To read about independent and dependent variables, see page R30.

4. Hypothesis: If I make a paper airplane with larger wings, then the airplane will glide farther, because the additional surface area of the wing will produce more lift.

> Notice that the part of the hypothesis after *because* adds an explanation of why the airplane will glide farther.

MORE ABOUT HYPOTHESES

- The results of an experiment cannot prove that a hypothesis is correct. Rather, the results either support or do not support the hypothesis.
- Valuable information is gained even when your hypothesis is not supported by your results. For example, it would be an important discovery to find that wing size is not related to how far an airplane glides.
- In science, a hypothesis is supported only after many scientists have conducted many experiments and produced consistent results.

Inferring

An **inference** is a logical conclusion drawn from the available evidence and prior knowledge. Inferences are often made from observations.

EXAMPLE

A student observing a set of acorns noticed something unexpected about one of them. He noticed a white, soft-bodied insect eating its way out of the acorn.

The student recorded these observations.

Observations
- There is a hole in the acorn, about 0.5 cm in diameter, where the insect crawled out.
- There is a second hole, which is about the size of a pinhole, on the other side of the acorn.
- The inside of the acorn is hollow.

Here are some inferences that can be made on the basis of the observations.

Inferences
- The insect formed from the material inside the acorn, grew to its present size, and ate its way out of the acorn.
- The insect crawled through the smaller hole, ate the inside of the acorn, grew to its present size, and ate its way out of the acorn.
- An egg was laid in the acorn through the smaller hole. The egg hatched into a larva that ate the inside of the acorn, grew to its present size, and ate its way out of the acorn.

When you make inferences, be sure to look at all of the evidence available and combine it with what you already know.

MORE ABOUT INFERENCES

Inferences depend both on observations and on the knowledge of the people making the inferences. Ancient people who did not know that organisms are produced only by similar organisms might have made an inference like the first one. A student today might look at the same observations and make the second inference. A third student might have knowledge about this particular insect and know that it is never small enough to fit through the smaller hole, leading her to the third inference.

Identifying Cause and Effect

In a **cause-and-effect relationship,** one event or characteristic is the result of another. Usually an effect follows its cause in time.

There are many examples of cause-and-effect relationships in everyday life.

Cause	Effect
Turn off a light.	Room gets dark.
Drop a glass.	Glass breaks.
Blow a whistle.	Sound is heard.

Scientists must be careful not to infer a cause-and-effect relationship just because one event happens after another event. When one event occurs after another, you cannot infer a cause-and-effect relationship on the basis of that information alone. You also cannot conclude that one event caused another if there are alternative ways to explain the second event. A scientist must demonstrate through experimentation or continued observation that an event was truly caused by another event.

EXAMPLE

Make an Observation

Suppose you have a few plants growing outside. When the weather starts getting colder, you bring one of the plants indoors. You notice that the plant you brought indoors is growing faster than the others are growing. You cannot conclude from your observation that the change in temperature was the cause of the increased plant growth, because there are alternative explanations for the observation. Some possible explanations are given below.

- The humidity indoors caused the plant to grow faster.

- The level of sunlight indoors caused the plant to grow faster.

- The indoor plant's being noticed more often and watered more often than the outdoor plants caused it to grow faster.

- The plant that was brought indoors was healthier than the other plants to begin with.

To determine which of these factors, if any, caused the indoor plant to grow faster than the outdoor plants, you would need to design and conduct an experiment.

See pages R28–R35 for information about designing experiments.

Recognizing Bias

Television, newspapers, and the Internet are full of experts claiming to have scientific evidence to back up their claims. How do you know whether the claims are really backed up by good science?

Bias is a slanted point of view, or personal prejudice. The goal of scientists is to be as objective as possible and to base their findings on facts instead of opinions. However, bias often affects the conclusions of researchers, and it is important to learn to recognize bias.

When scientific results are reported, you should consider the source of the information as well as the information itself. It is important to critically analyze the information that you see and read.

SOURCES OF BIAS

There are several ways in which a report of scientific information may be biased. Here are some questions that you can ask yourself:

1. **Who is sponsoring the research?**

   Sometimes, the results of an investigation are biased because an organization paying for the research is looking for a specific answer. This type of bias can affect how data are gathered and interpreted.

2. **Is the research sample large enough?**

   Sometimes research does not include enough data. The larger the sample size, the more likely that the results are accurate, assuming a truly random sample.

3. **In a survey, who is answering the questions?**

   The results of a survey or poll can be biased. The people taking part in the survey may have been specifically chosen because of how they would answer. They may have the same ideas or lifestyles. A survey or poll should make use of a random sample of people.

4. **Are the people who take part in a survey biased?**

   People who take part in surveys sometimes try to answer the questions the way they think the researcher wants them to answer. Also, in surveys or polls that ask for personal information, people may be unwilling to answer questions truthfully.

SCIENTIFIC BIAS

It is also important to realize that scientists have their own biases because of the types of research they do and because of their scientific viewpoints. Two scientists may look at the same set of data and come to completely different conclusions because of these biases. However, such disagreements are not necessarily bad. In fact, a critical analysis of disagreements is often responsible for moving science forward.

Identifying Faulty Reasoning

Faulty reasoning is wrong or incorrect thinking. It leads to mistakes and to wrong conclusions. Scientists are careful not to draw unreasonable conclusions from experimental data. Without such caution, the results of scientific investigations may be misleading.

EXAMPLE

Scientists try to make generalizations based on their data to explain as much about nature as possible. If only a small sample of data is looked at, however, a conclusion may be faulty. Suppose a scientist has studied the effects of the El Niño and La Niña weather patterns on flood damage in California from 1989 to 1995. The scientist organized the data in the bar graph below.

The scientist drew the following conclusions:

1. The La Niña weather pattern has no effect on flooding in California.

2. When neither weather pattern occurs, there is almost no flood damage.

3. A weak or moderate El Niño produces a small or moderate amount of flooding.

4. A strong El Niño produces a lot of flooding.

Flood and Storm Damage in California

Estimated damage (millions of dollars): 0, 500, 1000, 1500, 2000
Starting year of season (July 1–June 30): 1989, 1992, 1995

Legend:
- Weak–moderate El Niño
- Strong El Niño

SOURCE: *Governor's Office of Emergency Services, California*

For the six-year period of the scientist's investigation, these conclusions may seem to be reasonable. However, a six-year study of weather patterns may be too small of a sample for the conclusions to be supported. Consider the following graph, which shows information that was gathered from 1949 to 1997.

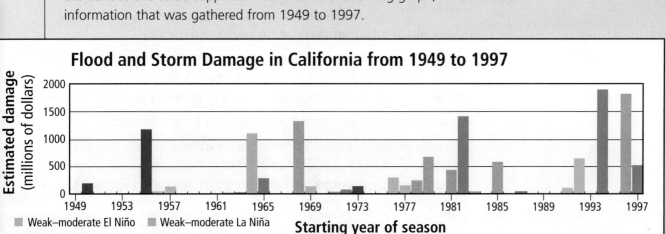

Flood and Storm Damage in California from 1949 to 1997

Estimated damage (millions of dollars): 0, 500, 1000, 1500, 2000
Starting year of season (July 1–June 30): 1949, 1953, 1957, 1961, 1965, 1969, 1973, 1977, 1981, 1985, 1989, 1993, 1997

Legend:
- Weak–moderate El Niño
- Strong El Niño
- Weak–moderate La Niña
- Strong La Niña
- Neither

SOURCE: *Governor's Office of Emergency Services, California*

The only one of the conclusions that all of this information supports is number 3: a weak or moderate El Niño produces a small or moderate amount of flooding. By collecting more data, scientists can be more certain of their conclusions and can avoid faulty reasoning.

Analyzing Statements

To **analyze** a statement is to examine its parts carefully. Scientific findings are often reported through media such as television or the Internet. A report that is made public often focuses on only a small part of research. As a result, it is important to question the sources of information.

Evaluate Media Claims

To **evaluate** a statement is to judge it on the basis of criteria you've established. Sometimes evaluating means deciding whether a statement is true.

Reports of scientific research and findings in the media may be misleading or incomplete. When you are exposed to this information, you should ask yourself some questions so that you can make informed judgments about the information.

1. **Does the information come from a credible source?**

   Suppose you learn about a new product and it is stated that scientific evidence proves that the product works. A report from a respected news source may be more believable than an advertisement paid for by the product's manufacturer.

2. **How much evidence supports the claim?**

   Often, it may seem that there is new evidence every day of something in the world that either causes or cures an illness. However, information that is the result of several years of work by several different scientists is more credible than an advertisement that does not even cite the subjects of the experiment.

3. **How much information is being presented?**

   Science cannot solve all questions, and scientific experiments often have flaws. A report that discusses problems in a scientific study may be more believable than a report that addresses only positive experimental findings.

4. **Is scientific evidence being presented by a specific source?**

   Sometimes scientific findings are reported by people who are called experts or leaders in a scientific field. But if their names are not given or their scientific credentials are not reported, their statements may be less credible than those of recognized experts.

Differentiate Between Fact and Opinion

Sometimes information is presented as a fact when it may be an opinion. When scientific conclusions are reported, it is important to recognize whether they are based on solid evidence. Again, you may find it helpful to ask yourself some questions.

1. **What is the difference between a fact and an opinion?**

   A **fact** is a piece of information that can be strictly defined and proved true. An **opinion** is a statement that expresses a belief, value, or feeling. An opinion cannot be proved true or false. For example, a person's age is a fact, but if someone is asked how old they feel, it is impossible to prove the person's answer to be true or false.

2. **Can opinions be measured?**

   Yes, opinions can be measured. In fact, surveys often ask for people's opinions on a topic. But there is no way to know whether or not an opinion is the truth.

HOW TO DIFFERENTIATE FACT FROM OPINION

Human Activities and the Environment

Opinions

Notice words or phrases that express beliefs or feelings. The words *unfortunately* and *careless* show that opinions are being expressed.

Unfortunately, human use of fossil fuels is one of the most significant developments of the past few centuries. Humans rely on fossil fuels, a non-renewable energy resource, for more than 90 percent of their energy needs.

Facts

Statements that contain statistics tend to be facts. Writers often use facts to support their opinions.

Opinion

Look for statements that speculate about events. These statements are opinions, because they cannot be proved.

This careless misuse of our planet's resources has resulted in pollution, global warming, and the destruction of fragile ecosystems. For example, oil pipelines carry more than one million barrels of oil each day across tundra regions. Transporting oil across such areas can only result in oil spills that poison the land for decades.

Lab Handbook

Safety Rules

Before you work in the laboratory, read these safety rules twice. Ask your teacher to explain any rules that you do not completely understand. Refer to these rules later on if you have questions about safety in the science classroom.

Directions

- Read all directions and make sure that you understand them before starting an investigation or lab activity. If you do not understand how to do a procedure or how to use a piece of equipment, ask your teacher.
- Do not begin any investigation or touch any equipment until your teacher has told you to start.
- Never experiment on your own. If you want to try a procedure that the directions do not call for, ask your teacher for permission first.
- If you are hurt or injured in any way, tell your teacher immediately.

Dress Code

goggles

apron

gloves

- Wear goggles when
  — using glassware, sharp objects, or chemicals
  — heating an object
  — working with anything that can easily fly up into the air and hurt someone's eye
- Tie back long hair or hair that hangs in front of your eyes.
- Remove any article of clothing—such as a loose sweater or a scarf—that hangs down and may touch a flame, chemical, or piece of equipment.
- Observe all safety icons calling for the wearing of eye protection, gloves, and aprons.

Heating and Fire Safety

fire safety

heating safety

- Keep your work area neat, clean, and free of extra materials.
- Never reach over a flame or heat source.
- Point objects being heated away from you and others.
- Never heat a substance or an object in a closed container.
- Never touch an object that has been heated. If you are unsure whether something is hot, treat it as though it is. Use oven mitts, clamps, tongs, or a test-tube holder.
- Know where the fire extinguisher and fire blanket are kept in your classroom.
- Do not throw hot substances into the trash. Wait for them to cool or use the container your teacher puts out for disposal.

Electrical Safety

electrical safety

- Never use lamps or other electrical equipment with frayed cords.
- Make sure no cord is lying on the floor where someone can trip over it.
- Do not let a cord hang over the side of a counter or table so that the equipment can easily be pulled or knocked to the floor.
- Never let cords hang into sinks or other places where water can be found.
- Never try to fix electrical problems. Inform your teacher of any problems immediately.
- Unplug an electrical cord by pulling on the plug, not the cord.

Chemical Safety

chemical safety

poison

fumes

- If you spill a chemical or get one on your skin or in your eyes, tell your teacher right away.
- Never touch, taste, or sniff any chemicals in the lab. If you need to determine odor, waft. Wafting consists of holding the chemical in its container 15 centimeters (6 in.) away from your nose, and using your fingers to bring fumes from the container to your nose.
- Keep lids on all chemicals you are not using.
- Never put unused chemicals back into the original containers. Throw away extra chemicals where your teacher tells you to.
- Pour chemicals over a sink or your work area, not over the floor.
- If you get a chemical in your eye, use the eyewash right away.
- Always wash your hands after handling chemicals, plants, or soil.

Wafting

Glassware and Sharp-Object Safety

sharp objects

- If you break glassware, tell your teacher right away.
- Do not use broken or chipped glassware. Give these to your teacher.
- Use knives and other cutting instruments carefully. Always wear eye protection and cut away from you.

Animal Safety

- Never hurt an animal.
- Touch animals only when necessary. Follow your teacher's instructions for handling animals.
- Always wash your hands after working with animals.

Cleanup

disposal

- Follow your teacher's instructions for throwing away or putting away supplies.
- Clean your work area and pick up anything that has dropped to the floor.
- Wash your hands.

Using Lab Equipment

Different experiments require different types of equipment. But even though experiments differ, the ways in which the equipment is used are the same.

Beakers

- Use beakers for holding and pouring liquids.
- Do not use a beaker to measure the volume of a liquid. Use a graduated cylinder instead. (See page R16.)
- Use a beaker that holds about twice as much liquid as you need. For example, if you need 100 milliliters of water, you should use a 200- or 250-milliliter beaker.

Test Tubes

- Use test tubes to hold small amounts of substances.
- Do not use a test tube to measure the volume of a liquid.
- Use a test tube when heating a substance over a flame. Aim the mouth of the tube away from yourself and other people.
- Liquids easily spill or splash from test tubes, so it is important to use only small amounts of liquids.

Test-Tube Holder

- Use a test-tube holder when heating a substance in a test tube.
- Use a test-tube holder if the substance in a test tube is dangerous to touch.
- Make sure the test-tube holder tightly grips the test tube so that the test tube will not slide out of the holder.
- Make sure that the test-tube holder is above the surface of the substance in the test tube so that you can observe the substance.

Test-Tube Rack

- Use a test-tube rack to organize test tubes before, during, and after an experiment.

- Use a test-tube rack to keep test tubes upright so that they do not fall over and spill their contents.

- Use a test-tube rack that is the correct size for the test tubes that you are using. If the rack is too small, a test tube may become stuck. If the rack is too large, a test tube may lean over, and some of its contents may spill or splash.

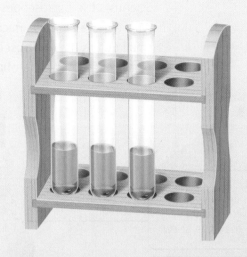

Forceps

- Use forceps when you need to pick up or hold a very small object that should not be touched with your hands.

- Do not use forceps to hold anything over a flame, because forceps are not long enough to keep your hand safely away from the flame. Plastic forceps will melt, and metal forceps will conduct heat and burn your hand.

Hot Plate

- Use a hot plate when a substance needs to be kept warmer than room temperature for a long period of time.

- Use a hot plate instead of a Bunsen burner or a candle when you need to carefully control temperature.

- Do not use a hot plate when a substance needs to be burned in an experiment.

- Always use "hot hands" safety mitts or oven mitts when handling anything that has been heated on a hot plate.

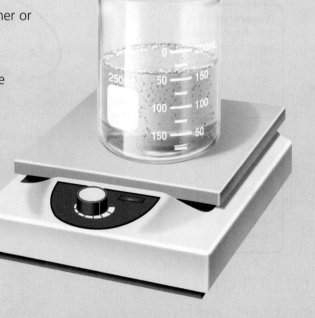

Microscope

Scientists use microscopes to see very small objects that cannot easily be seen with the eye alone. A microscope magnifies the image of an object so that small details may be observed. A microscope that you may use can magnify an object 400 times—the object will appear 400 times larger than its actual size.

Eyepiece Objects are viewed through the eyepiece. The eyepiece contains a lens that commonly magnifies an image 10 times.

Coarse Adjustment This knob is used to focus the image of an object when it is viewed through the low-power lens.

Fine Adjustment This knob is used to focus the image of an object when it is viewed through the high-power lens.

Low-Power Objective Lens This is the smallest lens on the nosepiece. It magnifies an image approximately 10 times.

Arm The arm supports the body above the stage. Always carry a microscope by the arm and base.

Stage Clip The stage clip holds a slide in place on the stage.

Base The base supports the microscope.

Body The body separates the lens in the eyepiece from the objective lenses below.

Nosepiece The nosepiece holds the objective lenses above the stage and rotates so that all lenses may be used.

High-Power Objective Lens This is the largest lens on the nosepiece. It magnifies an image approximately 40 times.

Stage The stage supports the object being viewed.

Diaphragm The diaphragm is used to adjust the amount of light passing through the slide and into an objective lens.

Mirror or Light Source Some microscopes use light that is reflected through the stage by a mirror. Other microscopes have their own light sources.

VIEWING AN OBJECT

1. Use the coarse adjustment knob to raise the body tube.

2. Adjust the diaphragm so that you can see a bright circle of light through the eyepiece.

3. Place the object or slide on the stage. Be sure that it is centered over the hole in the stage.

4. Turn the nosepiece to click the low-power lens into place.

5. Using the coarse adjustment knob, slowly lower the lens and focus on the specimen being viewed. Be sure not to touch the slide or object with the lens.

6. When switching from the low-power lens to the high-power lens, first raise the body tube with the coarse adjustment knob so that the high-power lens will not hit the slide.

7. Turn the nosepiece to click the high-power lens into place.

8. Use the fine adjustment knob to focus on the specimen being viewed. Again, be sure not to touch the slide or object with the lens.

MAKING A SLIDE, OR WET MOUNT

1 Place the specimen in the center of a clean slide.

2 Place a drop of water on the specimen.

3 Place a cover slip on the slide. Put one edge of the cover slip into the drop of water and slowly lower it over the specimen.

4 Remove any air bubbles from under the cover slip by gently tapping the cover slip.

5 Dry any excess water before placing the slide on the microscope stage for viewing.

Spring Scale (Force Meter)

- Use a spring scale to measure a force pulling on the scale.

- Use a spring scale to measure the force of gravity exerted on an object by Earth.

- To measure a force accurately, a spring scale must be zeroed before it is used. The scale is zeroed when no weight is attached and the indicator is positioned at zero.

- Do not attach a weight that is either too heavy or too light to a spring scale. A weight that is too heavy could break the scale or exert too great a force for the scale to measure. A weight that is too light may not exert enough force to be measured accurately.

Graduated Cylinder

- Use a graduated cylinder to measure the volume of a liquid.

- Be sure that the graduated cylinder is on a flat surface so that your measurement will be accurate.

- When reading the scale on a graduated cylinder, be sure to have your eyes at the level of the surface of the liquid.

- The surface of the liquid will be curved in the graduated cylinder. Read the volume of the liquid at the bottom of the curve, or meniscus (muh-NIHS-kuhs).

- You can use a graduated cylinder to find the volume of a solid object by measuring the increase in a liquid's level after you add the object to the cylinder.

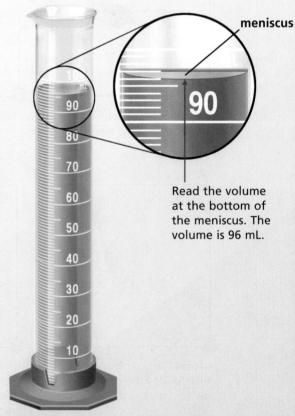

meniscus

Read the volume at the bottom of the meniscus. The volume is 96 mL.

Metric Rulers

- Use metric rulers or meter sticks to measure objects' lengths.

- Do not measure an object from the end of a metric ruler or meter stick, because the end is often imperfect. Instead, measure from the 1-centimeter mark, but remember to subtract a centimeter from the apparent measurement.

- Estimate any lengths that extend between marked units. For example, if a meter stick shows centimeters but not millimeters, you can estimate the length that an object extends between centimeter marks to measure it to the nearest millimeter.

- **Controlling Variables** If you are taking repeated measurements, always measure from the same point each time. For example, if you're measuring how high two different balls bounce when dropped from the same height, measure both bounces at the same point on the balls—either the top or the bottom. Do not measure at the top of one ball and the bottom of the other.

EXAMPLE

How to Measure a Leaf

1. Lay a ruler flat on top of the leaf so that the 1-centimeter mark lines up with one end. Make sure the ruler and the leaf do not move between the time you line them up and the time you take the measurement.

2. Look straight down on the ruler so that you can see exactly how the marks line up with the other end of the leaf.

3. Estimate the length by which the leaf extends beyond a marking. For example, the leaf below extends about halfway between the 4.2-centimeter and 4.3-centimeter marks, so the apparent measurement is about 4.25 centimeters.

4. Remember to subtract 1 centimeter from your apparent measurement, since you started at the 1-centimeter mark on the ruler and not at the end. The leaf is about 3.25 centimeters long (4.25 cm – 1 cm = 3.25 cm).

Triple-Beam Balance

This balance has a pan and three beams with sliding masses, called riders. At one end of the beams is a pointer that indicates whether the mass on the pan is equal to the masses shown on the beams.

1. Make sure the balance is zeroed before measuring the mass of an object. The balance is zeroed if the pointer is at zero when nothing is on the pan and the riders are at their zero points. Use the adjustment knob at the base of the balance to zero it.

2. Place the object to be measured on the pan.

3. Move the riders one notch at a time away from the pan. Begin with the largest rider. If moving the largest rider one notch brings the pointer below zero, begin measuring the mass of the object with the next smaller rider.

4. Change the positions of the riders until they balance the mass on the pan and the pointer is at zero. Then add the readings from the three beams to determine the mass of the object.

300 g	position of largest rider
90 g	position of middle rider
+ 3 g	position of smallest rider
393 g	mass of beaker

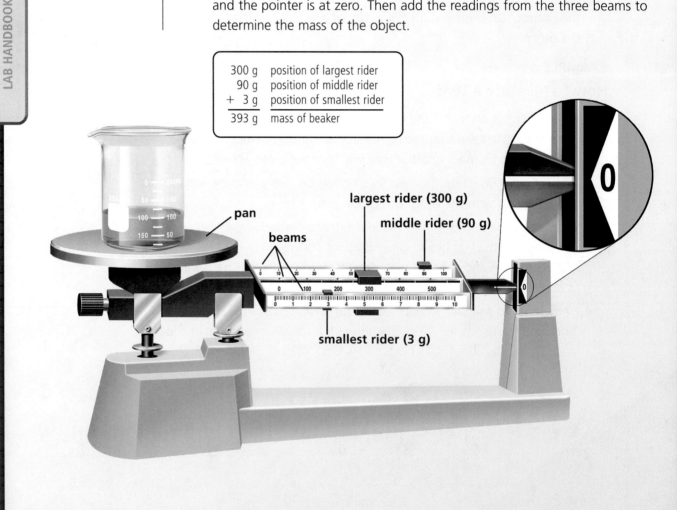

pan

beams

largest rider (300 g)

middle rider (90 g)

smallest rider (3 g)

Double-Pan Balance

This type of balance has two pans. Between the pans is a pointer that indicates whether the masses on the pans are equal.

1. Make sure the balance is zeroed before measuring the mass of an object. The balance is zeroed if the pointer is at zero when there is nothing on either of the pans. Many double-pan balances have sliding knobs that can be used to zero them.

2. Place the object to be measured on one of the pans.

3. Begin adding standard masses to the other pan. Begin with the largest standard mass. If this adds too much mass to the balance, begin measuring the mass of the object with the next smaller standard mass.

4. Add standard masses until the masses on both pans are balanced and the pointer is at zero. Then add the standard masses together to determine the mass of the object being measured.

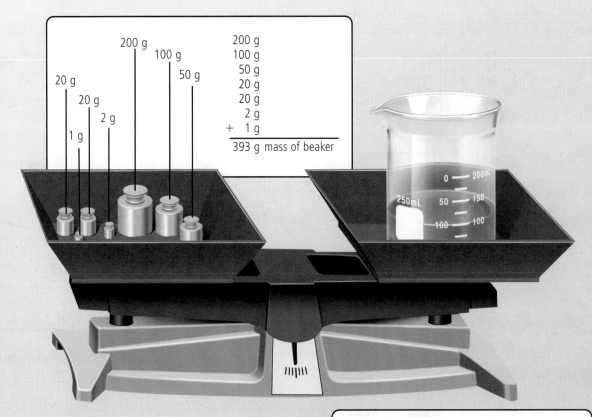

```
      200 g
    100 g
      50 g
      20 g
      20 g
       2 g
    +  1 g
  ─────────────
   393 g  mass of beaker
```

Never place chemicals or liquids directly on a pan. Instead, use the following procedure:

1 Determine the mass of an empty container, such as a beaker.

2 Pour the substance into the container, and measure the total mass of the substance and the container.

3 Subtract the mass of the empty container from the total mass to find the mass of the substance.

The Metric System and SI Units

Scientists use International System (SI) units for measurements of distance, volume, mass, and temperature. The International System is based on multiples of ten and the metric system of measurement.

Basic SI Units		
Property	Name	Symbol
length	meter	m
volume	liter	L
mass	kilogram	kg
temperature	kelvin	K

SI Prefixes		
Prefix	Symbol	Multiple of 10
kilo-	k	1000
hecto-	h	100
deca-	da	10
deci-	d	$0.1 \left(\frac{1}{10}\right)$
centi-	c	$0.01 \left(\frac{1}{100}\right)$
milli-	m	$0.001 \left(\frac{1}{1000}\right)$

Changing Metric Units

You can change from one unit to another in the metric system by multiplying or dividing by a power of 10.

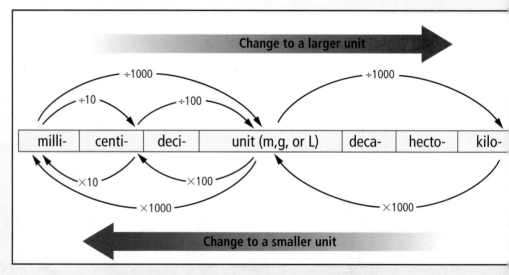

Example

Change 0.64 liters to milliliters.

(1) Decide whether to multiply or divide.

(2) Select the power of 10.

ANSWER 0.64 L = 640 mL

Change to a smaller unit by multiplying.

$$mL \longleftarrow \times 1000 \longrightarrow L$$

$$0.64 \times 1000 = 640.$$

Example

Change 23.6 grams to kilograms.

(1) Decide whether to multiply or divide.

(2) Select the power of 10.

ANSWER 23.6 g = 0.0236 kg

Change to a larger unit by dividing.

$$g \longrightarrow \div 1000 \longrightarrow kg$$

$$23.6 \div 1000 = 0.0236$$

Temperature Conversions

Even though the kelvin is the SI base unit of temperature, the degree Celsius will be the unit you use most often in your science studies. The formulas below show the relationships between temperatures in degrees Fahrenheit (°F), degrees Celsius (°C), and kelvins (K).

$$°C = \frac{5}{9}(°F - 32)$$

$$°F = \frac{9}{5}°C + 32$$

$$K = °C + 273$$

See page R42 for help with using formulas.

Examples of Temperature Conversions

Condition	Degrees Celsius	Degrees Fahrenheit
Freezing point of water	0	32
Cool day	10	50
Mild day	20	68
Warm day	30	86
Normal body temperature	37	98.6
Very hot day	40	104
Boiling point of water	100	212

Converting Between SI and U.S. Customary Units

Use the chart below when you need to convert between SI units and U.S. customary units.

SI Unit	From SI to U.S. Customary			From U.S. Customary to SI		
Length	When you know	multiply by	to find	When you know	multiply by	to find
kilometer (km) = 1000 m	kilometers	0.62	miles	miles	1.61	kilometers
meter (m) = 100 cm	meters	3.28	feet	feet	0.3048	meters
centimeter (cm) = 10 mm	centimeters	0.39	inches	inches	2.54	centimeters
millimeter (mm) = 0.1 cm	millimeters	0.04	inches	inches	25.4	millimeters
Area	When you know	multiply by	to find	When you know	multiply by	to find
square kilometer (km²)	square kilometers	0.39	square miles	square miles	2.59	square kilometers
square meter (m²)	square meters	1.2	square yards	square yards	0.84	square meters
square centimeter (cm²)	square centimeters	0.155	square inches	square inches	6.45	square centimeters
Volume	When you know	multiply by	to find	When you know	multiply by	to find
liter (L) = 1000 mL	liters	1.06	quarts	quarts	0.95	liters
	liters	0.26	gallons	gallons	3.79	liters
	liters	4.23	cups	cups	0.24	liters
	liters	2.12	pints	pints	0.47	liters
milliliter (mL) = 0.001 L	milliliters	0.20	teaspoons	teaspoons	4.93	milliliters
	milliliters	0.07	tablespoons	tablespoons	14.79	milliliters
	milliliters	0.03	fluid ounces	fluid ounces	29.57	milliliters
Mass	When you know	multiply by	to find	When you know	multiply by	to find
kilogram (kg) = 1000 g	kilograms	2.2	pounds	pounds	0.45	kilograms
gram (g) = 1000 mg	grams	0.035	ounces	ounces	28.35	grams

Precision and Accuracy

When you do an experiment, it is important that your methods, observations, and data be both precise and accurate.

low precision

precision,
but not accuracy

precision and
accuracy

Precision

In science, **precision** is the exactness and consistency of measurements. For example, measurements made with a ruler that has both centimeter and millimeter markings would be more precise than measurements made with a ruler that has only centimeter markings. Another indicator of precision is the care taken to make sure that methods and observations are as exact and consistent as possible. Every time a particular experiment is done, the same procedure should be used. Precision is necessary because experiments are repeated several times and if the procedure changes, the results will change.

EXAMPLE

Suppose you are measuring temperatures over a two-week period. Your precision will be greater if you measure each temperature at the same place, at the same time of day, and with the same thermometer than if you change any of these factors from one day to the next.

Accuracy

In science, it is possible to be precise but not accurate. **Accuracy** depends on the difference between a measurement and an actual value. The smaller the difference, the more accurate the measurement.

EXAMPLE

Suppose you look at a stream and estimate that it is about 1 meter wide at a particular place. You decide to check your estimate by measuring the stream with a meter stick, and you determine that the stream is 1.32 meters wide. However, because it is hard to measure the width of a stream with a meter stick, it turns out that you didn't do a very good job. The stream is actually 1.14 meters wide. Therefore, even though your estimate was less precise than your measurement, your estimate was actually more accurate.

LAB HANDBOOK

Making Data Tables and Graphs

Data tables and graphs are useful tools for both recording and communicating scientific data.

Making Data Tables

You can use a **data table** to organize and record the measurements that you make. Some examples of information that might be recorded in data tables are frequencies, times, and amounts.

EXAMPLE

Suppose you are investigating photosynthesis in two elodea plants. One sits in direct sunlight, and the other sits in a dimly lit room. You measure the rate of photosynthesis by counting the number of bubbles in the jar every ten minutes.

1. Title and number your data table.
2. Decide how you will organize the table into columns and rows.
3. Any units, such as seconds or degrees, should be included in column headings, not in the individual cells.

Table 1. Number of Bubbles from Elodea

Always number and title data tables.

Time (min)	Sunlight	Dim Light
0	0	0
10	15	5
20	25	8
30	32	7
40	41	10
50	47	9
60	42	9

The data in the table above could also be organized in a different way.

Table 1. Number of Bubbles from Elodea

Put units in column heading.

Light Condition	Time (min)						
	0	10	20	30	40	50	60
Sunlight	0	15	25	32	41	47	42
Dim light	0	5	8	7	10	9	9

Making Line Graphs

You can use a **line graph** to show a relationship between variables. Line graphs are particularly useful for showing changes in variables over time.

EXAMPLE

Suppose you are interested in graphing temperature data that you collected over the course of a day.

Table 1. Outside Temperature During the Day on March 7

	Time of Day						
	7:00 A.M.	9:00 A.M.	11:00 A.M.	1:00 P.M.	3:00 P.M.	5:00 P.M.	7:00 P.M.
Temp (°C)	8	9	11	14	12	10	6

1. Use the vertical axis of your line graph for the variable that you are measuring—temperature.

2. Choose scales for both the horizontal axis and the vertical axis of the graph. You should have two points more than you need on the vertical axis, and the horizontal axis should be long enough for all of the data points to fit.

3. Draw and label each axis.

4. Graph each value. First find the appropriate point on the scale of the horizontal axis. Imagine a line that rises vertically from that place on the scale. Then find the corresponding value on the vertical axis, and imagine a line that moves horizontally from that value. The point where these two imaginary lines intersect is where the value should be plotted.

5. Connect the points with straight lines.

Be sure to add a number and a title to your graph.

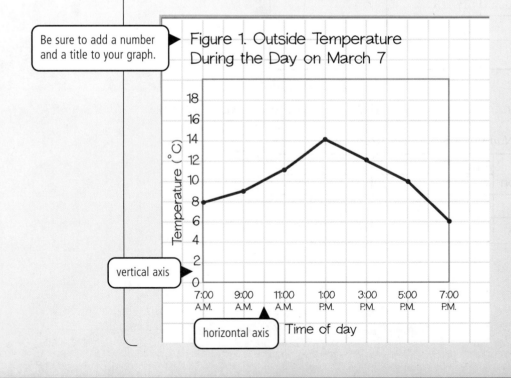

Figure 1. Outside Temperature During the Day on March 7

vertical axis

horizontal axis

LAB HANDBOOK

Making Circle Graphs

You can use a **circle graph,** sometimes called a pie chart, to represent data as parts of a circle. Circle graphs are used only when the data can be expressed as percentages of a whole. The entire circle shown in a circle graph is equal to 100 percent of the data.

EXAMPLE

Suppose you identified the species of each mature tree growing in a small wooded area. You organized your data in a table, but you also want to show the data in a circle graph.

1. To begin, find the total number of mature trees.

   $56 + 34 + 22 + 10 + 28 = 150$

2. To find the degree measure for each sector of the circle, write a fraction comparing the number of each tree species with the total number of trees. Then multiply the fraction by 360°.

   Oak: $\frac{56}{150} \times 360° = 134.4°$

3. Draw a circle. Use a protractor to draw the angle for each sector of the graph.

4. Color and label each sector of the graph.

5. Give the graph a number and title.

Table 1. Tree Species in Wooded Area

Species	Number of Specimens
Oak	56
Maple	34
Birch	22
Willow	10
Pine	28

LAB HANDBOOK

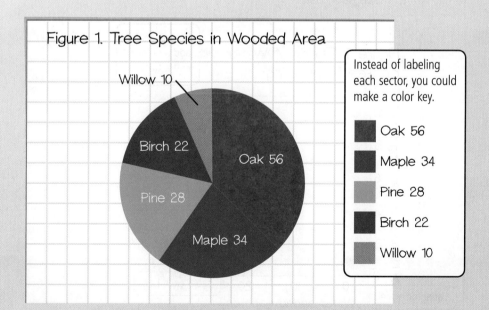

Figure 1. Tree Species in Wooded Area

Instead of labeling each sector, you could make a color key.

- Oak 56
- Maple 34
- Pine 28
- Birch 22
- Willow 10

Bar Graph

A **bar graph** is a type of graph in which the lengths of the bars are used to represent and compare data. A numerical scale is used to determine the lengths of the bars.

EXAMPLE

To determine the effect of water on seed sprouting, three cups were filled with sand, and ten seeds were planted in each. Different amounts of water were added to each cup over a three-day period.

Table 1. Effect of Water on Seed Sprouting

Daily Amount of Water (mL)	Number of Seeds That Sprouted After 3 Days in Sand
0	1
10	4
20	8

1. Choose a numerical scale. The greatest value is 8, so the end of the scale should have a value greater than 8, such as 10. Use equal increments along the scale, such as increments of 2.

2. Draw and label the axes. Mark intervals on the vertical axis according to the scale you chose.

3. Draw a bar for each data value. Use the scale to decide how long to make each bar.

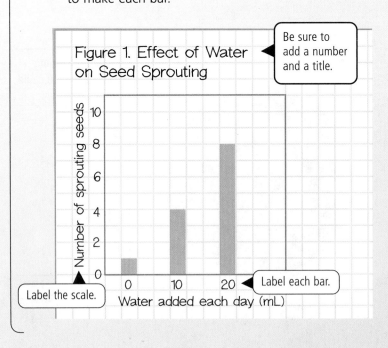

Figure 1. Effect of Water on Seed Sprouting

Be sure to add a number and a title.

Label the scale.

Label each bar.

Double Bar Graph

A **double bar graph** is a bar graph that shows two sets of data. The two bars for each measurement are drawn next to each other.

EXAMPLE

The same seed-sprouting experiment was repeated with potting soil. The data for sand and potting soil can be plotted on one graph.

1. Draw one set of bars, using the data for sand, as shown below.
2. Draw bars for the potting-soil data next to the bars for the sand data. Shade them a different color. Add a key.

Table 2. Effect of Water and Soil on Seed Sprouting

Daily Amount of Water (mL)	Number of Seeds That Sprouted After 3 Days in Sand	Number of Seeds That Sprouted After 3 Days in Potting Soil
0	1	2
10	4	5
20	8	9

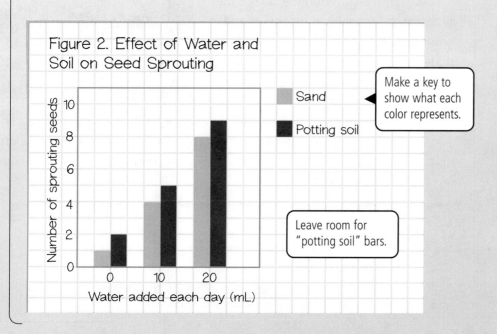

Figure 2. Effect of Water and Soil on Seed Sprouting

Make a key to show what each color represents.

Leave room for "potting soil" bars.

Designing an Experiment

Use this section when designing or conducting an experiment.

Determining a Purpose

Don't forget to learn as much as possible about your topic before you begin.

You can find a purpose for an experiment by doing research, by examining the results of a previous experiment, or by observing the world around you. An **experiment** is an organized procedure to study something under controlled conditions.

1. Write the purpose of your experiment as a question or problem that you want to investigate.

2. Write down research questions and begin searching for information that will help you design an experiment. Consult the library, the Internet, and other people as you conduct your research.

EXAMPLE

Middle school students observed an odor near the lake by their school. They also noticed that the water on the side of the lake near the school was greener than the water on the other side of the lake. The students did some research to learn more about their observations. They discovered that the odor and green color in the lake

came from algae. They also discovered that a new fertilizer was being used on a field nearby. The students inferred that the use of the fertilizer might be related to the presence of the algae and designed a controlled experiment to find out whether they were right.

> **Problem**
> How does fertilizer affect the presence of algae in a lake?
>
> **Research Questions**
> - Have other experiments been done on this problem? If so, what did those experiments show?
> - What kind of fertilizer is used on the field? How much?
> - How do algae grow?
> - How do people measure algae?
> - Can fertilizer and algae be used safely in a lab? How?

Research
As you research, you may find a topic that is more interesting to you than your original topic, or learn that a procedure you wanted to use is not practical or safe. It is OK to change your purpose as you research.

LAB HANDBOOK

Writing a Hypothesis

A **hypothesis** is a tentative explanation for an observation or scientific problem that can be tested by further investigation. You can write your hypothesis in the form of an "If . . . , then . . . , because . . ." statement.

Hypothesis

If the amount of fertilizer in lake water is increased, then the amount of algae will also increase, because fertilizers provide nutrients that algae need to grow.

Hypotheses
For help with hypotheses, refer to page R3.

Determining Materials

Make a list of all the materials you will need to do your experiment. Be specific, especially if someone else is helping you obtain the materials. Try to think of everything you will need.

Materials

- 1 large jar or container
- 4 identical smaller containers
- rubber gloves that also cover the arms
- sample of fertilizer-and-water solution
- eyedropper
- clear plastic wrap
- scissors
- masking tape
- marker
- ruler

Determining Variables and Constants

EXPERIMENTAL GROUP AND CONTROL GROUP

An experiment to determine how two factors are related always has two groups—a control group and an experimental group.

1. Design an experimental group. Include as many trials as possible in the experimental group in order to obtain reliable results.

2. Design a control group that is the same as the experimental group in every way possible, except for the factor you wish to test.

Experimental Group: two containers of lake water with one drop of fertilizer solution added to each

Control Group: two containers of lake water with no fertilizer solution added

Go back to your materials l and make sure you have enough items listed to cove both your experimental gro and your control group.

LAB HANDBOOK

VARIABLES AND CONSTANTS

Identify the variables and constants in your experiment. In a controlled experiment, a **variable** is any factor that can change. **Constants** are all of the factors that are the same in both the experimental group and the control group.

1. Read your hypothesis. The **independent variable** is the factor that you wish to test and that is manipulated or changed so that it can be tested. The independent variable is expressed in your hypothesis after the word *if*. Identify the independent variable in your laboratory report.

2. The **dependent variable** is the factor that you measure to gather results. It is expressed in your hypothesis after the word *then*. Identify the dependent variable in your laboratory report.

Hypothesis
If the amount of fertilizer in lake water is increased, then the amount of algae will also increase, because fertilizers provide nutrients that algae need to grow.

Table 1. Variables and Constants in Algae Experiment

Independent Variable	Dependent Variable	Constants
Amount of fertilizer in lake water	Amount of algae that grow	• Where the lake water is obtained • Type of container used • Light and temperature conditions where water will be stored

Set up your experiment so that you will test only one variable.

MEASURING THE DEPENDENT VARIABLE

Before starting your experiment, you need to define how you will measure the dependent variable. An **operational definition** is a description of the one particular way in which you will measure the dependent variable.

Your operational definition is important for several reasons. First, in any experiment there are several ways in which a dependent variable can be measured. Second, the procedure of the experiment depends on how you decide to measure the dependent variable. Third, your operational definition makes it possible for other people to evaluate and build on your experiment.

EXAMPLE 1

An operational definition of a dependent variable can be qualitative. That is, your measurement of the dependent variable can simply be an observation of whether a change occurs as a result of a change in the independent variable. This type of operational definition can be thought of as a "yes or no" measurement.

Table 2. Qualitative Operational Definition of Algae Growth

Independent Variable	Dependent Variable	Operational Definition
Amount of fertilizer in lake water	Amount of algae that grow	Algae grow in lake water

A qualitative measurement of a dependent variable is often easy to make and record. However, this type of information does not provide a great deal of detail in your experimental results.

EXAMPLE 2

An operational definition of a dependent variable can be quantitative. That is, your measurement of the dependent variable can be a number that shows how much change occurs as a result of a change in the independent variable.

Table 3. Quantitative Operational Definition of Algae Growth

Independent Variable	Dependent Variable	Operational Definition
Amount of fertilizer in lake water	Amount of algae that grow	Diameter of largest algal growth (in mm)

A quantitative measurement of a dependent variable can be more difficult to make and analyze than a qualitative measurement. However, this type of data provides much more information about your experiment and is often more useful.

Writing a Procedure

Write each step of your procedure. Start each step with a verb, or action word, and keep the steps short. Your procedure should be clear enough for someone else to use as instructions for repeating your experiment.

If necessary, go back to your materials list and add any materials that you left out.

Controlling Variables
The same amount of fertilizer solution must be added to two of the four containers.

Controlling Variables
All four containers must receive the same amount of light.

Procedure

1. Put on your gloves. Use the large container to obtain a sample of lake water.

2. Divide the sample of lake water equally among the four smaller containers.

3. Use the eyedropper to add one drop of fertilizer solution to two of the containers.

4. Use the masking tape and the marker to label the containers with your initials, the date, and the identifiers "Jar 1 with Fertilizer," "Jar 2 with Fertilizer," "Jar 1 without Fertilizer," and "Jar 2 without Fertilizer."

5. Cover the containers with clear plastic wrap. Use the scissors to punch ten holes in each of the covers.

6. Place all four containers on a window ledge. Make sure that they all receive the same amount of light.

7. Observe the containers every day for one week.

8. Use the ruler to measure the diameter of the largest clump of algae in each container, and record your measurements daily.

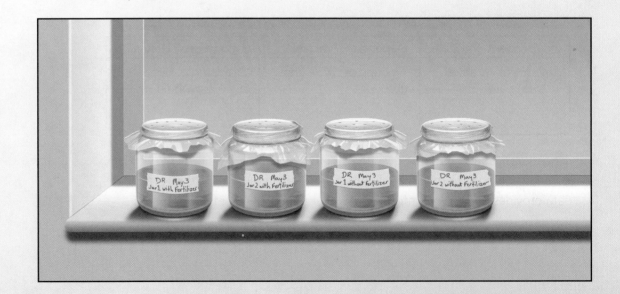

Recording Observations

Once you have obtained all of your materials and your procedure has been approved, you can begin making experimental observations. Gather both quantitative and qualitative data. If something goes wrong during your procedure, make sure you record that too.

> **Observations**
> For help with making qualitative and quantitative observations, refer to page R2.

> For more examples of data tables, see page R23.

Table 4. Fertilizer and Algae Growth

| Date and Time | Experimental Group | | Control Group | | |
	Jar 1 with Fertilizer (diameter of algae in mm)	Jar 2 with Fertilizer (diameter of algae in mm)	Jar 1 without Fertilizer (diameter of algae in mm)	Jar 2 without Fertilizer (diameter of algae in mm)	Observations
5/3 4:00 P.M.	0	0	0	0	condensation in all containers
5/4 4:00 P.M.	0	3	0	0	tiny green blobs in jar 2 with fertilizer
5/5 4:15 P.M.	4	5	0	3	green blobs in jars 1 and 2 with fertilizer and jar 2 without fertilizer
5/6 4:00 P.M.	5	6	0	4	water light green in jar 2 with fertilizer
5/7 4:00 P.M.	8	10	0	6	water light green in jars 1 and 2 with fertilizer and in jar 2 without fertilizer
5/8 3:30 P.M.	10	18	0	6	cover off jar 2 with fertilizer
5/9 3:30 P.M.	14	23	0	8	drew sketches of each container

> Notice that on the sixth day, the observer found that the cover was off one of the containers. It is important to record observations of unintended factors because they might affect the results of the experiment.

Drawings of Samples Viewed Under Microscope on 5/9 at 100x

> Use technology, such as a microscope, to help you make observations when possible.

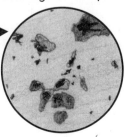

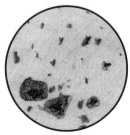

| Jar 1 with Fertilizer | Jar 2 with Fertilizer | Jar 1 without Fertilizer | Jar 2 without Fertilizer |

Summarizing Results

To summarize your data, look at all of your observations together. Look for meaningful ways to present your observations. For example, you might average your data or make a graph to look for patterns. When possible, use spreadsheet software to help you analyze and present your data. The two graphs below show the same data.

EXAMPLE 1

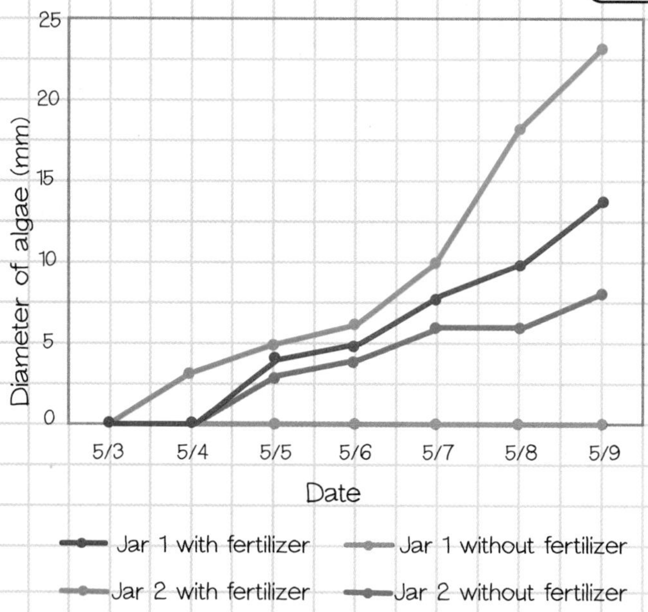

Always include a number and a title with a graph.

Figure 1. Fertilizer and Algae Growth

Line graphs are useful for showing changes over time. For help with line graphs, refer to page R24.

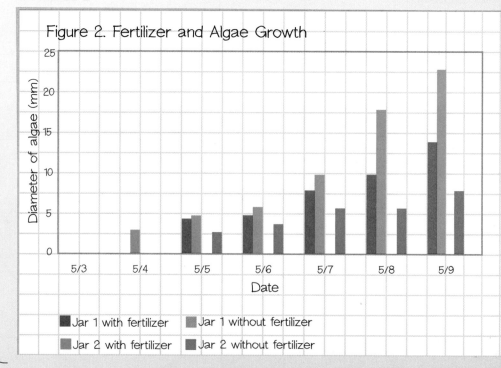

Figure 2. Fertilizer and Algae Growth

Bar graphs are useful for comparing different data sets. This bar graph has four bars for each day. Another way to present the data would be to calculate averages for the tests and the controls, and to show one test bar and one control bar for each day.

EXAMPLE 2

Drawing Conclusions

RESULTS AND INFERENCES

To draw conclusions from your experiment, first write your results. Then compare your results with your hypothesis. Do your results support your hypothesis? Be careful not to make inferences about factors that you did not test.

> For help with making inferences, see page R4.

Results and Inferences

The results of my experiment show that more algae grew in lake water to which fertilizer had been added than in lake water to which no fertilizer had been added. My hypothesis was supported. I infer that it is possible that the growth of algae in the lake was caused by the fertilizer used on the field.

> Notice that you cannot conclude from this experiment that the presence of algae in the lake was due only to the fertilizer.

QUESTIONS FOR FURTHER RESEARCH

Write a list of questions for further research and investigation. Your ideas may lead you to new experiments and discoveries.

Questions for Further Research

- What is the connection between the amount of fertilizer and algae growth?
- How do different brands of fertilizer affect algae growth?
- How would algae growth in the lake be affected if no fertilizer were used on the field?
- How do algae affect the lake and the other life in and around it?
- How does fertilizer affect the lake and the life in and around it?
- If fertilizer is getting into the lake, how is it getting there?

Math Handbook

Describing a Set of Data

Means, medians, modes, and ranges are important math tools for describing data sets such as the following widths of fossilized clamshells.

13 mm   25 mm   14 mm   21 mm   16 mm   23 mm   14 mm

Mean

The **mean** of a data set is the sum of the values divided by the number of values.

> **Example**
>
> To find the mean of the clamshell data, add the values and then divide the sum by the number of values.
>
> $$\frac{13 \text{ mm} + 25 \text{ mm} + 14 \text{ mm} + 21 \text{ mm} + 16 \text{ mm} + 23 \text{ mm} + 14 \text{ mm}}{7} = \frac{126 \text{ mm}}{7} = 18 \text{ mm}$$
>
> **ANSWER** The mean is 18 mm.

Median

The **median** of a data set is the middle value when the values are written in numerical order. If a data set has an even number of values, the median is the mean of the two middle values.

> **Example**
>
> To find the median of the clamshell data, arrange the values in order from least to greatest. The median is the middle value.
>
> 13 mm   14 mm   14 mm   16 mm   21 mm   23 mm   25 mm
>
> **ANSWER** The median is 16 mm.

Mode

The **mode** of a data set is the value that occurs most often.

Example

To find the mode of the clamshell data, arrange the values in order from least to greatest and determine the value that occurs most often.

13 mm   14 mm   14 mm   16 mm   21 mm   23 mm   25 mm

ANSWER The mode is 14 mm.

A data set can have more than one mode or no mode. For example, the following data set has modes of 2 mm and 4 mm:

2 mm   2 mm   3 mm   4 mm   4 mm

The data set below has no mode, because no value occurs more often than any other.

2 mm   3 mm   4 mm   5 mm

Range

The **range** of a data set is the difference between the greatest value and the least value.

Example

To find the range of the clamshell data, arrange the values in order from least to greatest.

13 mm   14 mm   14 mm   16 mm   21 mm   23 mm   25 mm

Subtract the least value from the greatest value.

13 mm is the least value.
25 mm is the greatest value.

25 mm − 13 mm = 12 mm

ANSWER The range is 12 mm.

Using Ratios, Rates, and Proportions

You can use ratios and rates to compare values in data sets. You can use proportions to find unknown values.

Ratios

A **ratio** uses division to compare two values. The ratio of a value a to a nonzero value b can be written as $\frac{a}{b}$.

> ### Example
>
> The height of one plant is 8 centimeters. The height of another plant is 6 centimeters. To find the ratio of the height of the first plant to the height of the second plant, write a fraction and simplify it.
>
> $$\frac{8 \text{ cm}}{6 \text{ cm}} = \frac{4 \times \overset{1}{\cancel{2}}}{3 \times \underset{1}{\cancel{2}}} = \frac{4}{3}$$
>
> **ANSWER** The ratio of the plant heights is $\frac{4}{3}$.

You can also write the ratio $\frac{a}{b}$ as "a to b" or as $a : b$. For example, you can write the ratio of the plant heights as "4 to 3" or as $4 : 3$.

Rates

A **rate** is a ratio of two values expressed in different units. A unit rate is a rate with a denominator of 1 unit.

> ### Example
>
> A plant grew 6 centimeters in 2 days. The plant's rate of growth was $\frac{6 \text{ cm}}{2 \text{ days}}$. To describe the plant's growth in centimeters per day, write a unit rate.
>
> *Divide numerator and denominator by 2:* $\quad \frac{6 \text{ cm}}{2 \text{ days}} = \frac{6 \text{ cm} \div 2}{2 \text{ days} \div 2}$
>
> *Simplify:* $\quad = \frac{3 \text{ cm}}{1 \text{ day}}$
>
> > You divide 2 days by 2 to get 1 day, so divide 6 cm by 2 also.
>
> **ANSWER** The plant's rate of growth is 3 centimeters per day.

Proportions

A **proportion** is an equation stating that two ratios are equivalent. To solve for an unknown value in a proportion, you can use cross products.

Example

If a plant grew 6 centimeters in 2 days, how many centimeters would it grow in 3 days (if its rate of growth is constant)?

Write a proportion:	$\dfrac{6 \text{ cm}}{2 \text{ days}} = \dfrac{x \text{ cm}}{3 \text{ days}}$
Set cross products:	$6 \cdot 3 = 2x$
Multiply 6 and 3:	$18 = 2x$
Divide each side by 2:	$\dfrac{18}{2} = \dfrac{2x}{2}$
Simplify:	$9 = x$

ANSWER The plant would grow 9 centimeters in 3 days.

Using Decimals, Fractions, and Percents

Decimals, fractions, and percentages are all ways of recording and representing data.

Decimals

A **decimal** is a number that is written in the base-ten place value system, in which a decimal point separates the ones and tenths digits. The values of each place is ten times that of the place to its right.

Example

A caterpillar traveled from point *A* to point *C* along the path shown.

A — 36.9 cm — B — 52.4 cm — C

ADDING DECIMALS To find the total distance traveled by the caterpillar, add the distance from *A* to *B* and the distance from *B* to *C*. Begin by lining up the decimal points. Then add the figures as you would whole numbers and bring down the decimal point.

```
   36.9 cm
 + 52.4 cm
   89.3 cm
```

ANSWER The caterpillar traveled a total distance of 89.3 centimeters.

Example *continued*

SUBTRACTING DECIMALS To find how much farther the caterpillar traveled on the second leg of the journey, subtract the distance from A to B from the distance from B to C.

$$\begin{array}{r} 52.4 \text{ cm} \\ -\,36.9 \text{ cm} \\ \hline 15.5 \text{ cm} \end{array}$$

ANSWER The caterpillar traveled 15.5 centimeters farther on the second leg of the journey.

Example

A caterpillar is traveling from point D to point F along the path shown. The caterpillar travels at a speed of 9.6 centimeters per minute.

MULTIPLYING DECIMALS You can multiply decimals as you would whole numbers. The number of decimal places in the product is equal to the sum of the number of decimal places in the factors.

For instance, suppose it takes the caterpillar 1.5 minutes to go from D to E. To find the distance from D to E, multiply the caterpillar's speed by the time it took.

> Align as shown. ▶

$$\begin{array}{rl} 9.6 & 1 \quad \text{decimal place} \\ \times\,1.5 & +\,1 \quad \text{decimal place} \\ \hline 480 & \\ 96 & \\ \hline 14.40 & 2 \quad \text{decimal places} \end{array}$$

ANSWER The distance from D to E is 14.4 centimeters.

DIVIDING DECIMALS When you divide by a decimal, move the decimal points the same number of places in the divisor and the dividend to make the divisor a whole number.

For instance, to find the time it will take the caterpillar to travel from E to F, divide the distance from E to F by the caterpillar's speed.

$$9.6\,\overline{)33.6}$$
◀ Move each decimal point one place to the right.

$$\begin{array}{r} 3.5 \\ 96\,\overline{)336.} \\ \underline{288} \\ 480 \\ \underline{480} \\ 0 \end{array}$$
◀ Line up decimal points.

ANSWER The caterpillar will travel from E to F in 3.5 minutes.

Fractions

A **fraction** is a number in the form $\frac{a}{b}$, where b is not equal to 0. A fraction is in **simplest form** if its numerator and denominator have a greatest common factor (GCF) of 1. To simplify a fraction, divide its numerator and denominator by their GCF.

Example

A caterpillar is 40 millimeters long. The head of the caterpillar is 6 millimeters long. To compare the length of the caterpillar's head with the caterpillar's total length, you can write and simplify a fraction that expresses the ratio of the two lengths.

Write the ratio of the two lengths:	$\dfrac{\text{Length of head}}{\text{Total length}} = \dfrac{6 \text{ mm}}{40 \text{ mm}}$
Write numerator and denominator as products of numbers and the GCF:	$= \dfrac{3 \times 2}{20 \times 2}$
Divide numerator and denominator by the GCF:	$= \dfrac{3 \times \overset{1}{\cancel{2}}}{20 \times \underset{1}{\cancel{2}}}$
Simplify:	$= \dfrac{3}{20}$

ANSWER In simplest form, the ratio of the lengths is $\frac{3}{20}$.

Percents

A **percent** is a ratio that compares a number to 100. The word *percent* means "per hundred" or "out of 100." The symbol for *percent* is %.

For instance, suppose 43 out of 100 caterpillars are female. You can represent this ratio as a percent, a decimal, or a fraction.

Percent	Decimal	Fraction
43%	0.43	$\dfrac{43}{100}$

Example

In the preceding example, the ratio of the length of the caterpillar's head to the caterpillar's total length is $\frac{3}{20}$. To write this ratio as a percent, write an equivalent fraction that has a denominator of 100.

Multiply numerator and denominator by 5:	$\dfrac{3}{20} = \dfrac{3 \times 5}{20 \times 5}$
	$= \dfrac{15}{100}$
Write as a percent:	$= 15\%$

ANSWER The caterpillar's head represents 15 percent of its total length.

Using Formulas

A mathematical **formula** is a statement of a fact, rule, or principle. It is usually expressed as an equation.

In science, a formula often has a word form and a symbolic form. The formula below expresses Ohm's law.

Word Form

$$\text{Current} = \frac{\text{voltage}}{\text{resistance}}$$

Symbolic Form

$$I = \frac{V}{R}$$

In this formula, I, V, and R are variables. A mathematical **variable** is a symbol or letter that is used to represent one or more numbers.

> The term *variable* is also used in science to refer to a factor that can change during an experiment.

Example

Suppose that you measure a voltage of 1.5 volts and a resistance of 15 ohms. You can use the formula for Ohm's law to find the current in amperes.

Write the formula for Ohm's law: $I = \dfrac{V}{R}$

Substitute 1.5 volts for V and 15 ohms for R: $I = \dfrac{1.5 \text{ volts}}{15 \text{ ohms}}$

Simplify: $I = 0.1 \text{ amp}$

ANSWER The current is 0.1 ampere.

If you know the values of all variables but one in a formula, you can solve for the value of the unknown variable. For instance, Ohm's law can be used to find a voltage if you know the current and the resistance.

Example

Suppose that you know that a current is 0.2 amperes and the resistance is 18 ohms. Use the formula for Ohm's law to find the voltage in volts.

Write the formula for Ohm's law: $I = \dfrac{V}{R}$

Substitute 0.2 amp for I and 18 ohms for R: $0.2 \text{ amp} = \dfrac{V}{18 \text{ ohms}}$

Multiply both sides by 18 ohms: $0.2 \text{ amp} \cdot 18 \text{ ohms} = V$

Simplify: $3.6 \text{ volts} = V$

ANSWER The voltage is 3.6 volts.

Finding Areas

The area of a figure is the amount of surface the figure covers.

Area is measured in square units, such as square meters (m²) or square centimeters (cm²). Formulas for the areas of three common geometric figures are shown below.

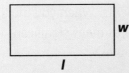

Area = (side length)²
$A = s^2$

Area = length × width
$A = lw$

Area = $\frac{1}{2}$ × base × height

$A = \frac{1}{2} bh$

Example

Each face of a halite crystal is a square like the one shown. You can find the area of the square by using the steps below.

Write the formula for the area of a square:	$A = s^2$
Substitute 3 mm for s:	$= (3 \text{ mm})^2$
Simplify:	$= 9 \text{ mm}^2$

3 mm

3 mm

ANSWER The area of the square is 9 square millimeters.

Finding Volumes

The volume of a solid is the amount of space contained by the solid.

Volume is measured in cubic units, such as cubic meters (m³) or cubic centimeters (cm³). The volume of a rectangular prism is given by the formula shown below.

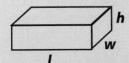

Volume = length × width × height
$V = lwh$

Example

A topaz crystal is a rectangular prism like the one shown. You can find the volume of the prism by using the steps below.

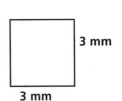

10 mm

12 mm

20 mm

Write the formula for the volume of a rectangular prism:	$V = lwh$
Substitute dimensions:	$= 20 \text{ mm} \times 12 \text{ mm} \times 10 \text{ mm}$
Simplify:	$= 2400 \text{ mm}^3$

ANSWER The volume of the rectangular prism is 2400 cubic millimeters.

Using Significant Figures

The **significant figures** in a decimal are the digits that are warranted by the accuracy of a measuring device.

When you perform a calculation with measurements, the number of significant figures to include in the result depends in part on the number of significant figures in the measurements. When you multiply or divide measurements, your answer should have only as many significant figures as the measurement with the fewest significant figures.

Example

Using a balance and a graduated cylinder filled with water, you determined that a marble has a mass of 8.0 grams and a volume of 3.5 cubic centimeters. To calculate the density of the marble, divide the mass by the volume.

Write the formula for density: $\quad$ Density $= \dfrac{\text{mass}}{\text{Volume}}$

Substitute measurements: $\qquad\qquad = \dfrac{8.0 \text{ g}}{3.5 \text{ cm}^3}$

Use a calculator to divide: $\qquad\quad \approx 2.285714286$ g/cm^3

ANSWER Because the mass and the volume have two significant figures each, give the density to two significant figures. The marble has a density of 2.3 grams per cubic centimeter.

Using Scientific Notation

Scientific notation is a shorthand way to write very large or very small numbers. For example, 73,500,000,000,000,000,000,000 kg is the mass of the Moon. In scientific notation, it is 7.35×10^{22} kg.

Example

You can convert from standard form to scientific notation.

Standard Form	Scientific Notation
720,000	7.2×10^5
5 decimal places left	Exponent is 5.
0.000291	2.91×10^{-4}
4 decimal places right	Exponent is −4.

You can convert from scientific notation to standard form.

Scientific Notation	Standard Form
4.63×10^7	46,300,000
Exponent is 7.	7 decimal places right
1.08×10^{-6}	0.00000108
Exponent is −6.	6 decimal places left

Note-Taking Handbook

Note-Taking Strategies

Taking notes as you read helps you understand the information. The notes you take can also be used as a study guide for later review. This handbook presents several ways to organize your notes.

Content Frame

1. Make a chart in which each column represents a category.

2. Give each column a heading.

3. Write details under the headings.

NAME	GROUP	CHARACTERISTICS	DRAWING
snail	mollusks	mantle, shell	
ant	arthropods	six legs, exoskeleton	
earthworm	segmented worms	segmented body, circulatory and digestive systems	
heartworm	roundworms	digestive system	
sea star	echinoderms	spiny skin, tube feet	
jellyfish	cnidarians	stinging cells	

categories

details

Combination Notes

1. For each new idea or concept, write an informal outline of the information.

2. Make a sketch to illustrate the concept, and label it.

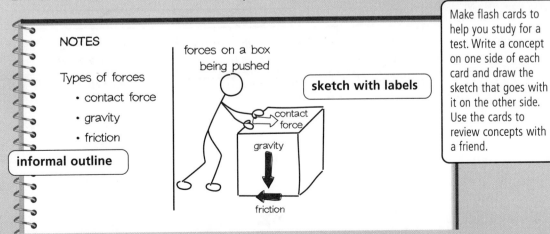

NOTES

Types of forces
- contact force
- gravity
- friction

informal outline

forces on a box being pushed

sketch with labels

contact force

gravity

friction

Make flash cards to help you study for a test. Write a concept on one side of each card and draw the sketch that goes with it on the other side. Use the cards to review concepts with a friend.

Main Idea and Detail Notes

1. In the left-hand column of a two-column chart, list main ideas. The blue headings express main ideas throughout this textbook.

2. In the right-hand column, write details that expand on each main idea.

You can shorten the headings in your chart. Be sure to use the most important words.

When studying for tests, cover up the detail notes column with a sheet of paper. Then use each main idea to form a question—such as "How does latitude affect climate?" Answer the question, and then uncover the detail notes column to check your answer.

MAIN IDEAS	DETAIL NOTES
1. Latitude affects climate. **main idea 1**	1. Places close to the equator are usually warmer than places close to the poles. 1. Latitude has the same effect in both hemispheres.
2. Altitude affects climate. **main idea 2**	2. Temperature decreases with altitude. 2. Altitude can overcome the effect of latitude on temperature.

**details ab[...]
main ide[...]**

**details ab[...]
main ide[...]**

Main Idea Web

1. Write a main idea in a box.

2. Add boxes around it with related vocabulary terms and important details.

You can find definitions near highlighted terms.

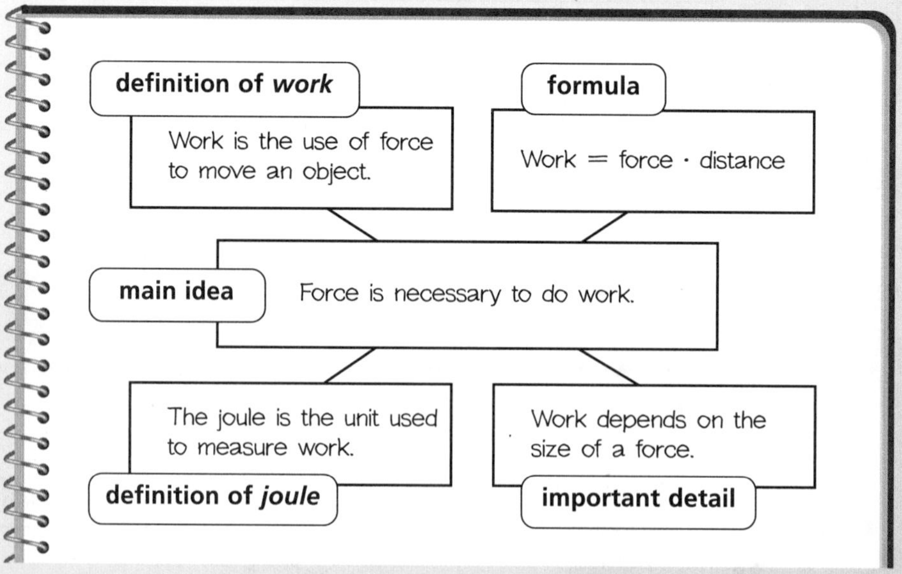

definition of *work*

Work is the use of force to move an object.

formula

Work = force · distance

main idea   Force is necessary to do work.

The joule is the unit used to measure work.

definition of *joule*

Work depends on the size of a force.

important detail

Mind Map

1. Write a main idea in the center.
2. Add details that relate to one another and to the main idea.

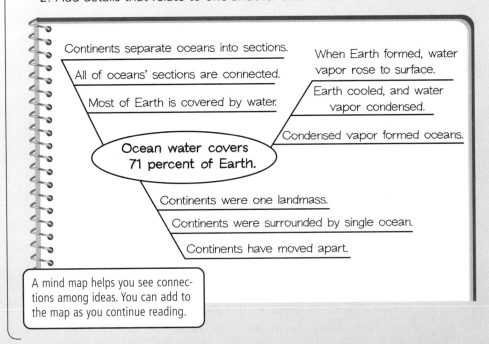

Continents separate oceans into sections.

All of oceans' sections are connected.

Most of Earth is covered by water.

When Earth formed, water vapor rose to surface.

Earth cooled, and water vapor condensed.

Condensed vapor formed oceans.

Ocean water covers 71 percent of Earth.

Continents were one landmass.

Continents were surrounded by single ocean.

Continents have moved apart.

A mind map helps you see connections among ideas. You can add to the map as you continue reading.

Supporting Main Ideas

1. Write a main idea in a box.
2. Add boxes underneath with information—such as reasons, explanations, and examples—that supports the main idea.

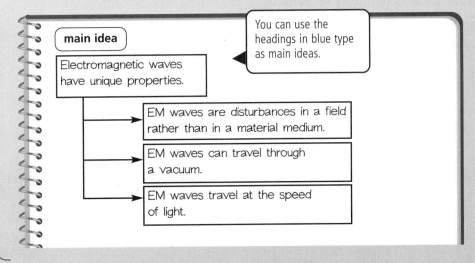

main idea

Electromagnetic waves have unique properties.

You can use the headings in blue type as main ideas.

EM waves are disturbances in a field rather than in a material medium.

EM waves can travel through a vacuum.

EM waves travel at the speed of light.

Outline

1. Copy the chapter title and headings from the book in the form of an outline.

2. Add notes that summarize in your own words what you read.

Cell Processes

1st key idea

I. Cells capture and release energy.

   A. All cells need energy. **1st subpoint of I**

   B. Some cells capture light energy. **2nd subpoint of I**

      1. Process of photosynthesis **1st detail about B**

      2. Chloroplasts (site of photosynthesis) **2nd detail about B**

      3. Carbon dioxide and water as raw materials

      4. Glucose and oxygen as products

   C. All cells release energy.

      1. Process of cellular respiration

      2. Fermentation of sugar to carbon dioxide

      3. Bacteria that carry out fermentation

II. Cells transport materials through membranes.

   A. Some materials move by diffusion.

      1. Particle movement from higher to lower concentrations

      2. Movement of water through membrane (osmosis)

   B. Some transport requires energy.

      1. Active transport

      2. Examples of active transport

Correct Outline Form

Include a title.

Arrange key ideas, subpoints, and details as shown.

Indent the divisions of the outline as shown.

Use the same grammatical form for items of the same rank. For example, if A is a sentence, B must also be a sentence.

You must have at least two main ideas or subpoints. That is, every A must be followed by a B, and every 1 must be followed by a 2.

Concept Map

1. Write an important concept in a large oval.
2. Add details related to the concept in smaller ovals.
3. Write linking words on arrows that connect the ovals.

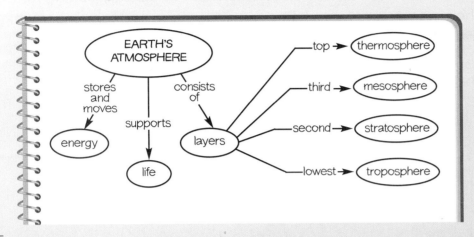

The main ideas or concepts can often be found in the blue headings. An example is "The atmosphere stores and moves energy." Use nouns from these concepts in the ovals, and use the verb or verbs on the lines.

Venn Diagram

1. Draw two overlapping circles, one for each item that you are comparing.
2. In the overlapping section, list the characteristics that are shared by both items.
3. In the outer sections, list the characteristics that are peculiar to each item.
4. Write a summary that describes the information in the Venn diagram.

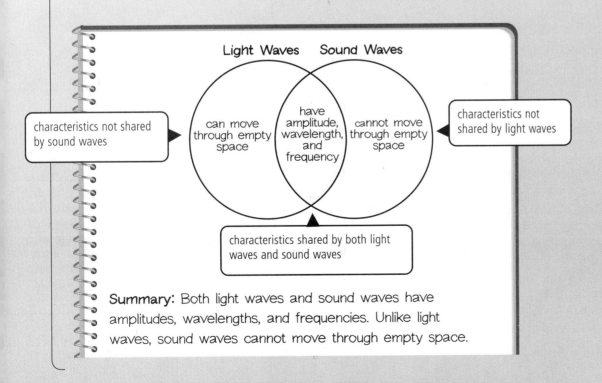

Summary: Both light waves and sound waves have amplitudes, wavelengths, and frequencies. Unlike light waves, sound waves cannot move through empty space.

Vocabulary Strategies

Important terms are highlighted in this book. A definition of each term can be found in the sentence or paragraph where the term appears. You can also find definitions in the Glossary. Taking notes about vocabulary terms helps you understand and remember what you read.

Description Wheel

1. Write a term inside a circle.
2. Write words that describe the term on "spokes" attached to the circle.

When studying for a test with a friend, read the phrases on the spokes one at a time until your friend identifies the correct term.

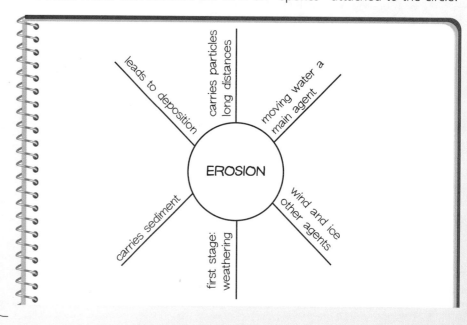

Description Wheel for EROSION with spokes: carries particles long distances, moving water a main agent, leads to deposition, wind and ice other agents, carries sediment, first stage: weathering

Four Square

1. Write a term in the center.
2. Write details in the four areas around the term.

Definition	Characteristics
any living thing	needs food, water, air; needs energy; grows, develops, reproduces
Examples	**Nonexamples**
dogs, cats, birds, insects, flowers, trees	rocks, water, dirt

ORGANISM

Include a definition, some characteristics, and examples. You may want to add a formula, a sketch, or examples of things that the term does *not* name.

Frame Game

1. Write a term in the center.

2. Frame the term with details.

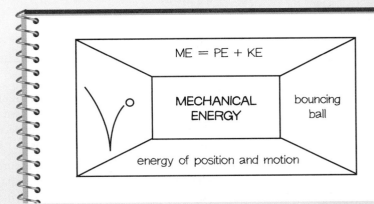

Include examples, descriptions, sketches, or sentences that use the term in context. Change the frame to fit each new term.

ME = PE + KE

MECHANICAL ENERGY

bouncing ball

energy of position and motion

Magnet Word

1. Write a term on the magnet.

2. On the lines, add details related to the term.

You can also use phrases or sentences on the lines.

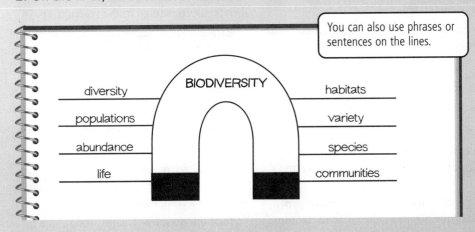

diversity

BIODIVERSITY

habitats

populations

variety

abundance

species

life

communities

Word Triangle

1. Write a term and its definition in the bottom section.

2. In the middle section, write a sentence in which the term is used correctly.

3. In the top section, draw a small picture to illustrate the term.

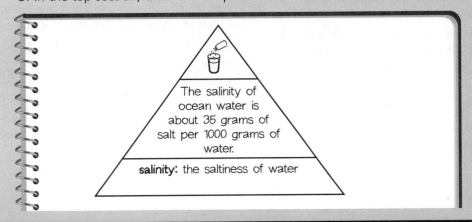

The salinity of ocean water is about 35 grams of salt per 1000 grams of water.

salinity: the saltiness of water

Glossary

A, B

acid
A substance that can donate a proton to another substance and has a pH below 7. (p. 126)

    ácido Una sustancia que puede donar un protón a otra sustancia y que tiene un pH menor a 7.

alloy
A solid mixture composed of a metal and one or more other substances. (p. 134)

    aleación Una mezcla sólida compuesta de un metal y una o más sustancias adicionales.

atom
The smallest particle of an element that has the chemical properties of that element. (p. xv)

    átomo La partícula más pequeña de un elemento que tiene las propiedades químicas de ese elemento.

atomic mass
The average mass of the atoms of an element. (p. 17)

    masa atómica La masa promedio de los átomos de un elemento.

atomic mass number
The total number of protons and neutrons in an atom's nucleus.

    número de masa atómica El número total de protones y neutrones que hay en el núcleo de un átomo.

atomic number
The number of protons in the nucleus of an atom. (p. 12)

    número atómico El número de protones en el núcleo de un átomo.

base
A substance that can accept a proton from another substance and has a pH above 7. (p. 126)

    base Una sustancia que puede aceptar un protón de otra sustancia y que tiene un pH superior a 7.

bond energy
The amount of energy in a chemical bond between atoms.

    energía de enlace La cantidad de energía que hay en un enlace químico entre átomos.

C, D

carbohydrate
A type of carbon-based molecule in living things. Carbohydrates include sugars and starches used for energy or as structural materials. Carbohydrate molecules contain carbon, hydrogen, and oxygen atoms.

    carbohidrato Un tipo de molécula de los organismos vivos basada en el carbono. Los carbohidratos incluyen los azúcares y los almidones usados como fuente de energía o como materiales estructurales. Las moléculas de los carbohidrato contienen átomos de carbono, hidrógeno y oxígeno.

catalyst
A substance that increases the rate of a chemical reaction but is not consumed in the reaction. (p. 76)

    catalizador Una sustancia que aumenta lel a ritmo velocidad de una reacción química pero que no es consumida en la reacción.

chemical change
A change of one substance into another substance.

    cambio químico Un cambio de una sustancia a otra sustancia.

chemical formula
An expression that shows the number and type of atoms joined in a compound. (p. 43)

    fórmula química Una expresión que muestra el número y el tipo de átomos unidos en un compuesto.

chemical reaction
The process by which chemical changes occur. In a chemical reaction, atoms are rearranged, and chemical bonds are broken and formed. (p. 69)

    reacción química El proceso mediante el cual ocurren cambios químicos. En una reacción química, los átomos se reorganizan y los enlaces químicos se rompen y se vuelven a formar.

coefficient
The number before a chemical formula that indicates how many molecules are involved in a chemical reaction.

    coeficiente El número anterior a una fórmula química que indica cuántas moléculas están involucradas en una reacción química.

compound

A substance made up of two or more different types of atoms bonded together.

compuesto Una sustancia formada por dos o más diferentes tipos de átomos enlazados.

concentration

The amount of solute dissolved in a solvent at a given temperature.

concentración La cantidad de soluto disuelta en un solvente a una temperatura determinada.

covalent bond

A pair of electrons shared by two atoms. (p. 50)

enlace covalente Un par de electrones compartidos por dos átomos.

cycle

n. A series of events or actions that repeat themselves regularly; a physical and/or chemical process in which one material continually changes locations and/or forms. Examples include the water cycle, the carbon cycle, and the rock cycle.

v. To move through a repeating series of events or actions.

ciclo *s.* Una serie de eventos o acciones que se repiten regularmente; un proceso físico y/o químico en el cual un material cambia continuamente de lugar y/o forma. Ejemplos: el ciclo del agua, el ciclo del carbono y el ciclo de las rocas.

data

Information gathered by observation or experimentation that can be used in calculating or reasoning. *Data* is a plural word; the singular is datum.

datos Información reunida mediante observación o experimentación y que se puede usar para calcular o para razonar.

density

A property of matter representing the mass per unit volume.

densidad Una propiedad de la materia que representa la masa por unidad de volumen.

dilute

adj. Having a low concentration of solute. (p. 118)

v. To add solvent in order to decrease the concentration of a solution.

diluido *adj.* Que tiene una baja concentración de soluto.

diluir *v.* Agregar solvente para disminuir la concentración de una solución.

E, F

electron

A negatively charged particle located outside an atom's nucleus. An electron is about 2000 times smaller than either a proton or neutron. (p. 11)

electrón Una partícula con carga negativa localizada fuera del núcleo de un átomo. Un electrón es como aproximadamente 2000 veces más pequeño que un protón o un neutrón.

element

A substance that cannot be broken down into a simpler substance by ordinary chemical changes. An element consists of atoms of only one type. (p. xv)

elemento Una sustancia que no puede descomponerse en otra sustancia más simple por medio de cambios químicos normales. Un elemento consta de átomos de un solo tipo.

endothermic reaction

A chemical reaction that absorbs energy. (p. 87)

reacción endotérmica Una reacción química que absorbe energía.

energy

The ability to do work or to cause a change. For example, the energy of a moving bowling ball knocks over pins; energy from food allows animals to move and to grow; and energy from the Sun heats Earth's surface and atmosphere, which causes air to move. (p. xix)

energía La capacidad para trabajar o causar un cambio. Por ejemplo, la energía de una bola de boliche en movimiento tumba los pinos; la energía proveniente de su alimento permite a los animales moverse y crecer; la energía del Sol calienta la superficie y la atmósfera de la Tierra, lo que ocasiona que el aire se mueva.

enzyme

A type of protein that is a catalyst for chemical reactions in living things. (p. 159)

enzima Un tipo de proteína que es un catalizador de reacciones químicas en organismos vivos.

exothermic reaction

A chemical reaction that releases energy. (p. 87)

reacción exotérmica Una reacción química que libera energía.

experiment

An organized procedure to study something under controlled conditions. (p. xxiv)

experimento Un procedimiento organizado para estudiar algo bajo condiciones controladas.

force
A push or a pull; something that changes the motion of an object. (p. xxi)

fuerza Un empuje o un jalón; algo que cambia el movimiento de un objeto.

friction
A force that resists the motion between two surfaces in contact. (p. xxi)

fricción Una fuerza que resiste el movimiento entre dos superficies en contacto.

G, H

gravity
The force that objects exert on each other because of their mass. (p. xxi)

gravedad La fuerza que los objetos ejercen entre sí debido a su masa.

group
A vertical column in the periodic table of the elements. Elements in a group have similar properties. (p. 22)

grupo Una columna vertical en la tabla periódica de los elementos. Los elementos en un grupo tienen propiedades similares.

half-life
The amount of time it takes for half of the nuclei of a radioactive isotope to decay into atoms of another element. (p. 32)

vida media La cantidad de tiempo que se necesita para que le toma a la mitad del núcleo de un isótopo radioactivo se en descomponganerse en átomos de otro elemento.

hydrocarbon
A compound that contains only carbon and hydrogen. (p. 163)

hidrocarburo Un compuesto que contiene solamente carbono e hidrógeno.

hypothesis
A tentative explanation for an observation or phenomenon. A hypothesis is used to make testable predictions. (p. xxiv)

hipótesi Una explicación provisional de una observación o de un fenómeno. Una hipótesis se usa para hacer predicciones que se pueden probar.

I, J

inorganic compound
A compound that is not considered organic. All compounds that do not contain carbon are inorganic, as are some types of carbon-containing compounds. (p. 148)

compuesto inorgánico Un compuesto que no se considera orgánico. Todos los compuestos que no contienen carbono son inorgánicos, al igual que algunos tipos de compuestos que contienen carbono.

ion
An atom or group of atoms that has a positive or negative electric charge. (p. 14)

ión Un átomo o un grupo de átomos que tiene una carga eléctrica positiva o negativa.

ionic bond
The electric attraction between a negative ion and a positive ion. (p. 48)

enlace iónico La atracción eléctrica entre un ión negativo y un ión positivo.

isomer
Any of two or more compounds that contain the same atoms but that have different structures. (p. 152)

isómero Cualquiera de dos o más compuestos que contienen los mismos átomos pero que tienen estructuras diferentes.

isotope
An atom of one element that has a different number of neutrons than another atom of the same element. (p. 12)

isótopo Un átomo de un elemento que tiene un número diferente de neutrones que otro átomo del mismo elemento.

K, L

law
In science, a rule or principle describing a physical relationship that always works in the same way under the same conditions. The law of conservation of energy is an example.

ley En las ciencias, una regla o un principio que describe una relación física que siempre funciona de la misma manera bajo las mismas condiciones. La ley de la conservación de la energía es un ejemplo.

law of conservation of energy
A law stating that no matter how energy is transferred or transformed, all of the energy is still present in one form or another. (p. 82)

ley de la conservación de la energía Una ley que establece que no importa cómo se transfiere o transforma la energía, toda la energía sigue presente en alguna forma u otra.

law of conservation of mass
A law stating that atoms are not created or destroyed in a chemical reaction. (p. 79)

ley de la conservación de la masa Una ley que establece que los átomos ni se crean ni se destruyen en una reacción química.

lipid
A type of carbon-based molecule in living things. Lipids include fats and oils used for energy or as structural materials. (p. 156)

lípido Un tipo de molecula de los organismos vivos basada en el carbono. Los lípidos incluyen las grasas y los aceites usados como fuente de energia o como materiales estructurales.

M, N

mass
A measure of how much matter an object is made of. (p. xv)

masa Una medida de la cantidad de materia de la que está compuesto un objeto.

matter
Anything that has mass and volume. Matter exists ordinarily as a solid, a liquid, or a gas. (p. xv)

materia Todo lo que tiene masa y volumen. Generalmente la materia existe como sólido, líquido o gas.

metal
An element that tends to be shiny, easily shaped, and a good conductor of electricity and heat. (p. 27)

metal Un elemento que tiende a ser brilloso, fácilmente deformable moldeado y buen conductor de electricidad y calor.

metallic bond
A certain type of bond in which nuclei float in a sea of electrons. (p. 56)

enlace metálico Cierto tipo de enlace en el cual los núcleos flotan en un mar de electrones.

metalloid
An element that has properties of both metals and non-metals. (p. 30)

metaloide Un elemento que tiene propiedades de los metales así como de los no metales.

mixture
A combination of two or more substances that do not combine chemically but remain the same individual substances. Mixtures can be separated by physical means.

mezcla Una combinación de dos o más sustancias que no se combinan químicamente sino que permanecen siendo las mismas sustancias individuales. Las mezclas se pueden separar por medios físicos.

molecule
A group of atoms that are held together by covalent bonds so that they move as a unit. (p. 51)

molécula Un grupo de átomos que se mantienen unidos por medio de enlaces covalentes de tal manera que se mueven como una sola unidad.

monomer
One of many small, repeating units linked together to form a polymer. (p. 166)

monómero Una de muchas unidades pequeñas que se repiten y están enlazadas unas con otras para formar un polímero.

neutral
Describing a solution that is neither an acid nor a base. A neutral solution has a pH of 7. (p. 129)

neutro Que describe una solución que no es un ácido ni una base. Una solución neutra tiene un pH de 7.

neutron
A particle that has no electric charge and is located in an atom's nucleus. (p. 11)

neutrón Una partícula que no tiene carga eléctrica y que se encuentra en el núcleo de un átomo.

nonmetal
An element that is not a metal and has properties generally opposite to those of a metal. (p. 29)

no metal Un elemento que no es un metal y que tiene propiedades generalmente opuestas a las de los metales.

nucleic acid
One of several carbon-based molecules that carry an organism's genetic code. One of the nucleic acids—DNA—contains the information needed to construct proteins. (p. 161)

ácido nucleico Una de varias moléculas basadas en el carbono que llevan el código genético de un organismo. Uno de los ácidos nucleicos, el ADN, contiene la información necesaria para construir proteínas.

nucleus

The central region of an atom where most of the atom's mass is found in protons and neutrons. (p. 11)

núcleo La región central de un átomo donde se encuentra la mayor parte de la masa del átomo en la forma de protones y neutrones.

O, P, Q

organic compound

A compound that is based on carbon. (p. 147)

compuesto orgánico Un compuesto basado en el carbono.

period

A horizontal row in the periodic table of the elements. Elements in a period have varying properties. (p. 22)

período Un renglón horizontal en la tabla periódica de los elementos. Los elementos en un período tienen distintas propiedades.

periodic table

A table of the elements, arranged by atomic number, that shows the patterns in their properties. (p. 18)

tabla periódica Una tabla de los elementos, organizada en base a número atómico, que muestra los patrones en sus propiedades.

pH

The concentration of hydrogen ions in a solution; a measurement of acidity. (p. 129)

pH La concentración de iones de hidrógeno en una solución;, una medida de acidez.

photosynthesis

In green plants, the endothermic process in which light is absorbed and used to change carbon dioxide and water into glucose and oxygen. (p. 90)

fotosíntesis En plantas verdes, el proceso endotérmico en el cual se absorbe luz y se usa para cambiar dióxido de carbono y agua a glucosa y oxígeno.

plastic

A polymer that can be molded or shaped. (p. 167)

plástico Un polímero que puede ser modelado o deformadomoldeado.

polar covalent bond

The unequal sharing of electrons between two atoms that gives rise to negative and positive regions of electric charge. (p. 51)

enlace polar covalente El compartir electrones desigualmente entre dos átomos y que lleva a la formación de regiones de carga eléctrica positiva y regiones de carga eléctrica negativa.

polymer

A very large carbon-based molecule made of smaller, repeating units. (p. 166)

polímero Una molécula muy grande basada en el carbono compuesta de unidades más pequeñas que se repiten.

precipitate

n. A solid substance that forms as a result of a reaction between chemicals in two liquids. (p. 72)

v. To come out of solution.

precipitado *s.* Una sustancia sólida que se forma como resultado de la reacción entre sustancias químicas en dos líquidos.

precipitar *v.* Salir de solución.

product

A substance formed by a chemical reaction. A product is made by the rearrangement of atoms and bonds in reactants. (p. 71)

producto Una sustancia formada por una reacción química. Un producto se hace mediante la reorganización de los átomos y los enlaces en los reactivos.

protein

A macromolecule in living things that is made of smaller molecules called amino acids. (p. 158)

proteína Una macromolécula en organismos vivos compuesta de moléculas más pequeñas llamadas aminoácidos.

proton

A positively charged particle located in an atom's nucleus. (p. 11)

protón Una partícula con cargada positivamente localizada en el núcleo de un átomo.

R, S

radioactivity

The process by which the nucleus of an atom of an element releases energy and particles. (p. 30)

radioactividad El proceso mediante el cual el núcleo de un átomo de un elemento libera energía y partículas.

reactant

A substance that is present at the beginning of a chemical reaction and is changed into a new substance. (p. 71)

reactivo Una sustancia que está presente en el comienzo de una reacción química y que se convierte en una nueva sustancia.

reactive
Likely to undergo a chemical change. (p. 26)

reactivo Que es probable que sufra un cambio químico.

respiration
The exothermic process by which living things release energy from glucose and oxygen and produce carbon dioxide and water. (p. 94)

respiración El proceso exotérmico mediante el cual los organismos vivos liberan energía de la glucosa y del oxígeno y producen dióxido de carbono y agua.

saturated
Containing the maximum amount of a solute that can be dissolved in a particular solvent at a given temperature and pressure. (p. 118)

saturado Que contiene la máxima cantidad de soluto que se puede disolver en un solvente en particular a determinada temperatura y presión.

solubility
The amount of solute that dissolves in a certain amount of a solvent at a given temperature and pressure to produce a saturated solution. (p. 119)

solubilidad La cantidad de soluto que se disuelve en cierta cantidad de solvente a determinada temperatura y presión para producir una solución saturada.

solute
In a solution, a substance that is dissolved in a solvent. (p. 112)

soluto En una solución, una sustancia que se disuelve en un solvente.

solution
A mixture of two or more substances that is identical throughout; a homogeneous mixture. (p. 111)

solución Una mezcla de dos o más sustancias que es idéntica en su totalidad;, una mezcla homogénea.

solvent
In a solution, the substance that dissolves a solute and makes up the largest percentage of a solution. (p. 112)

solvente En una solución, la sustancia que disuelve un soluto y que compone el porcentaje mayor de la una solución.

subscript
A number written slightly below and to the right of a chemical symbol that shows how many atoms of an element are in a compound. (p. 43)

subíndice Un número que se escribe en la parte inferior a la derecha de un símbolo químico y que muestra cuantos átomos de un elemento están en un compuesto.

suspension
A mixture in which the different parts are identifiable as separate substances; a heterogeneous mixture. (p. 113)

suspensión Una mezcla en la cual las diferentes partes son identificables como sustancias distintas; una mezcla heterogénea.

system
A group of objects or phenomena that interact. A system can be as simple as a rope, a pulley, and a mass. It also can be as complex as the interaction of energy and matter in the four spheres of the Earth system.

sistema Un grupo de objetos o fenómenos que interactúan. Un sistema puede ser algo tan sencillo como una cuerda, una polea y una masa. También puede ser algo tan complejo como la interacción de la energía y la materia en las cuatro esferas del sistema de la Tierra.

T, U

technology
The use of scientific knowledge to solve problems or to engineer new products, tools, or processes.

tecnología El uso de conocimientos científicos para resolver problemas o para diseñar nuevos productos, herramientas o procesos.

theory
In science, a set of widely accepted explanations of observations and phenomena. A theory is a well-tested explanation that is consistent with all available evidence.

teoría En las ciencias, un conjunto de explicaciones de observaciones y fenómenos que es ampliamente aceptado. Una teoría es una explicación bien probada que es consecuente con la evidencia disponible.

V, W, X, Y, Z

variable
Any factor that can change in a controlled experiment, observation, or model. (p. R30)

variable Cualquier factor que puede cambiar en un experimento controlado, en una observación o en un modelo.

volume
An amount of three-dimensional space, often used to describe the space that an object takes up. (p. xv)

volumen Una cantidad de espacio tridimensional; a menudo se usa este término para describir el espacio que ocupa un objeto.

Index

Page numbers for definitions are printed in **boldface** type.
Page numbers for illustrations, maps, and charts are printed in *italics*.

A, B

E, F

G, H

I, J

ice cream, 115
inert gases, 29, *29*
inference, **R4,** R35
infrared spectroscopy, 4
inorganic compounds, **148,** *148*
International Space Station, 138, *138*
International System of Units, R20–R21
Internet activity
    alloys, 109
    chemical bonding, 39
    chemical reactions, 67
    periodic table, 7
    polymers, 145
Investigations. *See* Chapter Investigations.
iodine, 22, 50, *50*
    molecular structure, 54, *54*
iodine clock reaction, 77
ionic bond, **48,** *48,* 48–49, *52,* 62
ionic compounds
    names of, 49
    properties of, 57–58
    in solutions, 114, *114,* 123, *123*
    structures of, 53
ions, **14**
    acids and bases, 126, 128–31, *128, 129*
    formation, *14,* 14–15, *15,* 48
    negative, 15, *15, 48,* 48–49
    periodic table and, 22–23, 48, *48*
    positive, 14, *14, 48,* 48–49
isobutane, 152
isomers, **152,** *152,* 169
isotopes, **12,** 12–13, *13*
    atomic mass number and, 17
    half-life, 32, *32*
    radioactive, 30, 31–32

K, L

keratin, 159, *159*
Kevlar, 169, *169*
laboratory equipment
    beaker, R12, *R12*
    double-pan balance, R19, *R19*
    force meter, R16, *R16*
    forceps, R13, *R13*
    graduated cylinder, R16, *R16*
    hot plate, R13, *R13*
    meniscus, **R16,** *R16*
    microscope, *xxiv, R14,* R14–R15
    ruler, metric, R17, *R17*
    spring scale, R16, *R16*
    test-tube holder, R12, *R12*
    test-tube rack, R13, *R13*
    test tube, R12, *R12*
    triple-beam balance, R18, *R18*
labs, R10–35. *See also* experiment.
    equipment, R12–R19
    safety, R10–R11
Langmuir, Irving, 106
lanthanides (rare earth elements), 21, 28, *28*
laser, chemical reaction and, *xxiv*
Lavoisier, Antoine, 78–79, *79*
law of conservation of energy, xix
law of conservation of mass, 78–79, **79,** 100
    in chemical equations, 80, 81, 84
lemon, pH, *130*
Lewis, G.N., 106

lipids, **156,** *156,* 156–57, 172
lithium, 10
litmus paper, 127, 129

M, N

magnesium, 10, 27
magnetic force, xxi
mass, **xv**
math skills
    area, **R43**
    bar graphs, 162
    decimal, **R39,** R40
    describing a set of data, R36–R37
    formulas, **R42**
    fractions, **R41**
    line graphs, 77
    mean, **R36**
    median, **R36**
    mode, **R37**
    percents, 139, **R41**
    proportions, **R39**
    range, **R37**
    rates, **R38**
    ratios, 46, **R38**
    scientific notation, 33, **R44**
    significant figures, **R44**
    volume, **R43, xv**
matter, xiv–xvii, **xv**
    conservation of, xvii
    forms of, xvi–xvii
    movement of, xvii
    particles and, xiv–xv
    physical forces and, xx
mean, **R36**
measurement
    area, **R43**
    mass, **xv**
    volume, **R43, xv**
median, **R36**
medicine
    alloys and, 137, *137*
    from nature, 2–5
    radioactivity in, 31
Mendeleev, Dmitri, 18, *18,* 19
mercury, 27
metallic bond, **56,** 56–57, *57,* 62
metalloids, 30, *30,* 34
metals, 22, **27,** 27–28
    alkali, 27
    alkaline earth, 27
    alloys, 28, 109, **134,** 134–38, 140
    properties and bonds, 56–57
    reactive, 27, *27*
    transition, *27,* 27–28
methane, *44*
    bonding of, *50,* 50–51
    combustion, 81, *81,* 82, *82,* 88, *88*
    molecular structure, 54, *54*
metric system, R20–R21
    changing metric units, R20, *R20*
    converting between U.S. customary units, R21, *R21*
    temperature conversion, R21, *R21*
microchips, 98–99, *99*
microscope, *R14,* R14–R15
    making a slide or wet mount, R15, *R15*
    scanning tunneling (STM), *xxiv*
    viewing an object, R15
milk, pH, *130*
mixtures, 111. *See also* solutions.

mode, **R37**
molecular structure, 54, *54*
   studying, 4
molecule, **xv, 51**
   carbon-based (*See* carbon)
monomers, **166,** 172
mosaic, *xiv*
nanotubes, carbon, 153, *153*
natural medicines, 2–5
negative ions, 15, *15, 48,* 48–49
Neptunium, 10
neutral, **129,** *130,* 131, 140
neutral atoms, 12
neutralization reaction, 131
neutron, **11,** *11,* 34, *34*
   number of, 12–13, *13*
Nitinol, 135
nitrogen, 16, 51, *51*
   compounds, 44
noble gases, 29, *29*
Nomex, 169, *169*
nonmetals, **29,** *29,* 34
nonpolar substances, 122–23, *123*
note-taking strategies, **R45–R49**
   combination notes, 68, *68,* R45, *R45*
   concept map, R49, *R49*
   content frame, R45, *R45*
   main idea and detail notes, 40, *40,* R46, *R46*
   main idea web, 8, *8,* R46, *R46*
   mind map, 110, *110,* R47, *R47*
   outline, R48, *R48*
   supporting main ideas, 146, *146,* R47, *R47*
   Venn diagram, R49, *R49*
nuclear magnetic resonance spectoscopy, 4
nucleic acids, **161,** *160,* 172
nucleus, *11,* **11,** 34, *34,* 106

O, P

observations, **R2,** R5, R33, **xxiv**
   qualitative, R2
   quantitative, R2
operational definition, **R31**
opinion, **R9**
   different from fact, R9
organic compounds, **147,** 147–48, *148*
osmium, 23
oxygen
   in combustion, 73, *73,* 81, *81*
   compounds, 44–45
   fuel cells and, xxvii
   in photosynthesis, 90, 45
   in respiration, 94–95
particle, xiv–xv
   atomic, *11,* 11–12
particle accelerators, 107
percents, 139, **R41**
period, *20, 22,* **22,** 34, *34*
periodic table, 17–30, **18,** *20–21,* 34, *34*
   atomic size and, 23, *23*
   density of elements and, 23
   group, *20, 22,* **22,** 34, *34*
   halogens, 22, *22,* 29, *29*
   how to read, 19, *19*
   Internet activity, 7
   Mendeleev's, 18, *18*
   metalloids, 30, *30,* 34
   metals, 27–28
   modern table, 19, *20–21*
   noble gases, 29, *29*

nonmetals, *29,* **29,** 34
   organization of, 19, 22–23
   period, *20, 22,* **22,** 34, *34*
   regions, 26, *26*
   trends and patterns, 22–23
petroleum, 164, *164*
pewter, *135*
pH, **129,** *130,* 140
phospholipids, 157, *157*
phosphorus, 16
photoresist, 98–99
photosynthesis, **90,** 90–91, 94
   equation, 95
physical science, xiii. *See also* science.
   unifying principles of, xiv–xxi
plastic, **167,** 167–68
   recycling, 168, *168*
pneumatic trough, 104
polar covalent bond, *51,* **51,** *52*
pollution, xxvii
polymers, **166,** 172
   Chapter Investigation, 170–71
   formation, 166–67, *167*
   Internet activity, 145
   monomers and, **166,** 167–69, *167, 168,* 172
   new materials, 168–69
   plastics, 167–68
polypropylene, 166, *167*
polystyrene, *151*
positive ions, 14, *14, 48,* 48–49
potassium, 16, 27
praseodymium, 28
precipitate, *72,* **72,** 118–19, *119*
precision, **R22**
prediction, **xxiv, R3**
pressure, solubility and, 122, *122*
products, **71**
propane, *44*
proportions, **R39**
propylene, 166, *167*
proteins, **158,** 158–59, *159,* 172
   DNA and, *160,* 161
   enzymes, 76, *76,* **159**
   structural, 159, *159*
   transport, 159, *159*
protons, *11,* **11,** 34, *34*
   in acids and bases, 126
   relation to element, and, 12

Q, R

quarks, 107
quartz, 98, *98*
radioactivity, **30,** 106
   detection, 30
   medical uses, 31, *31*
range, **R37**
rare earth elements (lanthanides), 21, 28, *28*
rates, **R38**
ratios, 46, **R38**
reactants, **71**
   changes in, 74–75, *75*
reactions. *See* chemical reactions.
reactive, **26**
reactive metals, 27, *27*
reasoning, faulty, **R7**
respiration, **94,** 94–95
   equation, 95
retinal, 152
RNA, 161

Acknowledgments

Photography

Cover © Photodisc/Getty Images; **i** © Photodisc/Getty Images; **iii** *left (top to bottom)* Photograph of James Trefil by Evan Cantwell; Photograph of Rita Ann Calvo by Joseph Calvo; Photograph of Linda Carnine by Amilcar Cifuentes; Photograph of Sam Miller by Samuel Miller; *right (top to bottom)* Photograph of Kenneth Cutler by Kenneth A. Cutler; Photograph of Donald Steely by Marni Stamm; Photograph of Vicky Vachon by Redfern Photographics; **vi** © Digital Vision/PictureQuest; **vii** From *General Chemistry* by P. W. Atkins, © 1989 by Peter Atkins. Used with permission of W. H. Freeman and Company; **ix** Photographs by Sharon Hoogstraten; **xiv–xv** © Larry Hamill/age fotostock america, inc.; **xvi–xvii** © Fritz Poelking/age fotostock america, inc.; **xviii–xix** © Galen Rowell/Corbis; **xx–xxi** © Jack Affleck/SuperStock; **xxii** AP/Wide World Photos; **xxiii** © David Parker/IMI/University of Birmingham High, TC Consortium/Photo Researchers; **xxiv** *left* AP/Wide World Photos; *right Washington University Record;* **xxv** *top* © Kim Steele/Getty Images; *bottom* Reprinted with permission from S. Zhou et al., *SCIENCE* 291:1944–47. © 2001 AAAS; **xxvi–xxvii** © Mike Fiala/Getty Images; **xxvii** *left* © Derek Trask/Corbis; *right* AP/Wide World Photos; **xxxii** © The Chedd-Angier Production Company; **2–3** © David Cavagnaro/Peter Arnold, Inc.; **3** Joel Sartore/National Geographic Image Collection; **4** © The Chedd-Angier Production Company; **5** © Colin Cuthbert/Photo Researchers; **6–7** IBM Research, Almaden Research Center; **7, 9** Photographs by Sharon Hoogstraten; **10** NASA; **12** © Pascal Goetgheluck/Photo Researchers; **13** Photograph by Sharon Hoogstraten; **16** © Cnri/Photo Researchers; **17** Photograph by Sharon Hoogstraten; **18** *left, right* The Granger Collection, New York; **24** *top* © A. Hart-Davis/Photo Researchers; *bottom* Photograph by Sharon Hoogstraten; **26** Photograph by Sharon Hoogstraten; **27** *left* © Charles D. Winters/Photo Researchers; *center* © Rich Treptow/Visuals Unlimited; *right* © Corbis Images/PictureQuest; **28** © Peter Christopher/Masterfile; **29** © M. Gibbon/Robertstock.com; **30** © Superstock; **31** *top* © Simon Fraser/Photo Researchers; *bottom* Photograph by Sharon Hoogstraten; **33** © Alfred Pasieka/Photo Researchers; *inset* © John Walsh/Photo Researchers; **38–39** © Digital Vision/PictureQuest; **39, 41** Photographs by Sharon Hoogstraten; **42** *left* © Rich Treptow/Visuals Unlimited; *center, right* © E. R. Degginger/Color-Pic, Inc.; **43, 45** Photograph by Sharon Hoogstraten; **46** © Lawrence M. Sawyer/Photodisc/PictureQuest; **47** © IFA/eStock Photography (PQ price control)/PictureQuest; **49** © Runk and Schoenberger/Grant Heilman Photography, Inc.; **52** © The Image Bank/Getty Images; **53** Photograph by Sharon Hoogstraten; **55** © Astrid & Hanns-Frieder Michler/Photo Researchers; *inset* © Volker Steger/Photo Researchers; **56** Photograph by Sharon Hoogstraten; **57** © David Wrobel/Visuals Unlimited; **58** © Rob Blakers/photolibrary/PictureQuest; **59** *left* © E. R. Degginger/Robertstock.com; *right* © C. Swartzell/Visuals Unlimited; **60** *top* © David Young-Wolff/Getty Images; *bottom* Photograph by Sharon Hoogstraten; **61** Photograph by Sharon Hoogstraten; **62** *left* © Rich Treptow/Visuals Unlimited; *center, right* © E. R. Degginger/Color-Pic, Inc.; **66–67** From *General Chemistry* by P. W. Atkins, © 1989 by Peter Atkins. Used with permission of W.H. Freeman and Company; **67, 69** Photographs by Sharon Hoogstraten; **70** © Daryl Benson/Masterfile; **72** *top left* © Science VU/Visuals Unlimited; *top right* © 1992 Richard Megna/Fundamental Photographs, NYC; *bottom left* © E. R. Degginger/Color-Pic, Inc.; *bottom right* © Larry Stepanowicz/Visuals Unlimited; **74** Photograph by Sharon Hoogstraten; **77** © Corbis Images/PictureQuest; *inset* © Andrew Lambert Photography/Photo Researchers; **78** © Wally Eberhart/Visuals Unlimited; **79** *top* The Granger Collection, New York; *bottom* Photograph by Sharon Hoogstraten; **80** © William Ervin/Photo Researchers; **82** © Maximilian Stock Ltd./Photo Researchers; **84** © Index Stock; **85** *left, inset* Courtesy of Chicago Fire Department; *center* Uline; *bottom right* Photograph by Sharon Hoogstraten; **86** Photograph by Sharon Hoogstraten; **87** *top* NASA; *bottom* © 1992 Richard Megna/Fundamental Photographs, NYC; **88** © Jeffrey L. Rotman/Corbis; **89** Thomas Eisner and Daniel Aneshansley, Cornell University; **91** © Harald Sund/Brand X Pictures/PictureQuest; **92** *top* AP/Wide World Photos; *bottom* Photographs by Sharon Hoogstraten; **93** Photograph by Sharon Hoogstraten; **94** © Runk and Schoenberger/Grant Heilman Photography, Inc.; **95** Photograph by Sharon Hoogstraten; **96** © Tom Yhlman/Visuals Unlimited; **97** *background* © Conor Caffrey/Photo Researchers; **98** © Arnold Fisher/Photo Researchers; **99** *left to right* © Bruce Forster/Getty Images; © Colin Cuthbert/Photo Researchers; © Fontarnau-Gutiérrez/age fotostock america, inc.; © D. Roberts/Photo Researchers; **100** © 1992 Richard Megna/Fundamental Photographs, NYC; **104** From Hales, *Vegetable Statiks* [1727]; **105** *top* The Granger Collection, New York; *bottom* Mary Evans Picture Library; **106** *top* AP/Wide World Photos; *bottom* © Dorling Kindersley; **107** *top, bottom* © David Parker/Photo Researchers; **108–109** © Stephen Frink/Index Stock; **109, 111** Photographs by Sharon Hoogstraten; **112** © Richard Cummins/Corbis; **113, 115** Photographs by Sharon Hoogstraten; **116** © Peter & Georgina Bowater/Stock Connection/PictureQuest; *inset* © 2001 Kim Fennema/Visuals Unlimited; **117, 118** Photographs by Sharon Hoogstraten; **119** *left, right* © 1990 Richard Megna/Fundamental Photographs, NYC; **120, 121** Photographs by Sharon Hoogstraten; **122** © Stephen Frink/StephenFrink.com; **123** Photograph by Sharon Hoogstraten; **124** © Thom Lang/Corbis; **125, 127, 129** Photographs by Sharon Hoogstraten; **130** *top left* © Martyn F. Chillmaid/Photo Researchers; *top right* © Chuck Swartzell/Visuals Unlimited; *center left* © E. R. Degginger/Color-Pic, Inc.; *center right* © Phil Degginger/Color-Pic, Inc.; *bottom left* © Stockbyte; *bottom right* © E. R. Degginger/Color-Pic, Inc.; **132** © Runk and Schoenberger/Grant Heilman Photography, Inc.; **132–133, 133, 134** Photographs by Sharon Hoogstraten; **135** *top to bottom* © Photodisc/Getty Images; © Greg Pease/Stock Connection/PictureQuest; © Stockbyte; © S. Feld/Robertstock.com; Jellinek & Sampson, London/Bridgeman Art Library; **136** © Joachim Messerschmidt/Bruce Coleman, Inc.; **137** *top* © Princess Margaret Rose Hospital/Photo Researchers; *inset* © Klaus Rose/Okapia/Photo Researchers; *bottom* Photograph by Sharon Hoogstraten; **138** NASA; **139** © IFA/eStock Photography (PQ price control)/PictureQuest; **140** © Joachim Messerschmidt/Bruce Coleman, Inc.; **144–145** © Jeff Greenberg/Index Stock/PictureQuest; **145, 147** Photographs by Sharon Hoogstraten; **148** *left* © E. R. Degginger/Color-Pic, Inc.; *right* © Charles D. Winters/Photo Researchers; **149** Photograph by Sharon Hoogstraten; **151** *top* © Claver Carroll/age fotostock america, inc.; *bottom left* © Fabio Cardoso/age fotostock america, inc.; *bottom right* Photograph by Sharon Hoogstraten; **153** *left* © S. J. Tans et al., Delft University of Technology/Photo Researchers; *right* Georgia Institute of Technology; **154** Photograph by Sharon Hoogstraten; **155** © Marcialis/StockFood; **156** © John Durham/Photo Researchers; **157** *top* © Meyer/StockFood; *bottom* © SPL/Photo Researchers; **158** Photograph by Sharon Hoogstraten; **159** *left* © Andrew Syred/Photo Researchers; *right* © SCIMAT 2000/Photo Researchers; **162** © Eising/StockFood; **163** Photograph by Sharon Hoogstraten; **165** *top* © Thomas Kitchin/Tom Stack & Associates; *center* © Superstock; *bottom left* © Bob Krist/Corbis; *bottom center* © Omni Photo Communications/Index Stock; *bottom right* © Gary Rhijnsburger/Masterfile; **167** *top* Image Club Graphics; *bottom* © 1994 CMCD, Inc.; **168** *bottom left* © J. Blank/Robertstock.com; **169** *left* © SuperStock; *right* © Cheryl A. Ertelt/Visuals Unlimited; **170** *top* © E. R. Degginger/Color-Pic, Inc.; *bottom* Photograph by Sharon Hoogstraten; **171** Photograph by Sharon Hoogstraten; **172** © SuperStock; **R28** © Photodisc/Getty Images.

Illustrations

Ampersand Design Group 85, 97; Stephen Durke 11, 12, 13, 14, 15, 34, 48, 49, 50, 51, 52, 54, 59, 62, 64, 70, 71, 73, 88, 89, 97, 102, 114, 142, 148, 158, 159, 160, 172, 174; Patrick Gnan 160; Gary Hincks 124; KO Studios 160; Debbie Maizels 160; Precision Graphics 165; Dan Stukenschneider R11–R19, R22, R32.

Content Standards: 5–8

A. Science as Inquiry

As a result of activities in grades 5–8, all students should develop

Abilities Necessary to do Scientific Inquiry

A.1   Identify questions that can be answered through scientific investigations. Students should develop the ability to refine and refocus broad and ill-defined questions. An important aspect of this ability consists of students' ability to clarify questions and inquiries and direct them toward objects and phenomena that can be described, explained, or predicted by scientific investigations. Students should develop the ability to identify their questions with scientific ideas, concepts, and quantitative relationships that guide investigation.

A.2   Design and conduct a scientific investigation. Students should develop general abilities, such as systematic observation, making accurate measurements, and identifying and controlling variables. They should also develop the ability to clarify their ideas that are influencing and guiding the inquiry, and to understand how those ideas compare with current scientific knowledge. Students can learn to formulate questions, design investigations, execute investigations, interpret data, use evidence to generate explanations, propose alternative explanations, and critique explanations and procedures.

A.3   Use appropriate tools and techniques to gather, analyze, and interpret data. The use of tools and techniques, including mathematics, will be guided by the question asked and the investigations students design. The use of computers for the collection, summary, and display of evidence is part of this standard. Students should be able to access, gather, store, retrieve, and organize data, using hardware and software designed for these purposes.

A.4   Develop descriptions, explanations, predictions, and models using evidence. Students should base their explanation on what they observed, and as they develop cognitive skills, they should be able to differentiate explanation from description—providing causes for effects and establishing relationships based on evidence and logical argument. This standard requires a subject matter knowledge base so the students can effectively conduct investigations, because developing explanations establishes connections between the content of science and the contexts within which students develop new knowledge.

A.5   Think critically and logically to make the relationships between evidence and explanations. Thinking critically about evidence includes deciding what evidence should be used and accounting for anomalous data. Specifically, students should be able to review data from a simple experiment, summarize the data, and form a logical argument about the cause-and-effect relationships in the experiment. Students should begin to state some explanations in terms of the relationship between two or more variables.

A.6   Recognize and analyze alternative explanations and predictions. Students should develop the ability to listen to and respect the explanations proposed by other students. They should remain open to and acknowledge different ideas and explanations, be able to accept the skepticism of others, and consider alternative explanations.

A.7   Communicate scientific procedures and explanations. With practice, students should become competent at communicating experimental methods, following instructions, describing observations, summarizing the results of other groups, and telling other students about investigations and explanations.

A.8   Use mathematics in all aspects of scientific inquiry. Mathematics is essential to asking and answering questions about the natural world. Mathematics can be used to ask questions; to gather, organize, and present data; and to structure convincing explanations.

Understandings about Scientific Inquiry

A.9.a   Different kinds of questions suggest different kinds of scientific investigations. Some investigations involve observing and describing objects, organisms, or events; some involve collecting specimens; some involve experiments; some involve seeking more information; some involve discovery of new objects and phenomena; and some involve making models.

A.9.b   Current scientific knowledge and understanding guide scientific investigations. Different scientific domains employ different methods, core theories, and standards to advance scientific knowledge and understanding.

A.9.c   Mathematics is important in all aspects of scientific inquiry.

A.9.d   Technology used to gather data enhances accuracy and allows scientists to analyze and quantify results of investigations.

A.9.e   Scientific explanations emphasize evidence, have logically consistent arguments, and use scientific principles, models, and theories. The scientific community accepts and uses such explanations until displaced by better scientific ones. When such displacement occurs, science advances.

A.9.f   Science advances through legitimate skepticism. Asking questions and querying other scientists' explanations is part of scientific inquiry. Scientists evaluate the explanations proposed by other scientists by examining evidence, comparing evidence, identifying faulty reasoning, pointing out statements that go beyond the evidence, and suggesting alternative explanations for the same observations.

A.9.g   Scientific investigations sometimes result in new ideas and phenomena for study, generate new methods or procedures for an investigation, or develop new technologies to improve the collection of data. All of these results can lead to new investigations.

B. Physical Science

As a result of their activities in grades 5–8, all students should develop an understanding of

Properties and Changes of Properties in Matter

B.1.a   A substance has characteristic properties, such as density, a boiling point, and solubility, all of which are independent of the amount of the sample. A mixture of substances often can be separated into the original substances using one or more of the characteristic properties.

B.1.b   Substances react chemically in characteristic ways with other substances to form new substances (compounds) with different characteristic properties. In chemical reactions, the total mass is conserved. Substances often are placed in categories or groups if they react in similar ways; metals is an example of such a group.

B.1.c   Chemical elements do not break down during normal laboratory reactions involving such treatments as heating, exposure to electric current, or reaction with acids. There are more than 100 known elements that combine in a multitude of ways to produce compounds, which account for the living and nonliving substances that we encounter.

Motions and Forces

B.2.a   The motion of an object can be described by its position, direction of motion, and speed. That motion can be measured and represented on a graph.

B.2.b   An object that is not being subjected to a force will continue to move at a constant speed and in a straight line.

B.2.c   If more than one force acts on an object along a straight line, then the forces will reinforce or cancel one another, depending on their direction and magnitude. Unbalanced forces will cause changes in the speed or direction of an object's motion.

Transfer of Energy

B.3.a   Energy is a property of many substances and is associated with heat, light, electricity, mechanical motion, sound, nuclei, and the nature of a chemical. Energy is transferred in many ways.

B.3.b   Heat moves in predictable ways, flowing from warmer objects to cooler ones, until both reach the same temperature.

B.3.c   Light interacts with matter by transmission (including refraction), absorption, or scattering (including reflection). To see an object, light from that object—emitted by or scattered from it—must enter the eye.

B.3.d   Electrical circuits provide a means of transferring electrical energy when heat, light, sound, and chemical changes are produced.

B.3.e   In most chemical and nuclear reactions, energy is transferred into or out of a system. Heat, light, mechanical motion, or electricity might all be involved in such transfers.

B.3.f   The sun is a major source of energy for changes on the earth's surface. The sun loses energy by emitting light. A tiny fraction of that light reaches the earth, transferring energy from the sun to the earth. The sun's energy arrives as light with a range of wavelengths, consisting of visible light, infrared, and ultraviolet radiation.

C. Life Science

As a result of their activities in grades 5–8, all students should develop understanding of

Structure and Function in Living Systems

C.1.a Living systems at all levels of organization demonstrate the complementary nature of structure and function. Important levels of organization for structure and function include cells, organs, tissues, organ systems, whole organisms, and ecosystems.

C.1.b All organisms are composed of cells—the fundamental unit of life. Most organisms are single cells; other organisms, including humans, are multicellular.

C.1.c Cells carry on the many functions needed to sustain life. They grow and divide, thereby producing more cells. This requires that they take in nutrients, which they use to provide energy for the work that cells do and to make the materials that a cell or an organism needs.

C.1.d Specialized cells perform specialized functions in multicellular organisms. Groups of specialized cells cooperate to form a tissue, such as a muscle. Different tissues are in turn grouped together to form larger functional units, called organs. Each type of cell, tissue, and organ has a distinct structure and set of functions that serve the organism as a whole.

C.1.e The human organism has systems for digestion, respiration, reproduction, circulation, excretion, movement, control, and coordination, and for protection from disease. These systems interact with one another.

C.1.f Disease is a breakdown in structures or functions of an organism. Some diseases are the result of intrinsic failures of the system. Others are the result of damage by infection by other organisms.

Reproduction and Heredity

C.2.a Reproduction is a characteristic of all living systems; because no individual organism lives forever, reproduction is essential to the continuation of every species. Some organisms reproduce asexually. Other organisms reproduce sexually.

C.2.b In many species, including humans, females produce eggs and males produce sperm. Plants also reproduce sexually—the egg and sperm are produced in the flowers of flowering plants. An egg and sperm unite to begin development of a new individual. That new individual receives genetic information from its mother (via the egg) and its father (via the sperm). Sexually produced offspring never are identical to either of their parents.

C.2.c Every organism requires a set of instructions for specifying its traits. Heredity is the passage of these instructions from one generation to another.

C.2.d Hereditary information is contained in genes, located in the chromosomes of each cell. Each gene carries a single unit of information. An inherited trait of an individual can be determined by one or by many genes, and a single gene can influence more than one trait. A human cell contains many thousands of different genes.

C.2.e The characteristics of an organism can be described in terms of a combination of traits. Some traits are inherited and others result from interactions with the environment.

Regulation and Behavior

C.3.a   All organisms must be able to obtain and use resources, grow, reproduce, and maintain stable internal conditions while living in a constantly changing external environment.

C.3.b   Regulation of an organism's internal environment involves sensing the internal environment and changing physiological activities to keep conditions within the range required to survive.

C.3.c   Behavior is one kind of response an organism can make to an internal or environmental stimulus. A behavioral response requires coordination and communication at many levels, including cells, organ systems, and whole organisms. Behavioral response is a set of actions determined in part by heredity and in part from experience.

C.3.d   An organism's behavior evolves through adaptation to its environment. How a species moves, obtains food, reproduces, and responds to danger are based in the species' evolutionary history.

Populations and Ecosystems

C.4.a   A population consists of all individuals of a species that occur together at a given place and time. All populations living together and the physical factors with which they interact compose an ecosystem.

C.4.b   Populations of organisms can be categorized by the function they serve in an ecosystem. Plants and some microorganisms are producers—they make their own food. All animals, including humans, are consumers, which obtain food by eating other organisms. Decomposers, primarily bacteria and fungi, are consumers that use waste materials and dead organisms for food. Food webs identify the relationships among producers, consumers, and decomposers in an ecosystem.

C.4.c   For ecosystems, the major source of energy is sunlight. Energy entering ecosystems as sunlight is transferred by producers into chemical energy through photosynthesis. That energy then passes from organism to organism in food webs.

C.4.d   The number of organisms an ecosystem can support depends on the resources available and abiotic factors, such as quantity of light and water, range of temperatures, and soil composition. Given adequate biotic and abiotic resources and no disease or predators, populations (including humans) increase at rapid rates. Lack of resources and other factors, such as predation and climate, limit the growth of populations in specific niches in the ecosystem.

Diversity and Adaptations of Organisms

C.5.a   Millions of species of animals, plants, and microorganisms are alive today. Although different species might look dissimilar, the unity among organisms becomes apparent from an analysis of internal structures, the similarity of their chemical processes, and the evidence of common ancestry.

C.5.b   Biological evolution accounts for the diversity of species developed through gradual processes over many generations. Species acquire many of their unique characteristics through biological adaptation, which involves the selection of naturally occurring variations in populations. Biological adaptations include changes in structures, behaviors, or physiology that enhance survival and reproductive success in a particular environment.

C.5.c   Extinction of a species occurs when the environment changes and the adaptive characteristics of a species are insufficient to allow its survival. Fossils indicate that many organisms that lived long ago are extinct. Extinction of species is common; most of the species that have lived on the earth no longer exist.

D. Earth and Space Science

As a result of their activities in grades 5–8, all students should develop an understanding of

Structure of the Earth System

D.1.a The solid earth is layered with a lithosphere; hot, convecting mantle; and dense, metallic core.

D.1.b Lithospheric plates on the scales of continents and oceans constantly move at rates of centimeters per year in response to movements in the mantle. Major geological events, such as earthquakes, volcanic eruptions, and mountain building, result from these plate motions.

D.1.c Land forms are the result of a combination of constructive and destructive forces. Constructive forces include crustal deformation, volcanic eruption, and deposition of sediment, while destructive forces include weathering and erosion.

D.1.d Some changes in the solid earth can be described as the "rock cycle." Old rocks at the earth's surface weather, forming sediments that are buried, then compacted, heated, and often recrystallized into new rock. Eventually, those new rocks may be brought to the surface by the forces that drive plate motions, and the rock cycle continues.

D.1.e Soil consists of weathered rocks and decomposed organic material from dead plants, animals, and bacteria. Soils are often found in layers, with each having a different chemical composition and texture.

D.1.f Water, which covers the majority of the earth's surface, circulates through the crust, oceans, and atmosphere in what is known as the "water cycle." Water evaporates from the earth's surface, rises and cools as it moves to higher elevations, condenses as rain or snow, and falls to the surface where it collects in lakes, oceans, soil, and in rocks underground.

D.1.g Water is a solvent. As it passes through the water cycle it dissolves minerals and gases and carries them to the oceans.

D.1.h The atmosphere is a mixture of nitrogen, oxygen, and trace gases that include water vapor. The atmosphere has different properties at different elevations.

D.1.i Clouds, formed by the condensation of water vapor, affect weather and climate.

D.1.j Global patterns of atmospheric movement influence local weather. Oceans have a major effect on climate, because water in the oceans holds a large amount of heat.

D.1.k Living organisms have played many roles in the earth system, including affecting the composition of the atmosphere, producing some types of rocks, and contributing to the weathering of rocks.

Earth's History

D.2.a The earth processes we see today, including erosion, movement of lithospheric plates, and changes in atmospheric composition, are similar to those that occurred in the past. Earth history is also influenced by occasional catastrophes, such as the impact of an asteroid or comet.

D.2.b Fossils provide important evidence of how life and environmental conditions have changed.

Earth in the Solar System

D.3.a   The earth is the third planet from the sun in a system that includes the moon, the sun, eight other planets and their moons, and smaller objects, such as asteroids and comets. The sun, an average star, is the central and largest body in the solar system.

D.3.b   Most objects in the solar system are in regular and predictable motion. Those motions explain such phenomena as the day, the year, phases of the moon, and eclipses.

D.3.c   Gravity is the force that keeps planets in orbit around the sun and governs the rest of the motion in the solar system. Gravity alone holds us to the earth's surface and explains the phenomena of the tides.

D.3.d   The sun is the major source of energy for phenomena on the earth's surface, such as growth of plants, winds, ocean currents, and the water cycle. Seasons result from variations in the amount of the sun's energy hitting the surface, due to the tilt of the earth's rotation on its axis and the length of the day.

E. Science and Technology

As a result of activities in grades 5–8, all students should develop

Abilities of Technological Design

E.1   Identify appropriate problems for technological design. Students should develop their abilities by identifying a specified need, considering its various aspects, and talking to different potential users or beneficiaries. They should appreciate that for some needs, the cultural backgrounds and beliefs of different groups can affect the criteria for a suitable product.

E.2   Design a solution or product. Students should make and compare different proposals in the light of the criteria they have selected. They must consider constraints—such as cost, time, trade-offs, and materials needed—and communicate ideas with drawings and simple models.

E.3   Implement a proposed design. Students should organize materials and other resources, plan their work, make good use of group collaboration where appropriate, choose suitable tools and techniques, and work with appropriate measurement methods to ensure adequate accuracy.

E.4   Evaluate completed technological designs or products. Students should use criteria relevant to the original purpose or need, consider a variety of factors that might affect acceptability and suitability for intended users or beneficiaries, and develop measures of quality with respect to such criteria and factors; they should also suggest improvements and, for their own products, try proposed modifications.

E.5   Communicate the process of technological design. Students should review and describe any completed piece of work and identify the stages of problem identification, solution design, implementation, and evaluation.

Understandings about Science and Technology

E.6.a   Scientific inquiry and technological design have similarities and differences. Scientists propose explanations for questions about the natural world, and engineers propose solutions relating to human problems, needs, and aspirations. Technological solutions are temporary; technologies exist within nature and so they cannot contravene physical or biological principles; technological solutions have side effects; and technologies cost, carry risks, and provide benefits.

E.6.b   Many different people in different cultures have made and continue to make contributions to science and technology.

E.6.c   Science and technology are reciprocal. Science helps drive technology, as it addresses questions that demand more sophisticated instruments and provides principles for better instrumentation and technique. Technology is essential to science, because it provides instruments and techniques that enable observations of objects and phenomena that are otherwise unobservable due to factors such as quantity, distance, location, size, and speed. Technology also provides tools for investigations, inquiry, and analysis.

E.6.d   Perfectly designed solutions do not exist. All technological solutions have trade-offs, such as safety, cost, efficiency, and appearance. Engineers often build in back-up systems to provide safety. Risk is part of living in a highly technological world. Reducing risk often results in new technology.

E.6.e   Technological designs have constraints. Some constraints are unavoidable, for example, properties of materials, or effects of weather and friction; other constraints limit choices in the design, for example, environmental protection, human safety, and aesthetics.

E.6.f   Technological solutions have intended benefits and unintended consequences. Some consequences can be predicted, others cannot.

F. Science in Personal and Social Perspectives

As a result of activities in grades 5–8, all students should develop understanding of

Personal Health

F.1.a   Regular exercise is important to the maintenance and improvement of health. The benefits of physical fitness include maintaining healthy weight, having energy and strength for routine activities, good muscle tone, bone strength, strong heart/lung systems, and improved mental health. Personal exercise, especially developing cardiovascular endurance, is the foundation of physical fitness.

F.1.b   The potential for accidents and the existence of hazards imposes the need for injury prevention. Safe living involves the development and use of safety precautions and the recognition of risk in personal decisions. Injury prevention has personal and social dimensions.

F.1.c   The use of tobacco increases the risk of illness. Students should understand the influence of short-term social and psychological factors that lead to tobacco use, and the possible long-term detrimental effects of smoking and chewing tobacco.

F.1.d   Alcohol and other drugs are often abused substances. Such drugs change how the body functions and can lead to addiction.

F.1.e   Food provides energy and nutrients for growth and development. Nutrition requirements vary with body weight, age, sex, activity, and body functioning.

F.1.f   Sex drive is a natural human function that requires understanding. Sex is also a prominent means of transmitting diseases. The diseases can be prevented through a variety of precautions.

F.1.g   Natural environments may contain substances (for example, radon and lead) that are harmful to human beings. Maintaining environmental health involves establishing or monitoring quality standards related to use of soil, water, and air.

Populations, Resources, and Environments

F.2.a   When an area becomes overpopulated, the environment will become degraded due to the increased use of resources.

F.2.b   Causes of environmental degradation and resource depletion vary from region to region and from country to country.

Natural Hazards

F.3.a   Internal and external processes of the earth system cause natural hazards, events that change or destroy human and wildlife habitats, damage property, and harm or kill humans. Natural hazards include earthquakes, landslides, wildfires, volcanic eruptions, floods, storms, and even possible impacts of asteroids.

F.3.b   Human activities also can induce hazards through resource acquisition, urban growth, land-use decisions, and waste disposal. Such activities can accelerate many natural changes.

F.3.c   Natural hazards can present personal and societal challenges because misidentifying the change or incorrectly estimating the rate and scale of change may result in either too little attention and significant human costs or too much cost for unneeded preventive measures.

Risks and Benefits

F.4.a   Risk analysis considers the type of hazard and estimates the number of people that might be exposed and the number likely to suffer consequences. The results are used to determine the options for reducing or eliminating risks.

F.4.b   Students should understand the risks associated with natural hazards (fires, floods, tornadoes, hurricanes, earthquakes, and volcanic eruptions), with chemical hazards (pollutants in air, water, soil, and food), with biological hazards (pollen, viruses, bacterial, and parasites), social hazards (occupational safety and transportation), and with personal hazards (smoking, dieting, and drinking).

F.4.c   Individuals can use a systematic approach to thinking critically about risks and benefits. Examples include applying probability estimates to risks and comparing them to estimated personal and social benefits.

F.4.d   Important personal and social decisions are made based on perceptions of benefits and risks.

Science and Technology in Society

F.5.a   Science influences society through its knowledge and world view. Scientific knowledge and the procedures used by scientists influence the way many individuals in society think about themselves, others, and the environment. The effect of science on society is neither entirely beneficial nor entirely detrimental.

F.5.b   Societal challenges often inspire questions for scientific research, and social priorities often influence research priorities through the availability of funding for research.

F.5.c   Technology influences society through its products and processes. Technology influences the quality of life and the ways people act and interact. Technological changes are often accompanied by social, political, and economic changes that can be beneficial or detrimental to individuals and to society. Social needs, attitudes, and values influence the direction of technological development.

F.5.d   Science and technology have advanced through contributions of many different people, in different cultures, at different times in history. Science and technology have contributed enormously to economic growth and productivity among societies and groups within societies.

F.5.e   Scientists and engineers work in many different settings, including colleges and universities, businesses and industries, specific research institutes, and government agencies.

F.5.f   Scientists and engineers have ethical codes requiring that human subjects involved with research be fully informed about risks and benefits associated with the research before the individuals choose to participate. This ethic extends to potential risks to communities and property. In short, prior knowledge and consent are required for research involving human subjects or potential damage to property.

F.5.g   Science cannot answer all questions and technology cannot solve all human problems or meet all human needs. Students should understand the difference between scientific and other questions. They should appreciate what science and technology can reasonably contribute to society and what they cannot do. For example, new technologies often will decrease some risks and increase others.

G. History and Nature of Science

As a result of activities in grades 5–8, all students should develop understanding of

Science as a Human Endeavor

G.1.a   Women and men of various social and ethnic backgrounds—and with diverse interests, talents, qualities, and motivations—engage in the activities of science, engineering, and related fields such as the health professions. Some scientists work in teams, and some work alone, but all communicate extensively with others.

G.1.b   Science requires different abilities, depending on such factors as the field of study and type of inquiry. Science is very much a human endeavor, and the work of science relies on basic human qualities, such as reasoning, insight, energy, skill, and creativity—as well as on scientific habits of mind, such as intellectual honesty, tolerance of ambiguity, skepticism, and openness to new ideas.

Nature of Science

G.2.a Scientists formulate and test their explanations of nature using observation, experiments, and theoretical and mathematical models. Although all scientific ideas are tentative and subject to change and improvement in principle, for most major ideas in science, there is much experimental and observational confirmation. Those ideas are not likely to change greatly in the future. Scientists do and have changed their ideas about nature when they encounter new experimental evidence that does not match their existing explanations.

G.2.b In areas where active research is being pursued and in which there is not a great deal of experimental or observational evidence and understanding, it is normal for scientists to differ with one another about the interpretation of the evidence or theory being considered. Different scientists might publish conflicting experimental results or might draw different conclusions from the same data. Ideally, scientists acknowledge such conflict and work towards finding evidence that will resolve their disagreement.

G.2.c It is part of scientific inquiry to evaluate the results of scientific investigations, experiments, observations, theoretical models, and the explanations proposed by other scientists. Evaluation includes reviewing the experimental procedures, examining the evidence, identifying faulty reasoning, pointing out statements that go beyond the evidence, and suggesting alternative explanations for the same observations. Although scientists may disagree about explanations of phenomena, about interpretations of data, or about the value of rival theories, they do agree that questioning, response to criticism, and open communication are integral to the process of science. As scientific knowledge evolves, major disagreements are eventually resolved through such interactions between scientists.

History of Science

G.3.a Many individuals have contributed to the traditions of science. Studying some of these individuals provides further understanding of scientific inquiry, science as a human endeavor, the nature of science, and the relationships between science and society.

G.3.b In historical perspective, science has been practiced by different individuals in different cultures. In looking at the history of many peoples, one finds that scientists and engineers of high achievement are considered to be among the most valued contributors to their culture.

G.3.c Tracing the history of science can show how difficult it was for scientific innovators to break through the accepted ideas of their time to reach the conclusions that we currently take for granted.

1. The Nature of Science

By the end of the 8th grade, students should know that

1.A The Scientific World View

1.A.1 When similar investigations give different results, the scientific challenge is to judge whether the differences are trivial or significant, and it often takes further studies to decide. Even with similar results, scientists may wait until an investigation has been repeated many times before accepting the results as correct.

1.A.2 Scientific knowledge is subject to modification as new information challenges prevailing theories and as a new theory leads to looking at old observations in a new way.

1.A.3 Some scientific knowledge is very old and yet is still applicable today.

1.A.4 Some matters cannot be examined usefully in a scientific way. Among them are matters that by their nature cannot be tested objectively and those that are essentially matters of morality. Science can sometimes be used to inform ethical decisions by identifying the likely consequences of particular actions but cannot be used to establish that some action is either moral or immoral.

1.B Scientific Inquiry

1.B.1 Scientists differ greatly in what phenomena they study and how they go about their work. Although there is no fixed set of steps that all scientists follow, scientific investigations usually involve the collection of relevant evidence, the use of logical reasoning, and the application of imagination in devising hypotheses and explanations to make sense of the collected evidence.

1.B.2 If more than one variable changes at the same time in an experiment, the outcome of the experiment may not be clearly attributable to any one of the variables. It may not always be possible to prevent outside variables from influencing the outcome of an investigation (or even to identify all of the variables), but collaboration among investigators can often lead to research designs that are able to deal with such situations.

1.B.3 What people expect to observe often affects what they actually do observe. Strong beliefs about what should happen in particular circumstances can prevent them from detecting other results. Scientists know about this danger to objectivity and take steps to try and avoid it when designing investigations and examining data. One safeguard is to have different investigators conduct independent studies of the same questions.

1.C The Scientific Enterprise

1.C.1 Important contributions to the advancement of science, mathematics, and technology have been made by different kinds of people, in different cultures, at different times.

1.C.2 Until recently, women and racial minorities, because of restrictions on their education and employment opportunities, were essentially left out of much of the formal work of the science establishment; the remarkable few who overcame those obstacles were even then likely to have their work disregarded by the science establishment.

1.C.3 No matter who does science and mathematics or invents things, or when or where they do it, the knowledge and technology that result can eventually become available to everyone in the world.

1.C.4 Scientists are employed by colleges and universities, business and industry, hospitals, and many government agencies. Their places of work include offices, classrooms, laboratories, farms, factories, and natural field settings ranging from space to the ocean floor.

1.C.5 In research involving human subjects, the ethics of science require that potential subjects be fully informed about the risks and benefits associated with the research and of their right to refuse to participate. Science ethics also demand that scientists must not knowingly subject coworkers, students, the neighborhood, or the community to health or property risks without their prior knowledge and consent. Because animals cannot make informed choices, special care must be taken in using them in scientific research.

1.C.6 Computers have become invaluable in science because they speed up and extend people's ability to collect, store, compile, and analyze data, prepare research reports, and share data and ideas with investigators all over the world.

1.C.7 Accurate record-keeping, openness, and replication are essential for maintaining an investigator's credibility with other scientists and society.

3. The Nature of Technology

By the end of the 8th grade, students should know that

3.A Technology and Science

3.A.1 In earlier times, the accumulated information and techniques of each generation of workers were taught on the job directly to the next generation of workers. Today, the knowledge base for technology can be found as well in libraries of print and electronic resources and is often taught in the classroom.

3.A.2 Technology is essential to science for such purposes as access to outer space and other remote locations, sample collection and treatment, measurement, data collection and storage, computation, and communication of information.

3.A.3 Engineers, architects, and others who engage in design and technology use scientific knowledge to solve practical problems. But they usually have to take human values and limitations into account as well.

3.B Design and Systems

3.B.1 Design usually requires taking constraints into account. Some constraints, such as gravity or the properties of the materials to be used, are unavoidable. Other constraints, including economic, political, social, ethical, and aesthetic ones, limit choices.

3.B.2 All technologies have effects other than those intended by the design, some of which may have been predictable and some not. In either case, these side effects may turn out to be unacceptable to some of the population and therefore lead to conflict between groups.

3.B.3 Almost all control systems have inputs, outputs, and feedback. The essence of control is comparing information about what is happening to what people want to happen and then making appropriate adjustments. This procedure requires sensing information, processing it, and making changes. In almost all modern machines, microprocessors serve as centers of performance control.

3.B.4 Systems fail because they have faulty or poorly matched parts, are used in ways that exceed what was intended by the design, or were poorly designed to begin with. The most common ways to prevent failure are pretesting parts and procedures, overdesign, and redundancy.

3.C Issues in Technology

3.C.1 The human ability to shape the future comes from a capacity for generating knowledge and developing new technologies—and for communicating ideas to others.

3.C.2 Technology cannot always provide successful solutions for problems or fulfill every human need.

3.C.3 Throughout history, people have carried out impressive technological feats, some of which would be hard to duplicate today even with modern tools. The purposes served by these achievements have sometimes been practical, sometimes ceremonial.

3.C.4 Technology has strongly influenced the course of history and continues to do so. It is largely responsible for the great revolutions in agriculture, manufacturing, sanitation and medicine, warfare, transportation, information processing, and communications that have radically changed how people live.

3.C.5 New technologies increase some risks and decrease others. Some of the same technologies that have improved the length and quality of life for many people have also brought new risks.

3.C.6 Rarely are technology issues simple and one-sided. Relevant facts alone, even when known and available, usually do not settle matters entirely in favor of one side or another. That is because the contending groups may have different values and priorities. They may stand to gain or lose in different degrees, or may make very different predictions about what the future consequences of the proposed action will be.

3.C.7 Societies influence what aspects of technology are developed and how these are used. People control technology (as well as science) and are responsible for its effects.

4. The Physical Setting

By the end of the 8th grade, students should know that

4.A The Universe

4.A.1   The sun is a medium-sized star located near the edge of a disk-shaped galaxy of stars, part of which can be seen as a glowing band of light that spans the sky on a very clear night. The universe contains many billions of galaxies, and each galaxy contains many billions of stars. To the naked eye, even the closest of these galaxies is no more than a dim, fuzzy spot.

4.A.2   The sun is many thousands of times closer to the earth than any other star. Light from the sun takes a few minutes to reach the earth, but light from the next nearest star takes a few years to arrive. The trip to that star would take the fastest rocket thousands of years. Some distant galaxies are so far away that their light takes several billion years to reach the earth. People on earth, therefore, see them as they were that long ago in the past.

4.A.3   Nine planets of very different size, composition, and surface features move around the sun in nearly circular orbits. Some planets have a great variety of moons and even flat rings of rock and ice particles orbiting around them. Some of these planets and moons show evidence of geologic activity. The earth is orbited by one moon, many artificial satellites, and debris.

4.A.4   Large numbers of chunks of rock orbit the sun. Some of those that the earth meets in its yearly orbit around the sun glow and disintegrate from friction as they plunge through the atmosphere—and sometimes impact the ground. Other chunks of rocks mixed with ice have long, off-center orbits that carry them close to the sun, where the sun's radiation (of light and particles) boils off frozen material from their surfaces and pushes it into a long, illuminated tail.

4.B The Earth

4.B.1   We live on a relatively small planet, the third from the sun in the only system of planets definitely known to exist (although other, similar systems may be discovered in the universe).

4.B.2   The earth is mostly rock. Three-fourths of its surface is covered by a relatively thin layer of water (some of it frozen), and the entire planet is surrounded by a relatively thin blanket of air. It is the only body in the solar system that appears able to support life. The other planets have compositions and conditions very different from the earth's.

4.B.3   Everything on or anywhere near the earth is pulled toward the earth's center by gravitational force.

4.B.4   Because the earth turns daily on an axis that is tilted relative to the plane of the earth's yearly orbit around the sun, sunlight falls more intensely on different parts of the earth during the year. The difference in heating of the earth's surface produces the planet's seasons and weather patterns.

4.B.5   The moon's orbit around the earth once in about 28 days changes what part of the moon is lighted by the sun and how much of that part can be seen from the earth—the phases of the moon.

4.B.6   Climates have sometimes changed abruptly in the past as a result of changes in the earth's crust, such as volcanic eruptions or impacts of huge rocks from space. Even relatively small changes in atmospheric or ocean content can have widespread effects on climate if the change lasts long enough.

4.B.7   The cycling of water in and out of the atmosphere plays an important role in determining climatic patterns. Water evaporates from the surface of the earth, rises and cools, condenses into rain or snow, and falls again to the surface. The water falling on land collects in rivers and lakes, soil, and porous layers of rock, and much of it flows back into the ocean.

4.B.8   Fresh water, limited in supply, is essential for life and also for most industrial processes. Rivers, lakes, and groundwater can be depleted or polluted, becoming unavailable or unsuitable for life.

4.B.9   Heat energy carried by ocean currents has a strong influence on climate around the world.

4.B.10 Some minerals are very rare and some exist in great quantities, but—for practical purposes—the ability to recover them is just as important as their abundance. As minerals are depleted, obtaining them becomes more difficult. Recycling and the development of substitutes can reduce the rate of depletion but may also be costly.

4.B.11 The benefits of the earth's resources—such as fresh water, air, soil, and trees—can be reduced by using them wastefully or by deliberately or inadvertently destroying them. The atmosphere and the oceans have a limited capacity to absorb wastes and recycle materials naturally. Cleaning up polluted air, water, or soil or restoring depleted soil, forests, or fishing grounds can be very difficult and costly.

4.C Processes that Shape the Earth

4.C.1   The interior of the earth is hot. Heat flow and movement of material within the earth cause earthquakes and volcanic eruptions and create mountains and ocean basins. Gas and dust from large volcanoes can change the atmosphere.

4.C.2   Some changes in the earth's surface are abrupt (such as earthquakes and volcanic eruptions) while other changes happen very slowly (such as uplift and wearing down of mountains). The earth's surface is shaped in part by the motion of water and wind over very long times, which act to level mountain ranges.

4.C.3   Sediments of sand and smaller particles (sometimes containing the remains of organisms) are gradually buried and are cemented together by dissolved minerals to form solid rock again.

4.C.4   Sedimentary rock buried deep enough may be reformed by pressure and heat, perhaps melting and recrystallizing into different kinds of rock. These re-formed rock layers may be forced up again to become land surface and even mountains. Subsequently, this new rock too will erode. Rock bears evidence of the minerals, temperatures, and forces that created it.

4.C.5   Thousands of layers of sedimentary rock confirm the long history of the changing surface of the earth and the changing life forms whose remains are found in successive layers. The youngest layers are not always found on top, because of folding, breaking, and uplift of layers.

4.C.6   Although weathered rock is the basic component of soil, the composition and texture of soil and its fertility and resistance to erosion are greatly influenced by plant roots and debris, bacteria, fungi, worms, insects, rodents, and other organisms.

4.C.7   Human activities, such as reducing the amount of forest cover, increasing the amount and variety of chemicals released into the atmosphere, and intensive farming, have changed the earth's land, oceans, and atmosphere. Some of these changes have decreased the capacity of the environment to support some life forms.

4.D Structure of Matter

4.D.1   All matter is made up of atoms, which are far too small to see directly through a micro-scope. The atoms of any element are alike but are different from atoms of other elements. Atoms may stick together in well-defined molecules or may be packed together in large arrays. Different arrangements of atoms into groups compose all substances.

4.D.2   Equal volumes of different substances usually have different weights.

4.D.3   Atoms and molecules are perpetually in motion. Increased temperature means greater average energy, so most substances expand when heated. In solids, the atoms are closely locked in position and can only vibrate. In liquids, the atoms or molecules have higher energy, are more loosely connected, and can slide past one another; some molecules may get enough energy to escape into a gas. In gases, the atoms or molecules have still more energy and are free of one another except during occasional collisions.

4.D.4   The temperature and acidity of a solution influence reaction rates. Many substances dissolve in water, which may greatly facilitate reactions between them.

4.D.5   Scientific ideas about elements were borrowed from some Greek philosophers of 2,000 years earlier, who believed that everything was made from four basic substances: air, earth, fire, and water. It was the combinations of these "elements" in different proportions that gave other substances their observable properties. The Greeks were wrong about those four, but now over 100 different elements have been identified, some rare and some plentiful, out of which everything is made. Because most elements tend to combine with others, few elements are found in their pure form.

4.D.6   There are groups of elements that have similar properties, including highly reactive metals, less-reactive metals, highly reactive nonmetals (such as chlorine, fluorine, and oxygen), and some almost completely nonreactive gases (such as helium and neon). An especially important kind of reaction between substances involves combination of oxygen with something else—as in burning or rusting. Some elements don't fit into any of the categories; among them are carbon and hydrogen, essential elements of living matter.

4.D.7   No matter how substances within a closed system interact with one another, or how they combine or break apart, the total weight of the system remains the same. The idea of atoms explains the conservation of matter: If the number of atoms stays the same no matter how they are rearranged, then their total mass stays the same.

4.E Energy Transformations

4.E.1   Energy cannot be created or destroyed, but only changed from one form into another.

4.E.2   Most of what goes on in the universe—from exploding stars and biological growth to the operation of machines and the motion of people—involves some form of energy being transformed into another. Energy in the form of heat is almost always one of the products of an energy transformation.

4.E.3   Heat can be transferred through materials by the collisions of atoms or across space by radiation. If the material is fluid, currents will be set up in it that aid the transfer of heat.

4.E.4   Energy appears in different forms. Heat energy is in the disorderly motion of molecules; chemical energy is in the arrangement of atoms; mechanical energy is in moving bodies or in elastically distorted shapes; gravitational energy is in the separation of mutually attracting masses.

4.F Motion

4.F.1 Light from the sun is made up of a mixture of many different colors of light, even though to the eye the light looks almost white. Other things that give off or reflect light have a different mix of colors.

4.F.2 Something can be "seen" when light waves emitted or reflected by it enter the eye—just as something can be "heard" when sound waves from it enter the ear.

4.F.3 An unbalanced force acting on an object changes its speed or direction of motion, or both. If the force acts toward a single center, the object's path may curve into an orbit around the center.

4.F.4 Vibrations in materials set up wavelike disturbances that spread away from the source. Sound and earthquake waves are examples. These and other waves move at different speeds in different materials.

4.F.5 Human eyes respond to only a narrow range of wavelengths of electromagnetic radiation—visible light. Differences of wavelength within that range are perceived as differences in color.

4.G Forces of Nature

4.G.1 Every object exerts gravitational force on every other object. The force depends on how much mass the objects have and on how far apart they are. The force is hard to detect unless at least one of the objects has a lot of mass.

4.G.2 The sun's gravitational pull holds the earth and other planets in their orbits, just as the planets' gravitational pull keeps their moons in orbit around them.

4.G.3 Electric currents and magnets can exert a force on each other.

5. The Living Environment

By the end of the 8th grade, students should know that

5.A Diversity of Life

5.A.1   One of the most general distinctions among organisms is between plants, which use sunlight to make their own food, and animals, which consume energy-rich foods. Some kinds of organisms, many of them microscopic, cannot be neatly classified as either plants or animals.

5.A.2   Animals and plants have a great variety of body plans and internal structures that contribute to their being able to make or find food and reproduce.

5.A.3   Similarities among organisms are found in internal anatomical features, which can be used to infer the degree of relatedness among organisms. In classifying organisms, biologists consider details of internal and external structures to be more important than behavior or general appearance.

5.A.4   For sexually reproducing organisms, a species comprises all organisms that can mate with one another to produce fertile offspring.

5.A.5   All organisms, including the human species, are part of and depend on two main interconnected global food webs. One includes microscopic ocean plants, the animals that feed on them, and finally the animals that feed on those animals. The other web includes land plants, the animals that feed on them, and so forth. The cycles continue indefinitely because organisms decompose after death to return food material to the environment.

5.B Heredity

5.B.1   In some kinds of organisms, all the genes come from a single parent, whereas in organisms that have sexes, typically half of the genes come from each parent.

5.B.2   In sexual reproduction, a single specialized cell from a female merges with a specialized cell from a male. As the fertilized egg, carrying genetic information from each parent, multiplies to form the complete organism with about a trillion cells, the same genetic information is copied in each cell.

5.B.3   New varieties of cultivated plants and domestic animals have resulted from selective breeding for particular traits.

5.C Cells

5.C.1   All living things are composed of cells, from just one to many millions, whose details usually are visible only through a microscope. Different body tissues and organs are made up of different kinds of cells. The cells in similar tissues and organs in other animals are similar to those in human beings but differ somewhat from cells found in plants.

5.C.2   Cells repeatedly divide to make more cells for growth and repair. Various organs and tissues function to serve the needs of cells for food, air, and waste removal.

5.C.3   Within cells, many of the basic functions of organisms—such as extracting energy from food and getting rid of waste—are carried out. The way in which cells function is similar in all living organisms.

5.C.4   About two-thirds of the weight of cells is accounted for by water, which gives cells many of their properties.

5.D Interdependence of Life

5.D.1 In all environments—freshwater, marine, forest, desert, grassland, mountain, and others—organisms with similar needs may compete with one another for resources, including food, space, water, air, and shelter. In any particular environment, the growth and survival of organisms depend on the physical conditions.

5.D.2 Two types of organisms may interact with one another in several ways: They may be in a producer/consumer, predator/prey, or parasite/host relationship. Or one organism may scavenge or decompose another. Relationships may be competitive or mutually beneficial. Some species have become so adapted to each other that neither could survive without the other.

5.E Flow of Matter and Energy

5.E.1 Food provides molecules that serve as fuel and building material for all organisms. Plants use the energy in light to make sugars out of carbon dioxide and water. This food can be used immediately for fuel or materials or it may be stored for later use. Organisms that eat plants break down the plant structures to produce the materials and energy they need to survive. Then they are consumed by other organisms.

5.E.2 Over a long time, matter is transferred from one organism to another repeatedly and between organisms and their physical environment. As in all material systems, the total amount of matter remains constant, even though its form and location change.

5.E.3 Energy can change from one form to another in living things. Animals get energy from oxidizing their food, releasing some of its energy as heat. Almost all food energy comes originally from sunlight.

5.F Evolution of Life

5.F.1 Small differences between parents and offspring can accumulate (through selective breeding) in successive generations so that descendants are very different from their ancestors.

5.F.2 Individual organisms with certain traits are more likely than others to survive and have offspring. Changes in environmental conditions can affect the survival of individual organisms and entire species.

5.F.3 Many thousands of layers of sedimentary rock provide evidence for the long history of the earth and for the long history of changing life forms whose remains are found in the rocks. More recently deposited rock layers are more likely to contain fossils resembling existing species.

6. The Human Organism

By the end of the 8th grade, students should know that

6.A Human Identity

6.A.1 Like other animals, human beings have body systems for obtaining and providing energy, defense, reproduction, and the coordination of body functions.

6.A.2 Human beings have many similarities and differences. The similarities make it possible for human beings to reproduce and to donate blood and organs to one another throughout the world. Their differences enable them to create diverse social and cultural arrangements and to solve problems in a variety of ways.

6.A.3 Fossil evidence is consistent with the idea that human beings evolved from earlier species.

6.A.4 Specialized roles of individuals within other species are genetically programmed, whereas human beings are able to invent and modify a wider range of social behavior.

6.A.5 Human beings use technology to match or excel many of the abilities of other species. Technology has helped people with disabilities survive and live more conventional lives.

6.A.6 Technologies having to do with food production, sanitation, and disease prevention have dramatically changed how people live and work and have resulted in rapid increases in the human population.

6.B Human Development

6.B.1 Fertilization occurs when sperm cells from a male's testes are deposited near an egg cell from the female ovary, and one of the sperm cells enters the egg cell. Most of the time, by chance or design, a sperm never arrives or an egg isn't available.

6.B.2 Contraception measures may incapacitate sperm, block their way to the egg, prevent the release of eggs, or prevent the fertilized egg from implanting successfully.

6.B.3 Following fertilization, cell division produces a small cluster of cells that then differentiate by appearance and function to form the basic tissues of an embryo. During the first three months of pregnancy, organs begin to form. During the second three months, all organs and body features develop. During the last three months, the organs and features mature enough to function well after birth. Patterns of human development are similar to those of other vertebrates.

6.B.4 The developing embryo—and later the newborn infant—encounters many risks from faults in its genes, its mother's inadequate diet, her cigarette smoking or use of alcohol or other drugs, or from infection. Inadequate child care may lead to lower physical and mental ability.

6.B.5 Various body changes occur as adults age. Muscles and joints become less flexible, bones and muscles lose mass, energy levels diminish, and the senses become less acute. Women stop releasing eggs and hence can no longer reproduce. The length and quality of human life are influenced by many factors, including sanitation, diet, medical care, sex, genes, environmental conditions, and personal health behaviors.

6.C Basic Functions

6.C.1 Organs and organ systems are composed of cells and help to provide all cells with basic needs.

6.C.2 For the body to use food for energy and building materials, the food must first be digested into molecules that are absorbed and transported to cells.

6.C.3 To burn food for the release of energy stored in it, oxygen must be supplied to cells, and carbon dioxide removed. Lungs take in oxygen for the combustion of food and they eliminate the carbon dioxide produced. The urinary system disposes of dissolved waste molecules, the intestinal tract removes solid wastes, and the skin and lungs rid the body of heat energy. The circulatory system moves all these substances to or from cells where they are needed or produced, responding to changing demands.

6.C.4 Specialized cells and the molecules they produce identify and destroy microbes that get inside the body.

6.C.5 Hormones are chemicals from glands that affect other body parts. They are involved in helping the body respond to danger and in regulating human growth, development, and reproduction.

6.C.6 Interactions among the senses, nerves, and brain make possible the learning that enables human beings to cope with changes in their environment.

6.D Learning

6.D.1 Some animal species are limited to a repertoire of genetically determined behaviors; others have more complex brains and can learn a wide variety of behaviors. All behavior is affected by both inheritance and experience.

6.D.2 The level of skill a person can reach in any particular activity depends on innate abilities, the amount of practice, and the use of appropriate learning technologies.

6.D.3 Human beings can detect a tremendous range of visual and olfactory stimuli. The strongest stimulus they can tolerate may be more than a trillion times as intense as the weakest they can detect. Still, there are many kinds of signals in the world that people cannot detect directly.

6.D.4 Attending closely to any one input of information usually reduces the ability to attend to others at the same time.

6.D.5 Learning often results from two perceptions or actions occurring at about the same time. The more often the same combination occurs, the stronger the mental connection between them is likely to be. Occasionally a single vivid experience will connect two things permanently in people's minds.

6.D.6 Language and tools enable human beings to learn complicated and varied things from others.

6.E Physical Health

6.E.1 The amount of food energy (calories) a person requires varies with body weight, age, sex, activity level, and natural body efficiency. Regular exercise is important to maintain a healthy heart/lung system, good muscle tone, and bone strength.

6.E.2 Toxic substances, some dietary habits, and personal behavior may be bad for one's health. Some effects show up right away, others may not show up for many years. Avoiding toxic substances, such as tobacco, and changing dietary habits to reduce the intake of such things as animal fat increases the chances of living longer.

6.E.3 Viruses, bacteria, fungi, and parasites may infect the human body and interfere with normal body functions. A person can catch a cold many times because there are many varieties of cold viruses that cause similar symptoms.

6.E.4 White blood cells engulf invaders or produce antibodies that attack them or mark them for killing by other white cells. The antibodies produced will remain and can fight off subsequent invaders of the same kind.

6.E.5 The environment may contain dangerous levels of substances that are harmful to human beings. Therefore, the good health of individuals requires monitoring the soil, air, and water and taking steps to keep them safe.

6.F Mental Health

6.F.1 Individuals differ greatly in their ability to cope with stressful situations. Both external and internal conditions (chemistry, personal history, values) influence how people behave.

6.F.2 Often people react to mental distress by denying that they have any problem. Sometimes they don't know why they feel the way they do, but with help they can sometimes uncover the reasons.

8. The Designed World

By the end of the 8th grade, students should know that

8.A Agriculture

8.A.1 Early in human history, there was an agricultural revolution in which people changed from hunting and gathering to farming. This allowed changes in the division of labor between men and women and between children and adults, and the development of new patterns of government.

8.A.2 People control the characteristics of plants and animals they raise by selective breeding and by preserving varieties of seeds (old and new) to use if growing conditions change.

8.A.3 In agriculture, as in all technologies, there are always trade-offs to be made. Getting food from many different places makes people less dependent on weather in any one place, yet more dependent on transportation and communication among far-flung markets. Specializing in one crop may risk disaster if changes in weather or increases in pest populations wipe out that crop. Also, the soil may be exhausted of some nutrients, which can be replenished by rotating the right crops.

8.A.4 Many people work to bring food, fiber, and fuel to U.S. markets. With improved technology, only a small fraction of workers in the United States actually plant and harvest the products that people use. Most workers are engaged in processing, packaging, transporting, and selling what is produced.

8.B Materials and Manufacturing

8.B.1   The choice of materials for a job depends on their properties and on how they interact with other materials. Similarly, the usefulness of some manufactured parts of an object depends on how well they fit together with the other parts.

8.B.2   Manufacturing usually involves a series of steps, such as designing a product, obtaining and preparing raw materials, processing the materials mechanically or chemically, and assembling, testing, inspecting, and packaging. The sequence of these steps is also often important.

8.B.3   Modern technology reduces manufacturing costs, produces more uniform products, and creates new synthetic materials that can help reduce the depletion of some natural resources.

8.B.4   Automation, including the use of robots, has changed the nature of work in most fields, including manufacturing. As a result, high-skill, high-knowledge jobs in engineering, computer programming, quality control, supervision, and maintenance are replacing many routine, manual-labor jobs. Workers therefore need better learning skills and flexibility to take on new and rapidly changing jobs.

8.C Energy Sources and Use

8.C.1   Energy can change from one form to another, although in the process some energy is always converted to heat. Some systems transform energy with less loss of heat than others.

8.C.2   Different ways of obtaining, transforming, and distributing energy have different environmental consequences.

8.C.3   In many instances, manufacturing and other technological activities are performed at a site close to an energy source. Some forms of energy are transported easily, others are not.

8.C.4   Electrical energy can be produced from a variety of energy sources and can be transformed into almost any other form of energy. Moreover, electricity is used to distribute energy quickly and conveniently to distant locations.

8.C.5   Energy from the sun (and the wind and water energy derived from it) is available indefinitely. Because the flow of energy is weak and variable, very large collection systems are needed. Other sources don't renew or renew only slowly.

8.C.6   Different parts of the world have different amounts and kinds of energy resources to use and use them for different purposes.

8.D Communication

8.D.1   Errors can occur in coding, transmitting, or decoding information, and some means of checking for accuracy is needed. Repeating the message is a frequently used method.

8.D.2   Information can be carried by many media, including sound, light, and objects. In this century, the ability to code information as electric currents in wires, electromagnetic waves in space, and light in glass fibers has made communication millions of times faster than is possible by mail or sound.

8.E Information Processing

8.E.1   Most computers use digital codes containing only two symbols, 0 and 1, to perform all operations. Continuous signals (analog) must be transformed into digital codes before they can be processed by a computer.

8.E.2   What use can be made of a large collection of information depends upon how it is organized. One of the values of computers is that they are able, on command, to reorganize information in a variety of ways, thereby enabling people to make more and better uses of the collection.

8.E.3   Computer control of mechanical systems can be much quicker than human control. In situations where events happen faster than people can react, there is little choice but to rely on computers. Most complex systems still require human oversight, however, to make certain kinds of judgments about the readiness of the parts of the system (including the computers) and the system as a whole to operate properly, to react to unexpected failures, and to evaluate how well the system is serving its intended purposes.

8.E.4   An increasing number of people work at jobs that involve processing or distributing information. Because computers can do these tasks faster and more reliably, they have become standard tools both in the workplace and at home.

8.F Health Technology

8.F.1   Sanitation measures such as the use of sewers, landfills, quarantines, and safe food handling are important in controlling the spread of organisms that cause disease. Improving sanitation to prevent disease has contributed more to saving human life than any advance in medical treatment.

8.F.2   The ability to measure the level of substances in body fluids has made it possible for physicians to make comparisons with normal levels, make very sophisticated diagnoses, and monitor the effects of the treatments they prescribe.

8.F.3   It is becoming increasingly possible to manufacture chemical substances such as insulin and hormones that are normally found in the body. They can be used by individuals whose own bodies cannot produce the amounts required for good health.

9. The Mathematical World

By the end of the 8th grade, students should know that

9.A Numbers

9.A.1   There have been systems for writing numbers other than the Arabic system of place values based on tens. The very old Roman numerals are now used only for dates, clock faces, or ordering chapters in a book. Numbers based on 60 are still used for describing time and angles.

9.A.2   A number line can be extended on the other side of zero to represent negative numbers. Negative numbers allow subtraction of a bigger number from a smaller number to make sense, and are often used when something can be measured on either side of some reference point (time, ground level, temperature, budget).

9.A.3   Numbers can be written in different forms, depending on how they are being used. How fractions or decimals based on measured quantities should be written depends on how precise the measurements are and how precise an answer is needed.

9.A.4   The operations + and − are inverses of each other—one undoes what the other does; likewise x and ÷ .

9.A.5   The expression a/b can mean different things: a parts of size 1/b each, a divided by b, or a compared to b.

9.A.6   Numbers can be represented by using sequences of only two symbols (such as 1 and 0, on and off); computers work this way.

9.A.7   Computations (as on calculators) can give more digits than make sense or are useful.

9.B Symbolic Relationships

9.B.1   An equation containing a variable may be true for just one value of the variable.

9.B.2   Mathematical statements can be used to describe how one quantity changes when another changes. Rates of change can be computed from differences in magnitudes and vice versa.

9.B.3   Graphs can show a variety of possible relationships between two variables. As one variable increases uniformly, the other may do one of the following: increase or decrease steadily, increase or decrease faster and faster, get closer and closer to some limiting value, reach some intermediate maximum or minimum, alternately increase and decrease indefinitely, increase or decrease in steps, or do something different from any of these.

9.C Shapes

9.C.1   Some shapes have special properties: triangular shapes tend to make structures rigid, and round shapes give the least possible boundary for a given amount of interior area. Shapes can match exactly or have the same shape in different sizes.

9.C.2   Lines can be parallel, perpendicular, or oblique.

9.C.3   Shapes on a sphere like the earth cannot be depicted on a flat surface without some distortion.

9.C.4   The graphic display of numbers may help to show patterns such as trends, varying rates of change, gaps, or clusters. Such patterns sometimes can be used to make predictions about the phenomena being graphed.

9.C.5   It takes two numbers to locate a point on a map or any other flat surface. The numbers may be two perpendicular distances from a point, or an angle and a distance from a point.

9.C.6   The scale chosen for a graph or drawing makes a big difference in how useful it is.

9.D Uncertainty

9.D.1   How probability is estimated depends on what is known about the situation. Estimates can be based on data from similar conditions in the past or on the assumption that all the possibilities are known.

9.D.2   Probabilities are ratios and can be expressed as fractions, percentages, or odds.

9.D.3   The mean, median, and mode tell different things about the middle of a data set.

9.D.4   Comparison of data from two groups should involve comparing both their middles and the spreads around them.

9.D.5   The larger a well-chosen sample is, the more accurately it is likely to represent the whole. But there are many ways of choosing a sample that can make it unrepresentative of the whole.

9.D.6   Events can be described in terms of being more or less likely, impossible, or certain.

9.E Reasoning

9.E.1   Some aspects of reasoning have fairly rigid rules for what makes sense; other aspects don't. If people have rules that always hold, and good information about a particular situation, then logic can help them to figure out what is true about it. This kind of reasoning requires care in the use of key words such as if, and, not, or, all, and some. Reasoning by similarities can suggest ideas but can't prove them one way or the other.

9.E.2   Practical reasoning, such as diagnosing or troubleshooting almost anything, may require many-step, branching logic. Because computers can keep track of complicated logic, as well as a lot of information, they are useful in a lot of problem-solving situations.

9.E.3   Sometimes people invent a general rule to explain how something works by summarizing observations. But people tend to overgeneralize, imagining general rules on the basis of only a few observations.

9.E.4   People are using incorrect logic when they make a statement such as "If A is true, then B is true; but A isn't true, therefore B isn't true either."

9.E.5   A single example can never prove that something is always true, but sometimes a single example can prove that something is not always true.

9.E.6   An analogy has some likenesses to but also some differences from the real thing.

10. Historical Perspectives

By the end of the 8th grade, students should know that

10.A Displacing the Earth from the Center of the Universe

10.A.1 The motion of an object is always judged with respect to some other object or point and so the idea of absolute motion or rest is misleading.

10.A.2 Telescopes reveal that there are many more stars in the night sky than are evident to the unaided eye, the surface of the moon has many craters and mountains, the sun has dark spots, and Jupiter and some other planets have their own moons.

10.F Understanding Fire

10.F.1 From the earliest times until now, people have believed that even though millions of different kinds of material seem to exist in the world, most things must be made up of combinations of just a few basic kinds of things. There has not always been agreement, however, on what those basic kinds of things are. One theory long ago was that the basic substances were earth, water, air, and fire. Scientists now know that these are not the basic substances. But the old theory seemed to explain many observations about the world.

10.F.2 Today, scientists are still working out the details of what the basic kinds of matter are and of how they combine, or can be made to combine, to make other substances.

10.F.3 Experimental and theoretical work done by French scientist Antoine Lavoisier in the decade between the American and French revolutions led to the modern science of chemistry.

10.F.4 Lavoisier's work was based on the idea that when materials react with each other many changes can take place but that in every case the total amount of matter afterward is the same as before. He successfully tested the concept of conservation of matter by conducting a series of experiments in which he carefully measured all the substances involved in burning, including the gases used and those given off.

10.F.5 Alchemy was chiefly an effort to change base metals like lead into gold and to produce an elixir that would enable people to live forever. It failed to do that or to create much knowledge of how substances react with each other. The more scientific study of chemistry that began in Lavoisier's time has gone far beyond alchemy in understanding reactions and producing new materials.

10.G Splitting the Atom

10.G.1 The accidental discovery that minerals containing uranium darken photographic film, as light does, led to the idea of radioactivity.

10.G.2 In their laboratory in France, Marie Curie and her husband, Pierre Curie, isolated two new elements that caused most of the radioactivity of the uranium mineral. They named one radium because it gave off powerful, invisible rays, and the other polonium in honor of Madame Curie's country of birth. Marie Curie was the first scientist ever to win the Nobel prize in two different fields—in physics, shared with her husband, and later in chemistry.

10.I Discovering Germs

10.I.1 Throughout history, people have created explanations for disease. Some have held that disease has spiritual causes, but the most persistent biological theory over the centuries was that illness resulted from an imbalance in the body fluids. The introduction of germ theory by Louis Pasteur and others in the 19th century led to the modern belief that many diseases are caused by microorganisms—bacteria, viruses, yeasts, and parasites.

10.I.2 Pasteur wanted to find out what causes milk and wine to spoil. He demonstrated that spoilage and fermentation occur when microorganisms enter from the air, multiply rapidly, and produce waste products. After showing that spoilage could be avoided by keeping germs out or by destroying them with heat, he investigated animal diseases and showed that microorganisms were involved. Other investigators later showed that specific kinds of germs caused specific diseases.

10.I.3 Pasteur found that infection by disease organisms—germs—caused the body to build up an immunity against subsequent infection by the same organisms. He then demonstrated that it was possible to produce vaccines that would induce the body to build immunity to a disease without actually causing the disease itself.

10.I.4 Changes in health practices have resulted from the acceptance of the germ theory of disease. Before germ theory, illness was treated by appeals to supernatural powers or by trying to adjust body fluids through induced vomiting, bleeding, or purging. The modern approach emphasizes sanitation, the safe handling of food and water, the pasteurization of milk, quarantine, and aseptic surgical techniques to keep germs out of the body; vaccinations to strengthen the body's immune system against subsequent infection by the same kind of microorganisms; and antibiotics and other chemicals and processes to destroy microorganisms.

10.I.5 In medicine, as in other fields of science, discoveries are sometimes made unexpectedly, even by accident. But knowledge and creative insight are usually required to recognize the meaning of the unexpected.

10.J Harnessing Power

10.J.1 Until the 1800s, most manufacturing was done in homes, using small, handmade machines that were powered by muscle, wind, or running water. New machinery and steam engines to drive them made it possible to replace craftsmanship with factories, using fuels as a source of energy. In the factory system, workers, materials, and energy could be brought together efficiently.

10.J.2 The invention of the steam engine was at the center of the Industrial Revolution. It converted the chemical energy stored in wood and coal, which were plentiful, into mechanical work. The steam engine was invented to solve the urgent problem of pumping water out of coal mines. As improved by James Watt, it was soon used to move coal, drive manufacturing machinery, and power locomotives, ships, and even the first automobiles.

11. Common Themes

By the end of the 8th grade, students should know that

11.A Systems

11.A.1 A system can include processes as well as things.

11.A.2 Thinking about things as systems means looking for how every part relates to others. The output from one part of a system (which can include material, energy, or information) can become the input to other parts. Such feedback can serve to control what goes on in the system as a whole.

11.A.3 Any system is usually connected to other systems, both internally and externally. Thus a system may be thought of as containing subsystems and as being a subsystem of a larger system.

11.B Models

11.B.1 Models are often used to think about processes that happen too slowly, too quickly, or on too small a scale to observe directly, or that are too vast to be changed deliberately, or that are potentially dangerous.

11.B.2 Mathematical models can be displayed on a computer and then modified to see what happens.

11.B.3 Different models can be used to represent the same thing. What kind of a model to use and how complex it should be depends on its purpose. The usefulness of a model may be limited if it is too simple or if it is needlessly complicated. Choosing a useful model is one of the instances in which intuition and creativity come into play in science, mathematics, and engineering.

11.C Constancy and Change

11.C.1 Physical and biological systems tend to change until they become stable and then remain that way unless their surroundings change.

11.C.2 A system may stay the same because nothing is happening or because things are happening but exactly counterbalance one another.

11.C.3 Many systems contain feedback mechanisms that serve to keep changes within specified limits.

11.C.4 Symbolic equations can be used to summarize how the quantity of something changes over time or in response to other changes.

11.C.5 Symmetry (or the lack of it) may determine properties of many objects, from molecules and crystals to organisms and designed structures.

11.C.6 Cycles, such as the seasons or body temperature, can be described by their cycle length or frequency, what their highest and lowest values are, and when these values occur. Different cycles range from many thousands of years down to less than a billionth of a second.

11.D Scale

11.D.1 Properties of systems that depend on volume, such as capacity and weight, change out of proportion to properties that depend on area, such as strength or surface processes.

11.D.2 As the complexity of any system increases, gaining an understanding of it depends increasingly on summaries, such as averages and ranges, and on descriptions of typical examples of that system.

12. Habits of Mind

By the end of the 8th grade, students should know that

12.A Values and Attitudes

12.A.1 Know why it is important in science to keep honest, clear, and accurate records.

12.A.2 Know that hypotheses are valuable, even if they turn out not to be true, if they lead to fruitful investigations.

12.A.3 Know that often different explanations can be given for the same evidence, and it is not always possible to tell which one is correct.

12.B Computation and Estimation

12.B.1 Find what percentage one number is of another and figure any percentage of any number.

12.B.2 Use, interpret, and compare numbers in several equivalent forms such as integers, fractions, decimals, and percents.

12.B.3 Calculate the circumferences and areas of rectangles, triangles, and circles, and the volumes of rectangular solids.

12.B.4 Find the mean and median of a set of data.

12.B.5 Estimate distances and travel times from maps and the actual size of objects from scale drawings.

12.B.6 Insert instructions into computer spreadsheet cells to program arithmetic calculations.

12.B.7 Determine what unit (such as seconds, square inches, or dollars per tankful) an answer should be expressed in from the units of the inputs to the calculation, and be able to convert compound units (such as yen per dollar into dollar per yen, or miles per hour into feet per second).

12.B.8 Decide what degree of precision is adequate and round off the result of calculator operations to enough significant figures to reasonably reflect those of the inputs.

12.B.9 Express numbers like 100, 1,000, and 1,000,000 as powers of 10.

12.B.10 Estimate probabilities of outcomes in familiar situations, on the basis of history or the number of possible outcomes.

12.C Manipulation and Observation

12.C.1 Use calculators to compare amounts proportionally.

12.C.2 Use computers to store and retrieve information in topical, alphabetical, numerical, and key-word files, and create simple files of their own devising.

12.C.3 Read analog and digital meters on instruments used to make direct measurements of length, volume, weight, elapsed time, rates, and temperature, and choose appropriate units for reporting various magnitudes.

12.C.4 Use cameras and tape recorders for capturing information.

12.C.5 Inspect, disassemble, and reassemble simple mechanical devices and describe what the various parts are for; estimate what the effect that making a change in one part of a system is likely to have on the system as a whole.

12.D Communication Skills

12.D.1 Organize information in simple tables and graphs and identify relationships they reveal.

12.D.2 Read simple tables and graphs produced by others and describe in words what they show.

12.D.3 Locate information in reference books, back issues of newspapers and magazines, compact disks, and computer databases.

12.D.4 Understand writing that incorporates circle charts, bar and line graphs, two-way data tables, diagrams, and symbols.

12.D.5 Find and describe locations on maps with rectangular and polar coordinates.

12.E Critical-Response Skills

12.E.1 Question claims based on vague attributions (such as "Leading doctors say...") or on statements made by celebrities or others outside the area of their particular expertise.

12.E.2 Compare consumer products and consider reasonable personal trade-offs among them on the basis of features, performance, durability, and cost.

12.E.3 Be skeptical of arguments based on very small samples of data, biased samples, or samples for which there was no control sample.

12.E.4 Be aware that there may be more than one good way to interpret a given set of findings.

12.E.5 Notice and criticize the reasoning in arguments in which (1) fact and opinion are intermingled or the conclusions do not follow logically from the evidence given, (2) an analogy is not apt, (3) no mention is made of whether the control groups are very much like the experimental group, or (4) all members of a group (such as teenagers or chemists) are implied to have nearly identical characteristics that differ from those of other groups.

The Periodic Table of the Elements

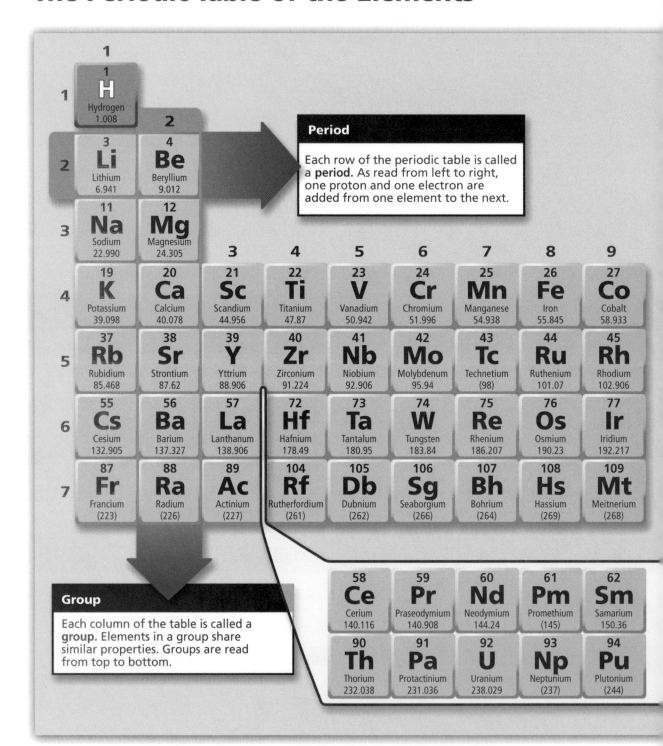

Period

Each row of the periodic table is called a **period.** As read from left to right, one proton and one electron are added from one element to the next.

Group

Each column of the table is called a **group.** Elements in a group share similar properties. Groups are read from top to bottom.

 Metal Metalloid Nonmetal **Fe** Solid **Hg** Liquid Gas